DISCARDED

AUTOMATIC DIATOM IDENTIFICATION

SERIES IN MACHINE PERCEPTION AND ARTIFICIAL INTELLIGENCE*

Editors: **H. Bunke** (Univ. Bern, Switzerland)
P. S. P. Wang (Northeastern Univ., USA)

Vol. 34: Advances in Handwriting Recognition
(Ed. *S.-W. Lee*)

Vol. 35: Vision Interface — Real World Applications of Computer Vision
(Eds. *M. Cheriet and Y.-H. Yang*)

Vol. 36: Wavelet Theory and Its Application to Pattern Recognition
(*Y. Y. Tang, L. H. Yang, J. Liu and H. Ma*)

Vol. 37: Image Processing for the Food Industry
(*E. R. Davies*)

Vol. 38: New Approaches to Fuzzy Modeling and Control — Design and Analysis
(*M. Margaliot and G. Langholz*)

Vol. 39: Artificial Intelligence Techniques in Breast Cancer Diagnosis and Prognosis
(Eds. *A. Jain, A. Jain, S. Jain and L. Jain*)

Vol. 40: Texture Analysis in Machine Vision
(Ed. *M. K. Pietikäinen*)

Vol. 41: Neuro-Fuzzy Pattern Recognition
(Eds. *H. Bunke and A. Kandel*)

Vol. 42: Invariants for Pattern Recognition and Classification
(Ed. *M. A. Rodrigues*)

Vol. 43: Agent Engineering
(Eds. *Jiming Liu, Ning Zhong, Yuan Y. Tang and Patrick S. P. Wang*)

Vol. 44: Multispectral Image Processing and Pattern Recognition
(Eds. *J. Shen, P. S. P. Wang and T. Zhang*)

Vol. 45: Hidden Markov Models: Applications in Computer Vision
(Eds. *H. Bunke and T. Caelli*)

Vol. 46: Syntactic Pattern Recognition for Seismic Oil Exploration
(*K. Y. Huang*)

Vol. 47: Hybrid Methods in Pattern Recognition
(Eds. *H. Bunke and A. Kandel*)

Vol. 48: Multimodal Interface for Human-Machine Communications
(Eds. *P. C. Yuen, Y. Y. Tang and P. S. P. Wang*)

Vol. 49: Neural Networks and Systolic Array Design
(Eds. *D. Zhang and S. K. Pal*)

*For the complete list of titles in this series, please write to the Publisher.

Series in Machine Perception and Artificial Intelligence – Vol. 51

AUTOMATIC DIATOM IDENTIFICATION

Editors

Hans du Buf
University of Algarve, Portugal

Micha M. Bayer
Royal Botanic Garden Edinburgh, UK

World Scientific
New Jersey • London • Singapore • Hong Kong

Published by

World Scientific Publishing Co. Pte. Ltd.
P O Box 128, Farrer Road, Singapore 912805
USA office: Suite 1B, 1060 Main Street, River Edge, NJ 07661
UK office: 57 Shelton Street, Covent Garden, London WC2H 9HE

British Library Cataloguing-in-Publication Data
A catalogue record for this book is available from the British Library.

AUTOMATIC DIATOM IDENTIFICATION

Copyright © 2002 by World Scientific Publishing Co. Pte. Ltd.

All rights reserved. This book, or parts thereof, may not be reproduced in any form or by any means, electronic or mechanical, including photocopying, recording or any information storage and retrieval system now known or to be invented, without written permission from the Publisher.

For photocopying of material in this volume, please pay a copying fee through the Copyright Clearance Center, Inc., 222 Rosewood Drive, Danvers, MA 01923, USA. In this case permission to photocopy is not required from the publisher.

ISBN 981-02-4886-5

Printed in Singapore by Uto-Print

Preface

Computer processing of diatom images used to be an area reserved for a very few specialists. With the advent of the ADIAC project (Automatic Diatom Identification And Classification), funded by the European Marine Science And Technology (MAST) programme, we may say that diatom research has entered a new era. In the future, computers could assist diatom researchers in many different ways: online electronic atlasses could largely replace printed ones, automatic slide scanning could free diatomists from routine microscopy tasks, and the identification of taxa could be carried out automatically or semi-automatically.

This book is actually the final scientific report of ADIAC, the first project ever to attempt diatom identification on a large scale by means of image analysis, using both contour and valve-face features. It also allowed the creation of large image databases, and explored the automatic scanning of microscope slides.

ADIAC was designed as a pilot study, that is, it was intended to explore the subject of automatic diatom identification and to pinpoint problems. Success cannot be bought with research grants, especially when a relatively new area is being explored, when there are only three years available to do the work, and when much of it is done by fresh PhD students. Now, at the end of the project, it is fair to say that our results have exceeded all expectations, and that the project's participants have done a magnificent job. However, much remains to be done before our techniques can be applied in routine practice. We hope that investigators and practitioners in both diatom research and pattern recognition will find this book useful, as it points the way to future research.

Hans du Buf and Micha Bayer

For public databases and online demo see:
http://www.ualg.pt/adiac/
http://www.rbge.org.uk/ADIAC/
http://www.rbge.org.uk/ADIAC/db/adiacdb.htm
http://www.cs.rug.nl/hpci/demos/adiac/

Acknowledgments

The ADIAC project was funded by the European MAST (MArine Science and Technology) programme, contract MAS3-CT97-0122. Additional funding came from the Swiss Federal Office for Education and Science (BBW 98.00.48), the CSIC in Spain and the Spanish programmes III Pricit CAM as well as MAR98-1537CE.

We would also like to thank Annemarie Clarke, Richie Head, Mary Lewis and Kaarina Weckström (Univ. of Newcastle) for providing diatom slides and for their expertise in selecting, identifying and scanning a substantial number diatom specimens for the image databases.

We would like to thank Stephen Droop, formerly Royal Botanic Garden Edinburgh, for his help in planning and developing the ADIAC project in the early stages, and for his participation in the first half of the project. We are also grateful to Dr. Martin Pullan, Royal Botanic Garden Edinburgh, for help in developing the ADIAC diatom database and the web-based search engine, and for continued advice and support in data management issues.

David Mann is grateful to the Royal Society for an equipment grant enabling the purchase of a photomicroscope.

The authors of Chapter 5 would like to thank all the participants that took part in the ring-test and the Internet-based experiments, and so generously gave up much of their precious time; without their contribution these valuable data could not have been generated. Many thanks also to AWEL (Amt für Abfall, Wasser, Energie and Luft), Kanton Zürich (Switzerland), for supporting the ring-test, to Dr. Pius Niederhauser (AWEL) who performed field and laboratory work, and to Kathryn Lyttle (Univ. of Newcastle) who kindly helped analyze the data from the Internet-based experiments.

Gabriel Cristóbal would like to thank J. Chamorro, J. Fernandez-Valdivia, B. Prados and D. Alvarez of Univ. Granada (Spain), and J. I. Uzquiano of CSIC (Spain) for their contributions and support of the work described in Chapter 13.

The editors also gratefully acknowledge the series editor Horst Bunke, who encouraged us to prepare this book, the editors of World Scientific for their valuable help, as well as Luís Santos for LaTeXing Chapters 2 to 5. Last but not least, to all authors: thanks for a timely production.

Authors' affiliations

Authors for correspondence are marked with a †.

Micha M. Bayer, David G. Mann[†]
Royal Botanic Garden Edinburgh (RBGE)
20A Inverleith Row, Edinburgh EH3 5LR, Scotland, UK

Richard J. Telford, Steve Juggins[†]
University of Newcastle-upon-Tyne, Dept. of Geography
Daysh Building, Claremont Road, Newcastle-upon-Tyne
Tyne and Wear NE1 7RU, UK

Bertrand Ludes
Université Louis Pasteur, Inst. de Médecine Légale
11 rue Humann, F-67085, Strasbourg Cedex, France

**Michael H.F. Wilkinson, Andrei C. Jalba, Erik R. Urbach,
Michel A. Westenberg, Jos B.T.M. Roerdink**[†]
Inst. for Mathematics and Computing Science
P.O. Box 800, 9700 AV Groningen, The Netherlands

Stefan Fischer, Horst Bunke[†]
Universität Bern, Inst. für Informatik und angewandte Mathematik
Neubrückstrasse 10, CH-3012 Bern, Switzerland

Jose L. Pech-Pacheco, Gabriel Cristóbal[†]
Instituto de Optica – CSIC
Serrano 121, 28006 Madrid, Spain

**Robert E. Loke, Adrian Ciobanu, Luís M. Santos,
Hamid R. Shahbazkia, Hans du Buf**[†]
Vision Laboratory, Faculty of Sciences and Technology
University of Algarve, Campus de Gambelas, 8000-810 Faro, Portugal

Martyn G. Kelly
Bowburn Consultancy
11 Monteigne Drive, Bowburn, Durham DH6 5QB, UK

Joachim Hürlimann
AquaPlus
Bundesstrasse 6, CH-6300 Zug, Switzerland

Authors' affiliations

Authors for correspondence are marked with a *.

Michla M. Boyer*, David C. Mann
Royal Botanic Garden Edinburgh (RBGE)
20A Inverleith Row, Edinburgh EH3 5LR, Scotland, UK

Richard J. Telford, Steve Juggins
University of Newcastle-upon-Tyne, Dept. of Geography
Daysh building, Claremont Road, Newcastle-upon-Tyne
NE1 7RU, UK

Bertrand Ludes
Institute home Ivoirien, Inst. de Médecine légale
11, rue Humann, F-67085 Strasbourg Cedex, France

Michael H.P. Wilkinson*, Arnaud C. Jalba, Erik R. Urbach,
Michael A. Westenberg, Jos B.T.M. Roerdink,
Inst. Wiskunde en Inform., and Computing Science,
P.O. Box 800, 9700AV Groningen, The Netherlands

Stefan Fischer, Horst Bunke,
Universität Bern, Inst. für Informatik und angewandte Mathematik,
Neubruckstrasse 10, CH-3012 Bern, Switzerland

Jose L. Fech-Pacheco, Gabriel Cristobal*,
Instituto de Óptica – CSIC,
Serrano 121, 28006 Madrid, Spain

Herbert F. bake, Adrian Ciobanu, Luis F. Sánchez,
Harald R. Skjalbsekia, Hans du Buf*,
Vision Laboratory, Faculty of Science and Technology,
University of Algarve, Campus de Gambelas, 8000 SIR Faro, Portugal

Martyn G. Kelly,
Bowburn Consultancy,
11 Monteigne Drive, Bowburn, Durham DH6 5QB, UK

Jonathan Hüfeman,
AquaPlus,
Ruepenstrasse 6, CH-6300 Zug, Switzerland

Contents

Preface		v
Acknowledgments		vii
Authors' Affiliations		ix

1 **Introduction to ADIAC and This Book**   1
   Hans du Buf and Micha M. Bayer

2 **Diatoms: Organism and Image**   9
   David G. Mann

3 **Diatom Applications**   41
   Richard J. Telford, Steve Juggins, Martyn G. Kelly, and Bertrand Ludes

4 **ADIAC Imaging Techniques and Databases**   55
   Micha M. Bayer and Steve Juggins

5 **Human Error and Quality Assurance in Diatom Analysis**   75
   Martyn G. Kelly, Micha M. Bayer, Joachim Hürlimann, and Richard J. Telford

6 **Contour Extraction**   93
   Stefan Fischer, Hamid R. Shahbazkia, and Horst Bunke

7 **Identification Using Classical and New Features in Combination with Decision Tree Ensembles**   109
   Stefan Fischer and Horst Bunke

8 **Identification by Curvature of Convex and Concave Segments**   141
   Robert E. Loke and Hans du Buf

9 **Identification by Contour Profiling and Legendre Polynomials**   167
   Adrian Ciobanu and Hans du Buf

10 **Identification by Gabor Features**   187
   Luís M. Santos and Hans du Buf

11 **Identification by Mathematical Morphology**   221
   Michael H. F. Wilkinson, Andrei C. Jalba, Erik R. Urbach, and Jos B. T. M. Roerdink

12 **Mixed-Method Identifications**   245
   Michel A. Westenberg and Jos B. T. M. Roerdink

13 **Automatic Slide Scanning**   259
   José L. Pech-Pacheco and Gabriel Cristóbal

14 **ADIAC Achievements and Future Work**   289
   Hans du Buf and Micha M. Bayer

Appendix: The Mixed Genera Data Set   299

CHAPTER 1

INTRODUCTION TO ADIAC AND THIS BOOK

HANS DU BUF AND MICHA M. BAYER

1 History and structure of the project

Pattern recognition is as old as image processing. Right from the beginning researchers were motivated by potential applications. Medical research is now unthinkable without computers; pattern recognition plays an important role in, for example, cell analysis for cancer. Police forces around the world use sophisticated software for fingerprint analysis and face recognition. There are numerous other applications of pattern recognition that impact on our daily lives in a more or less direct fashion.

Some eight years ago, when talking to Dr. Simone Servant of the Musée National de l'Histoire Naturelle in Paris, one of us (HdB) learned that shape analysis of diatoms was something vaguely known but only used by very few diatomists. The idea of creating a project on diatoms and image processing then steadily became reality, after establishing contacts with diatomists, among others Stephen Droop at the Royal Botanic Garden Edinburgh. He agreed that computer processing of diatom images was indeed a very interesting topic, and he pointed out application areas and how diatom research could benefit from it. We drafted together two groups of scientists to make up a European consortium of the necessary calibre and geographical spread: one consisting of diatomists from a variety of backgrounds; the other consisting of researchers in pattern recognition and image processing. The partnership of the approved ADIAC project was:

- Stephen Droop, Royal Botanic Garden Edinburgh, after leaving RBGE replaced by David Mann (taxonomy, morphology)
- Steve Juggins, Univ. of Newcastle (climatology, ecology)
- Bertrand Ludes, Univ. Louis Pasteur, Strasbourg (forensics)
- Horst Bunke, Univ. of Bern (character recognition, graph matching)
- Jos Roerdink, Univ. of Groningen (mathematical morphology)
- Gabriel Cristóbal, Inst. de Optica, Madrid (texture analysis, microscopy)
- Hans du Buf, Univ. of Algarve, Faro, project coordinator (texture analysis, visual perception)

These were the "scientific officers" responsible for the work; the full team included, of course, many postdocs and PhD students, most of them now co-author of chapters of this book.

ADIAC, which stands for Automatic Diatom Identification and Classification, started in May 1998 and lasted for three years. The reasoning behind the project title was that, once good algorithms for diatom identification had been developed, these could then be applied in order to corroborate existing taxonomic concepts (n.b. there is a potentially confusing issue of terminology regarding the term "classification," which in biology means the establishing of class-forming rules, but in pattern recognition is used in the same sense as "identification" in biology; in this book, the readership should interpret both of the terms in the context of the respective chapters where they occur). Ultimately, this could not be a realistic goal for ADIAC, because three years is insufficient time for resolving a problem as complex as automatic identification, *and* then applying the solutions to another, even more demanding problem. However, diatomists have been using feature extraction and statistical analyses for some time already, to discriminate between morphologically similar taxa, for example, different demes of *Sellaphora pupula* (see Chapter 4). Some of the methods described in this book have produced extremely good results with a specially created test set of *Sellaphora pupula* specimens, which is encouraging and suggests that some of the new methods presented here are indeed suitable for generating class-forming rules.

One of the facts that helped in obtaining funding for the project were the clearly focused goals, that are directly related to the way diatomists work: in all diatom applications (e.g. water quality monitoring, paleoecology, forensics etc.), microscope slides must be scanned for diatoms and the diatoms must be identified. This is very tedious and repetitive work, and any degree of automation would be a vast improvement over the present state of affairs. Most identifications are done using identification keys and/or comparing specimens on slides to photographs or drawings of diatoms in books and atlasses. This is not a trivial task by any standard: taxonomists estimate that there may be as many as 200,000 different diatom species, half of them still undiscovered (Mann and Droop, 1996), and many of these are extremely hard to distinguish on the basis of morphology. Because of this, diatom identification can involve significant error rates (see Chapter 5), and computer algorithms could be very useful here in order to improve results and to free up time that could be spent doing more important things.

In summary, ADIAC's main objectives were:

1. To create large image databases, including a general database with taxonomic and ecological information (this was implemented as a public WWW database at RBGE), and specific databases for testing the algorithms developed in ADIAC. The latter were intended to contain sufficient images to train classifiers and apply the trained classifiers to unseen samples.

2. To implement and test existing (classical) as well as new feature extraction methods for the characterization of valve contours and ornamentations (e.g. the striation pattern), if possible also classifiers developed specifically for use with diatoms (graph matching). The reason for using classical methods (e.g. morphometric features, Fourier descriptors) in parallel with new methods is that there are numerous feature sets that are already being used in other computer vision applications, and it would be wasteful to not use these, but equally, diatoms are distictive enough from other organisms or objects to require new methods in order to optimize identification.

Chapter 1: Introduction to ADIAC and this book

3. To explore all aspects of the automatic scanning of microscope slides, i.e. particle detection at low/medium magnification, autofocusing, and image capture at high magnification. An important part of this is the discrimination of usable diatom valves and debris. This needs to be done at medium magnification because of the huge increase in processing (CPU) time at higher magnifications. This part of the project was rather specialized: nowadays, most research laboratories can easily afford a scientific-grade CCD camera for microscopy, but equipping a microscope with an accurate XYZ-stage controller may cost as much as a complete set of microscope lenses. However, once slide scanning software has been fully developed, it may be possible that an institute makes one or several systems available centrally for general access by different groups.

There was another topic on the wish-list, that was not included officially in the project, but that has nevertheless been realized before the end of the project: "groundtruthing" by diatomists. This involved experiments in which experts identified the same diatom taxa that were also used for the testing of ADIAC software. This is a very important issue, because the computer results need to be validated by comparing identification rates with those obtained by experts. Computer results need to be comparable, preferably even better, before computer processing can be applied in practice; see Chapter 5.

Although ADIAC was the first project on this particular topic, the results were rather impressive: most methods produced identification rates in excess of 90%, and even outperformed most human experts (see Chapter 5). However, ADIAC was intended as a pilot project, i.e. aimed at pinpointing problems rather than providing final solutions, and there were several problems that, for one reason or another, could not be addressed or fully resolved:

(1) The complexity of automatic contour extraction was clearly underestimated. Automatic contour extraction is critical when large numbers of images need to be processed, and when identification is unsupervised (Chapter 6). However, many diatom images contain inorganic debris or diatom fragments, which may touch or even partly obscure the specimen shown, and therefore automatically extracted contours may be faulty and must be checked visually, and, if necessary, corrected by user intervention. An automatic method for this is still unavailable. A dramatic example for this are images of diatoms from forensic samples (Chapter 4), which are usually badly contaminated with ash from the burning of organic material; here, existing contour extraction methods regularly produce deformed contours. This problem also extends to samples from sediment cores which may show a similar extent of contamination with debris. Here, much additional work is necessary, and in Chapter 10 we show that methods for the detection of striae on a valve could possibly be used to guide the contour extraction in the future.

(2) The suggested work on graph matching could not be carried out because the implementation of feature extraction methods took up more time than expected, but, regardless of this, the results obtained with normal classifiers were more than adequate. However, graph matching is now one of the priorities of ongoing PhD projects.

(3) The slide scanning and autofocusing work has improved on existing methods (Chapter 13), but one problem turned out to be far more difficult than anticipated:

particle detection at low magnification works well, but following this, usable diatoms must be selected, and debris and broken or slanting valves rejected. Currently about 80% of usable valves can be detected, and further work in this area is therefore necessary to improve the result.

Overall, ADIAC's achievements may look impressive, but the results presented in this book are initial, not final results. There is still much room for improvements, and other groups could—and should—contribute to this since most of the data are publicly available.

2 Research in biology and diatoms

ADIAC was, of course, not the first project on image analysis in biology. Automatic identification by means of image analysis has been explored for a wide variety of major taxonomic groups, including bacteria (Dubuisson et al., 1994), blue-green algae (Thiel et al., 1995), fungi (Dorge et al., 2000), dinoflagellates (Ishii et al., 1987; Estep and Macintyre, 1989; Simpson et al., 1992; Culverhouse et al., 1996; Pech-Pacheco et al., 1999), coccoliths (Brechner and Thierstein, 1999), foraminifera (Yu et al., 1996), mixed plankton samples (Jeffries et al., 1984; Gorsky et al., 1989), nematodes (Sommer, 1996; Theodoropoulos et al., 2000), insects (Yu et al., 1992; Dietrich and Pooley, 1994; Batra, 2001), as well as pollen (France et al., 1997), seeds (Cullen-Refai et al., 1988; Petersen and Krutz, 1992) and whole plants (Gerhards et al., 1993; Guyer et al., 1986). This list is not exhaustive, and there are many other studies that address this problem. However, most of the studies mentioned above were limited to a small number of taxa, from a few to several tens, and generally addressed specific problems where only the few taxa under consideration were critical, for example the identification of toxic plankters (Culverhouse et al., 1996), or agricultural weeds (Gerhards et al., 1993). Unfortunately, the general pattern with work carried out on image analysis in biology appears to be that solid, promising work has often been abandoned once the funding for a particular project ran out. To the best of our knowledge, none of the systems described in the literature has been developed into a finished, distributable computer program that other researchers can actually use.

Diatoms, too, have been the subject of research in this area. The automation of diatom analysis has been explored as early as the 1970s and '80s (Cairns et al., 1977; 1979; 1982), but unfortunately this work has failed to make a lasting impact on diatom research. There could have been a number of reasons for this, but the most likely is that the hardware involved was too specialized and possibly too expensive for the wider diatomist community. This prototype system was based on laser holography and optical correlation, and matched diatom template images to those generated by a laser beam on a microscope.

The combined potential of image analysis and multivariate statistics was recognized in the 1980s, and in a number of studies this methodology has been used for diatom taxonomy. All of these studies used shape analysis, a powerful method for the discrimination of diatom taxa, and employed as shape descriptors either Legendre polynomials (Stoermer and Ladewski, 1982; Stoermer et al., 1986; Theriot and Ladewski, 1986; Steinman and Ladewski, 1987; Goldman et al., 1990; Rhode et al., 2001), Fourier coefficients (Mou and Stoermer, 1992), or simpler shape measures such as rectangularity, combined with "traditional" diatom characters such as stria density and size measurements (Droop, 1994, 2000). All of the studies show that the extraction of outline/shape features, combined with multivariate analysis methods such as principal component analysis, is a very powerful tool for resolving the subtle morphological variation in diatoms, at the level of species and beyond. This is an important pre-requisite for automating identification.

However, all of the previous studies were concerned with shape analysis of very few taxa. ADIAC has taken this work a step further, in that (1) it used ornamentation (striation) characteristics as well as shape features, (2) the goal was primarily identification, rather than classification, and (3) large scale tests involving many taxa were carried out. Furthermore, software for specimen detection on microscope slides is novel, and even a rather limited (semi-)automatic tool could save a very significant amount of labor.

3 The chapters

This book has two very different audiences: (1) diatomists and other phycologists interested in state-of-the-art computer processing, and (2) researchers in pattern recognition interested in a novel application. We tried to do both audiences justice, and to make it possible for each to grasp what the other group is talking about. This was not an easy task, but we tried to include sufficient information so that both groups will find this book useful. We also hope that the book will foster future collaborations to help advance this challenging new topic. There is a definite need to specify procedures in detail so that future researchers can continue the work, and this was one of the main reasons for writing this book.

The core chapters in this book describe the work done during the project, and several additional chapters introduce the necessary diatom "background." The structure of the book is as follows:

Chapter 2 gives a general introduction to diatoms, and covers their biology, morphology, taxonomy, ecology, as well as classification (class-forming rules) and identification.

Chapter 3 presents a general introduction to diatom applications, such as paleoecology, water quality monitoring, and forensics.

Chapter 4 describes imaging techniques (image acquisition, preprocessing) and image databases, including image sets for the testing of identification algorithms. It also describes the online diatom web browser at RBGE.

Chapter 5 looks at human error in diatom identification, and presents results from two unique studies which were aimed at quantifying error rates in identification. The data from these experiments are intended as a benchmark for computer identification rates in the future.

Chapter 6 focuses on automatic contour extraction, the first important step towards unsupervised identification, and also a requirement for the processing of large numbers of images.

Chapter 7 is the first of the chapters on automatic identification. It describes some "classical" contour feature sets (morphometric features, moment invariants,

Fourier descriptors), as well as texture features for the characterization of striation patterns. It introduces decision-tree classifiers, which are also used in other chapters. The identification rates presented here can be viewed as baseline data that challenge more recent feature sets described in other chapters.

Chapter 8 describes an identification method in which a contour is segmented into convex, concave and straight parts, and in which length as well as curvature features are computed from the segments. In this chapter one of the simplest classifiers is used (Bayes minimum distance, or nearest mean).

Chapter 9 introduces an improved version of the characteristic profile of contours, based on dynamic ellipse fitting, and it also deals with Legendre polynomials. This is the only instance in the book in which a hand-optimized syntactical classifier is used, along with other classifiers.

Chapter 10 is devoted to Gabor analysis, which was originally inspired by image processing in the visual system of primates. Using the 1D characteristic profile of contours introduced in Chapter 9, an almost continuous Gabor filter scaling is applied, and events which are stable in scale space are extracted. The striation analysis applies computational models of cortical grating and bar cells, and density and orientation statistics are computed.

Chapter 11 is about the state-of-the-art in mathematical morphology. Curvature scale space analysis is used for diatom contours, whereas striation patterns are described by size and shape distributions.

Chapter 12 describes protocols and results for mixed-method identifications, in which the individual feature sets from the previous chapters are combined. It shows that good results can be obtained by carefully selecting a small subset of robust features from the entire set. This chapter also introduces the interactive identification web-demo based at the University of Groningen. (N.B. Because this chapter was prepared using preliminary data sets, i.e. before the development work on algorithms was completely finished, the results may differ from those presented in other chapters.)

Chapter 13 is about automatic slide scanning and autofocusing, techniques which aim at automating the process of specimen location and image capture.

Chapter 14 summarizes ADIAC's results and achievements, and presents ideas for future improvements and developments.

References

Batra, S.W.T. (2001) Automatic image analysis for rapid identification of africanized honey bees. In: Africanized honey bees and bee mites, G.R. Needham (ed.), Ellis Horwood, Chichester, pp. 260-263.

Brechner, S. and Thierstein, H.R. (1999) Classifying microfossils: detecting symmetry versus neural networks. Schriftenreihe Österr. Computergesellschaft, Vol. 130, pp. 181-192.

Cairns, J.J., Almeida, S.P. and Fujii, H. (1982) Automated identification of diatoms. Bioscience, Vol. 32, pp. 98-102.

Cairns, J.J., Dickson, K.L. and Slocomb, J. (1977) The ABC's of diatom identification using laser holography. Hydrobiologia, Vol. 54, pp. 7-16.

Cairns, J.J., Dickson, K.L., Pryfogle, P., Almeida, S.P., Case, S.K., Fournier, J.M. and Fujii, H. (1979). Determining the accuracy of coherent optical identification of diatoms. Water Resources Bulletin, Vol. 15, pp. 1770-1775.

Cullen-Refai, A., Faubion, J. and Hoseney, R. (1988) Identification of wheat-varieties using digital image-analysis. Cereal Foods World, Vol. 33, pp. 669-669.

Culverhouse, P.F., Simpson, R.G., Ellis, R., Lindley, J.A., Williams, R., Parisini, T., Reguera, B., Bravo, I., Zoppoli, R., Earnshaw, G., McCall, H. and Smith, G. (1996) Automatic classification of field-collected dinoflagellates by artificial neural network. Marine Ecology Progress Series, Vol. 138, pp. 281- 287.

Dietrich, C.H. and Pooley, C.D. (1994) Automated identification of leafhoppers (Homoptera, Cicadellidae, Draeculacephala Ball). Annals Entomological Society of America, Vol. 87, pp. 412-423.

Dorge, T., Carstensen, J.M. and Frisvad, J.C. (2000) Direct identification of pure *Penicillium* species using image analysis. J. Microbiological Methods, Vol. 41, pp. 121-133.

Droop, S.J.M. (1994) Morphological variation in *Diploneis smithii* and *D. fusca* (Bacillariophyceae). Archiv für Protistenkunde, Vol. 144, pp. 249-270.

Droop, S.J.M. (2000) Spatial and temporal stability of demes in *Diploneis smithii/D.fusca* (Bacillariophyta) supports a narrow species concept. Phycologia, Vol. 39, pp. 527-546.

Dubuisson, M.-P., Jain, A.K. and Jain, M.K. (1994) Segmentation and classification of bacterial culture images. J. Microbiological Methods, Vol. 19, pp. 279-295.

Estep, K.W. and MacIntyre, F. (1989) Counting, sizing, and identification of algae using image-analysis. Sarsia, Vol. 74, pp. 261-268.

France, I., Duller, A.W.G., Lamb, H.F. and Duller, G.A.T. (1997) A comparative study of approaches to automatic pollen identification. Proc. British Machine Vision Conf., pp. 340-349.

Gerhards, R., Nabout, A., Sokefeld, M., Kuhbauch, W. and Eldin, H. (1993) Automatic identification of 10 weed species in digital images using Fourier descriptors and shape-parameters. J. Agronomy and Crop Science / Zeitschrift für Acker und Pflanzenbau, Vol. 171, pp. 321-328.

Goldman, N., Paddock, T.B.B. and Shaw, K.M. (1990) Quantitative analysis of shape variation in populations of *Surirella fastuosa*. Diatom Research, Vol. 5, pp. 25-42.

Gorsky, G., Guilbert, P. and Valenta, E. (1989) The autonomous image analyzer – enumeration, measurement and identification of marine-phytoplankton. Marine Ecology Progress Series, Vol. 58, pp. 133-142.

Guyer, D., Miles, G., Schreiber, M., Mitchell, O. and Vanderbilt, V. (1986) Machine vision and image-processing for plant-identification. Trans. of the ASAE, Vol. 29, pp. 1500-1507.

Ishii, T., Adachi, R., Omori, M., Shimizu, U. and Irie, H. (1987) The identification, counting, and measurement of phytoplankton by an image-processing system. J. du Conseil, Vol. 43, pp. 253-260.

Jeffries, H.P., Berman, M.S., Poularikas, A.D., Katsinis, C., Melas, I., Sherman, K. and Bivins, L. (1984) Automated sizing, counting, and identification of zooplankton by pattern recognition. Marine Biology, Vol. 78, pp. 329-334.

Mann, D.G. and Droop, S.J.M. (1996) Biodiversity, biogeography and conservation of diatoms. Hydrobiologia, Vol. 336, pp. 19-32.

Mou, D. and Stoermer, E.F. (1992) Separating *Tabellaria* (Bacillariophyceae) shape groups: A large sample approach based on Fourier descriptor analysis. J. Phycology, Vol. 28, pp. 386-395.

Pech-Pacheco, J.L., Alvarez-Borrego, J., Orellana-Cepeda, E. and Cortes-Altamirano, R. (1999) Diffraction pattern applicability in the identification of *Ceratium* species. J. Plankton Research, Vol. 21, 1455-1474.

Petersen, P.E.H. and Krutz, G.W. (1992) Automatic identification of weed seeds by color machine vision. Seed Science and Technology, Vol. 20, pp. 193-208.

Rhode, K.M., Pappas, J.L. and Stoermer, E.F. (2001) Quantitative analysis of shape variation in type and modern populations of *Meridion* (Bacillariophyceae). J. Phycology, Vol. 37, pp. 175-183.

Simpson, R., Williams, R., Ellis, R. and Culverhouse, P.F. (1992) Biological pattern recognition by neural networks. Marine Ecology Progress Series, Vol. 79, pp. 303-308.

Sommer, C. (1996) Digital image analysis and identification of eggs from bovine parasitic nematodes. J. Helminthology, Vol. 70, pp. 143-151.

Steinman, A.D. and Ladewski, T.B. (1987) Quantitative shape analysis of *Eunotia pectinalis* (Bacillariophyceae) and its application to seasonal distribution patterns. Phycologia, Vol. 26, pp. 467-477.

Stoermer, E.F. and Ladewski, T.B. (1982) Quantitative analysis of shape variation in type and modern populations of *Gomphoneis herculeana*. Nova Hedwigia, Beiheft 73, pp. 347-386.

Stoermer, E.F., Qi, Y.-Z. and Ladewski, T.B. (1986) A quantitative investigation of shape variation in *Didymosphenia* (Lyngbye) M. Schmidt (Bacillariophyta). Phycologia, Vol. 25, pp. 494-502.

Theodoropoulos, G., Loumos, V., Anagnostopoulos, C., Kayafas, E. and Martinez-Gonzales, B. (2000) A digital image analysis and neural network based system for identification of third-stage parasitic strongyle larvae from domestic animals. Computer Methods and Programs in Biomedicine, Vol. 62, pp. 69-76.

Theriot, E. and Ladewski, T.B. (1986) Morphometric analysis of shape of specimens from the neotype of *Tabellaria flocculosa* (Bacillariophyceae). American J. Botany, Vol. 73, pp. 224-229.

Thiel, S.U., Wiltshire, R.J. and Davies, L.A. (1995) Automated object recognition of blue-green algae for measuring water quality – a preliminary study. Water Research, Vol. 29, pp. 2398-2404.

Yu, D., Kokko, E., Barron, J., Schaalje, G. and Gowen, B. (1992) Identification of ichneumonid wasps using image-analysis of wings. Systematic Entomology, Vol. 17, pp. 389-395.

Yu, S., Saintmarc, P., Thonnat, M. and Berthod, M. (1996) Feasibility study of automatic identification of planktic foraminifera by computer vision. J. Foraminiferal Research, Vol. 26, pp. 113-123.

CHAPTER 2

DIATOMS: ORGANISM AND IMAGE

DAVID G. MANN

Diatoms are a large and ecologically important group of unicellular or colonial algae. They are characterized by their unique, bipartite, highly patterned and perforate cell wall, which is composed largely of hydrated amorphous silica. Each half of the cell wall consists of a valve and a number of girdle bands. One half is slightly larger than the other and overlaps it. Together, the halves make a cylinder, with the two valves at the ends. The cross section of the cylinder, and hence the outline of the valve, varies greatly in shape between species and genera. This, together with the pattern of pores and other markings on the valve, provides the information needed for species identification, although it is only part of the information used by diatomists to produce evolutionary classifications. The identification and classification are made more difficult by the curious life cycle of diatoms, in which a slow reduction in size, accompanied by changes in shape, alternates with a much more rapid expansion.

1 General introduction

There are two major groups of plants, one dominating the land, the other the water. On land, seed plants generally prevail. The principal group of seed plants—the angiosperms (flowering plants)—is very diverse, with over 270,000 species already discovered and named (Groombridge and Jenkins, 2000). Aquatic habitats, on the other hand, are dominated by algae. The most familiar of these are the large red and brown seaweeds found along rocky shores. However, in terms of the contribution they make to global cycles of carbon, oxygen, nitrogen, phosphorus, silicon and other biologically significant elements, these large seaweeds are insignificant, compared to the microscopic algae (phytoplankton) growing suspended in the upper layers of the open ocean. Several groups of microscopic algae are abundant in marine phytoplankton, including coccolithophorids, dinoflagellates and picoplanktonic cyanobacteria, but the diatoms dominate in the more nutrient-rich and productive parts of the ocean. Diatoms may be responsible for nearly 20% or more of the net primary carbon production worldwide, fixing 20 Pg per year out of a global total of 105 Pg (Mann, 1999), which is more than all the world's rainforests combined (Field et al., 1998).

This massive contribution to the carbon cycle is achieved by relatively few species (Hasle and Syvertsen, 1996): the phytoplankton comprises a small group of specialized species, adapted to cope with the special problems of living in suspension in the surface veneer (10^2–10^3 m) of the oceans. There are probably only a few thousand planktonic diatom species, represented by vast numbers of individuals. In the benthos, however—both in freshwater and the sea—diatom evolution has run riot, producing a bewildering array of species. There are probably about 200,000 in total, most of which have yet to be described or named (Mann and Droop, 1996; Mann, 1999). Many of them are relatively unimportant in terms of the global carbon cycle, but they can be of great significance for detecting environmental change over timescales of 10s to millions of years (Stoermer and Smol, 1999; see also Chapter 3). It is the mismatch between the plethora of diatom species, on the one hand, and the very small and dispersed group of people capable of identifying them, on the other (together with the fact that the huge literature is uncollated and of very uneven quality, and with the tedium of identification by eye), that prompts projects like ADIAC.

1.1 What is a diatom?

Diatoms are unicellular or colonial algae; they are regarded as a class (Bacillariophyceae) or phylum (Bacillariophyta) of plants (Round et al., 1990; van den Hoek et al., 1995). The special feature of the diatoms, which makes them easily recognizable, is the unique structure of their cell wall, which is composed of pieces of highly patterned and perforated hydrated silica ($SiO_2.nH_2O$). Other organisms metabolize silicon and produce silica structures—some chrysophyte algae produce ornate scales; radiolaria and silicoflagellates produce silica endoskeletons (van den Hoek et al., 1995)—but none produce a bipartite, perforate silica wall. Diatoms construct the wall internally, through condensation of orthosilicic acid [$Si(OH)_4$] within a special vesicle lying just beneath the cell membrane. Wall formation has been studied in detail in several diatoms (Pickett-Heaps et al., 1990) and there has recently been exciting progress in our understanding of the biochemistry of silica precipitation: polyamines and lysine-rich polypeptides (silaffins) promote silica precipitation and control the morphology of the silica produced (e.g. Kröger et al., 2000, 2001). Diatom silicification mechanisms are interesting to materials chemists, since they may lead to new industrial processes for synthesizing ceramics and new composite materials at ambient temperatures (e.g. Brott et al., 2001). Nevertheless, we are still very far from understanding how the intricate micropatterning of the silica shell is controlled, in which repeating elements are spaced at intervals of about 0.1–2 μm.

Individual diatom cells are usually invisible to the naked eye or may appear as tiny brown spots. Most species measure 10–200 μm in their largest dimension and their discovery was therefore delayed until the invention of microscopes. Ford (1991) has shown that material collected by Leeuwenhoek in the 1680s contains diatoms, but although Leeuwenhoek discovered many types of aquatic organisms, including bacteria, and must have seen diatoms (the capacity of his lenses to resolve the detail of diatom cell walls is demonstrated conclusively by Ford), there is nothing in Leeuwenhoek's publications that can be proved to refer to diatoms. Credit for the discovery of diatoms therefore goes to an English gentleman (Anonymous, 1703), who made unmistakable drawings of the freshwater diatom *Tabellaria* (Round et al., 1990). Initially, misled by the ability of many benthic diatoms to move, some biologists, notably Ehrenberg (e.g. 1838), considered that diatoms were animals, with digestive systems and sex organs, but later there was (e.g. Pritchard, 1861) a consensus that diatoms are plants. Since then, especially since the acceptance of the endosymbiosis theory, which postulates that the organelles responsible for photosynthesis in eukaryotic plants (chloroplasts) arose through the ingestion and modification of photosynthetic bacteria—it has become clear that the boundaries between "plant," "animal" and "fungus" are almost meaningless in terms of phylogeny (e.g. van den Hoek et al., 1995): these categories indicate how organisms function physiologically, but they do not reflect evolutionary relationships. This reopened the question of where diatoms should be classified.

The relationship between diatoms and other living organisms has been clarified by electron microscopy and by molecular biological techniques that allow the direct analysis of genes. Comparisons between different algae and protozoa on the basis of nucleotide sequences in genes coding for proteins or ribosome subunits (e.g. Saunders et al., 1995; Medlin et al., 1997) reveal that diatoms belong to a group variously called "Heterokonta" (Cavalier-Smith, 1986), "stramenopiles" (Patterson, 1989) or "Heterokontophyta" (van den Hoek et al., 1995). This large group includes organisms as different as the giant kelps of rocky seashores (brown algae, such as *Laminaria* and *Sargassum*); fungal pathogens such as *Phytophthora infestans* (the cause of late blight of potatoes); toxic flagellates like *Chattonella*, occurring in the marine phytoplankton; the dark-green velvety growths of *Vaucheria* on salt marsh muds; and diatoms (van den Hoek et al., 1995; Graham and Wilcox, 2000). The heterokonts form a group that is quite distinct from other eukaryote groups, such as the green plants (including the green algae, as well as mosses, ferns, flowers and trees), red algae (Rhodophyta), animals (Metazoa) and the alveolates (a complex, diverse group containing the dinoflagellates, ciliate protozoa, and apicomplexans). Thus, although it is convenient to refer to diatoms as "algae," indicating that they are simple photosynthetic organisms that use chlorophyll to harvest light energy to drive the fixation of carbon dioxide into sugars, generating oxygen in the process, it must be remembered that this does not imply that diatoms are closely related to other "algae;" for the most part they are not. "Algae" have evolved many times, quite independently, through the formation of novel chimaeric organisms via the symbiosis of a variety of photosynthetic cells within different heterotrophic host cells (van den Hoek et al., 1995; Graham and Wilcox, 2000).

Few morphological or structural characteristics are common to all Heterokontophyta: the adaptation of heterokonts to many, very different types of environment has disguised the phylogenetic relationships between them. When motile cells are formed in Heterokontophyta, they often have two flagella and, if so, one flagellum is smooth whereas the other bears complex stiff tubular hairs ("hetero-kont" = possessing different punting poles!). The hairs function to reverse the thrust of the flagellum (Holwill and Taylor, 2000). In diatoms, the only flagellate cells are the sperm. Other characteristics shared by photosynthetic Heterokontophyta, including the diatoms, are the presence of four membranes around each chloroplast; the presence of chlorophyll c (c_1, c_2 or c_3); a tendency for the chlorophylls (which are green) to be masked by relatively large amounts of yellow or orange carotenoid pigments, which, like the chlorophylls, are involved in harvesting light energy; and grouping of the thylakoids into triplets (Jeffrey, 1989; Björnland and Liaaen-Jensen, 1989; van den Hoek et al., 1995).

1.2 Ecology

Diatoms are almost ubiquitous in aquatic habitats, both in freshwater and in the sea. Diatoms are abundant in the plankton, particularly when the water column is turbulent and nutrient-rich. If the water column becomes stratified, so that only a shallow surface layer is mixed by convection and wind stress, diatoms are frequently at a disadvantage because of their relatively high density (e.g. Reynolds, 1984), caused by the silica cell wall.

Most diatom species live in the benthos, i.e. on rock, gravel, sand, silt, plant and animal surfaces at the bottom of lakes, rivers and seas. The vast majority of diatoms are photosynthetic organisms and they are therefore restricted to relatively shallow habitats (at most a few hundred meters' depth). Many grow attached to solid substrata by stalks or pads of polysaccharide, and some form polysaccharide tubes, within which they move and multiply. Attached algae can be classified by the kind of substratum: the community of algae attached to rocks is termed "epilithon," to plants "epiphyton," and to sand grains "epipsammon." Sediments are often colonized by very diverse communities of highly mobile diatoms and other algae, which are usually referred to as "epipelon." Epipelic diatoms often exhibit rhythms of vertical migration and are important in stabilizing sediments, protecting them from erosion (Paterson et al., 1998; Paterson and Black, 1999; Paterson and Hagerthey, 2001). Some diatoms live on land, on soils or mosses or rock faces, but they probably do not grow unless covered by a film of water. An overview of microalgal ecology is given by Round (1981).

Individual species have particular habitat requirements. Marine species cannot generally grow in freshwater, nor *vice versa*; some species are restricted to acid habitats, some to alkaline; and so on. Knowing these preferences is essential if diatoms are to be used in ecological and palaeoecological monitoring (see Chapter 3).

2 Introduction to the diatom cell

2.1 The protoplast

The diatom cell consists of the living protoplast itself and the cell wall that surrounds it (Fig. 1a). The protoplast contains all the organelles that are characteristic of eukaryotic plants, including a nucleus, Golgi apparatus, mitochondria and chloroplasts. The cell is generally highly vacuolate and the cytoplasm and most organelles are often restricted to the periphery of the cell; the nucleus generally lies at or near the center (Fig. 1b). Diatom chloroplasts are brown, gold or greenish yellow, with large amounts of carotenoids (principally β,β-carotene, fucoxanthin, diatoxanthin and diadinoxanthin) disguising the chlorophylls, which are chls a and c (Björnland and Liaaen-Jensen, 1989; Jeffrey, 1989).

In planktonic species, there are usually many small chloroplasts. Benthic species, on the other hand, often have fewer (1, 2 or 4) chloroplasts, which are much larger and more elaborate, and undergo complex rearrangements and shape changes during the cell cycle (Mann, 1996); in these diatoms, chloroplast morphology is generally constant enough within species or genera for its characteristics to be useful both for classification (e.g. Round et al., 1990; Mann, 1996) and identification (Cox, 1996). The genus *Sellaphora*, for example, is defined largely by its protoplast structure and also by its mode of sexual reproduction (Mann, 1989). One or more pyrenoids are usually present within each chloroplast and these often have taxon-specific positions and morphology.

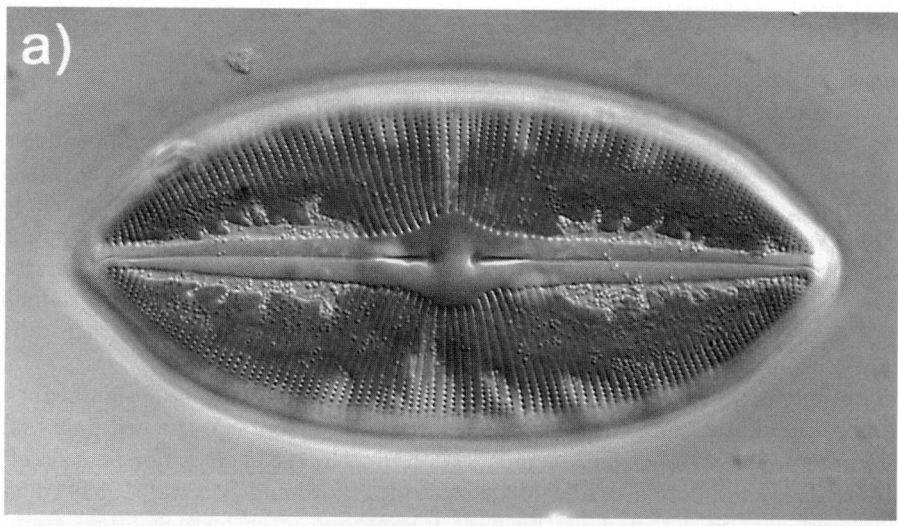

Figure 1. *Petroneis latissima*, light microscopy of a live cell. a: Focus close to one of the valves, showing the slightly radiate striae and the two longitudinal slits of the central raphe system, all underlain by a complex lobed chloroplast (darker areas). b: Median focus, showing the central nucleus surrounded by a large vacuole, which fills the cell lumen apart from a thin peripheral layer of cytoplasm containing organelles: lobes of the chloroplast are visible in section at the top and bottom.

2.2 The cell wall

The cell wall of diatoms has a unique structure and a special, complex terminology has evolved to deal with it (e.g. von Stosch, 1975, 1981; Ross et al., 1979). Only the basic terms will be used here.

2.2.1 Organic components of the cell wall

The cell wall is usually composed predominantly of silica but it also contains organic material, in the form of discrete layers coating or underlying the silica elements, and components enclosed within the silica itself, e.g. silaffins and polyamines involved in promoting silica precipitation. The organic components include cell-wall–specific proteins, e.g. the frustulins (Kröger et al., 1996, 1997), which may play a role in wall morphogenesis, as well as lipids and polysaccharides (data summarized by Round et al., 1990). The functions of most organic components are poorly understood. Some probably protect the silica elements from dissolution, others link the silica elements together, and some may be involved in cell–cell recognition, e.g. during sexual reproduction.

The only organic component with any load-bearing structural significance is probably the diatotepum, which forms a discrete layer outside the cell membrane, immediately beneath the silica elements (von Stosch, 1981). It is not present in all diatoms. It is composed of acidic polysaccharides (Ford and Percival, 1965; Volcani, 1978), and can be detected in the light microscope after histochemical staining (e.g. Liebisch, 1928, 1929). In a few diatoms, the diatotepic layer is the principal turgor-resisting component of the cell wall (Borowitzka et al., 1977; Borowitzka and Volcani, 1978; von Stosch, 1981; von Stosch and Reimann, 1970). In some diatoms, the silica envelope is partly separated from the diatotepum, and hence from the protoplast, by wide spaces (von Stosch, 1977; Schnepf et al., 1980).

2.2.2 Silica components of the cell wall: the frustule

The silica components of the cell wall are collectively called the "frustule"; it is this that is left when diatom cells are cleaned with concentrated acids or other oxidizing agents to remove all organic matter, which is the traditional first step in any study that requires diatoms to be identified to species level. From Section 2.2.1 it will be clear that the frustule is *not* the whole of the cell wall, and in a few cases it is not even the predominant part of the cell wall. Furthermore, the cell wall is now only one of several important sources of characters for use in classification and identification (Section 2.1): the form, position and replication of the chloroplasts (Mann 1996), the position and behavior of the nucleus (Mann and Stickle, 1988; Round et al., 1990), the development of the cell wall (Mann, 1984; Cox, 1999), the method of sexual reproduction (Mann, 1989; Mann and Stickle, 1995), and, recently, molecular genetic data (Medlin et al., 1996; Lundholm et al., 2001; Mann et al., 2001) are all important in diatom systematics. However, from around 1850 until 1980, diatom taxonomy was almost wholly dependent on the characteristics of the frustule, and in many important diatom applications, especially in palaeoecological studies (e.g. of climate change or environmental degradation) or stratigraphy, the frustule is the only part of the diatom available for study.

The frustule is a composite structure and can be very complex in shape and organization; in extreme cases, it can comprise tens of separate elements. However, it is basically a cylinder, which can be circular in Section, elliptical, almost linear (Fig. 2), triangular or square, or so complex as to be beyond verbal description. The two ends of the cylinder, which may be flat-topped or highly elaborate, are each made of a single piece of silica. These end-pieces, called "valves," are usually the largest of the silica elements of the frustule. The sides of the cylinder are formed by the down-turned edges of the valves (referred to as the "mantles," to distinguish them from the "valve face") and by several to many strips of silica, called "girdle bands," which together form the "girdle." Closer examination reveals that the cylinder is in fact bipartite, consisting of two halves that overlap at or near the center of the girdle: the girdle bands fall into two discrete, overlapping series ("cingula"), each associated with one of the valves (Fig. 3). We can therefore distinguish an outer half-frustule, referred to as the "epitheca," consisting of epivalve and epicingulum, and an inner half-frustule, the "hypotheca," consisting of hypovalve and hypocingulum. Like any cylinder, the frustule presents two principal aspects—its end view ("valve view") and side view ("girdle view"). If the valves are not circular (which is often the case: Fig. 2), there will be more than one "girdle view." The girdle cylinder is sometimes curved, so that the valves lie at an angle to each other.

The bipartite structure of the frustule is related to the way diatoms grow and divide (Section 2.3). The morphology of diatom valves is explored in Section 3, in connection with identification.

2.3 The cell cycle and cell division

In eukaryotic cells, there is a series of fairly well differentiated phases between one cell division and the next, which comprise the "cell cycle." Typically, after cell division, there is initially a phase of growth in volume and mass (the "G1 phase"), which is taken to end when the DNA present in the nucleus is replicated (the "S phase"). Then there is another growth phase ("G2"), which is ended by the "M phase," comprising nuclear division (mitosis) and generally also cell division (cytokinesis). During the G1, S and G2 phases, the volume and content of the cell must roughly double, if cell size is not to decrease or increase in successive cell generations. The cell cycle follows this pattern in diatoms (e.g. Olson et al., 1986a), although very few species have been studied in detail.

Increases in cell volume during the G1, S and G2 phases cannot be accommodated by stretching the frustule elements: unlike cell walls that are made of cellulose microfibrils or other polysaccharides, the silica valves and girdle bands cannot be stretched, though they can be bent (further discussion by Mann, 1994). Instead, expansion is made possible through the sliding apart of the epitheca and hypotheca and the coordinated addition of new girdle bands to the edge of the hypotheca (the protoplast at all times remains covered by the cell wall). The frustule cylinder thus grows at its center, not at the ends, and only through the growth of the hypotheca. Growth may be steady throughout the cell cycle, or may take place in one or more

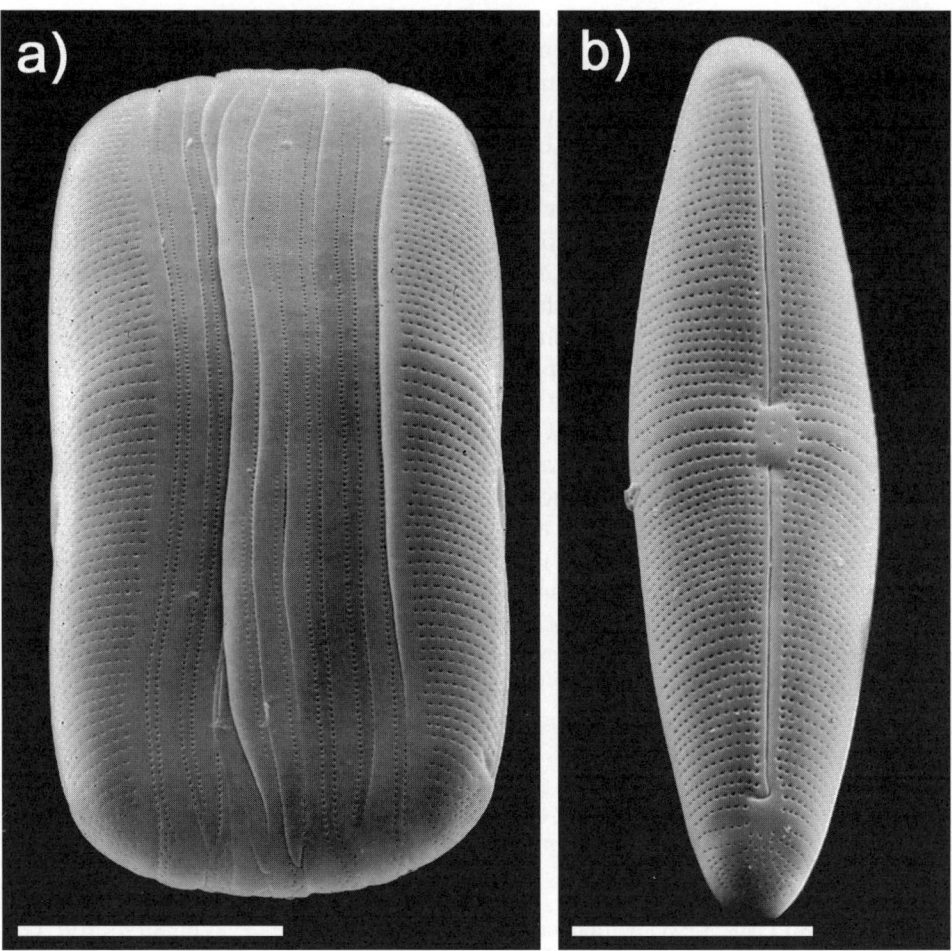

Figure 2. The raphid pennate diatom *Parlibellus*, scanning electron micrographs (SEM). a: Girdle view of a whole frustule, showing the two valves (in profile, left and right), linked by girdle bands. The slightly larger epitheca (see Fig. 3) is to the right. b: A single valve in face view. In this case the valve outline, corresponding to the cross Section of the girdle cylinder, is highly elongate. The valve is pierced by numerous pores in transverse rows, and by two longitudinal raphe slits, which are involved in cell movement. Scale bars 10 μm.

discrete phases (Olson et al., 1986b; see also Round et al., 1990); the amount of expansion during the cell cycle appears to be under strict control.

During the M phase, the nucleus divides and then the protoplast constricts into two within the intact cell wall (Fig. 4c, d). Each daughter cell then produces a new hypotheca, while it is still contained within the wall of the parent cell (Fig. 4a, b). This process has been studied in detail (e.g. Pickett-Heaps et al., 1990), especially the development of the new valves. The formation of each valve—including both the production of the pattern and the deposition of silica—occurs outwards from a pattern center (Round et al., 1990; Pickett-Heaps et al., 1990) and is complete

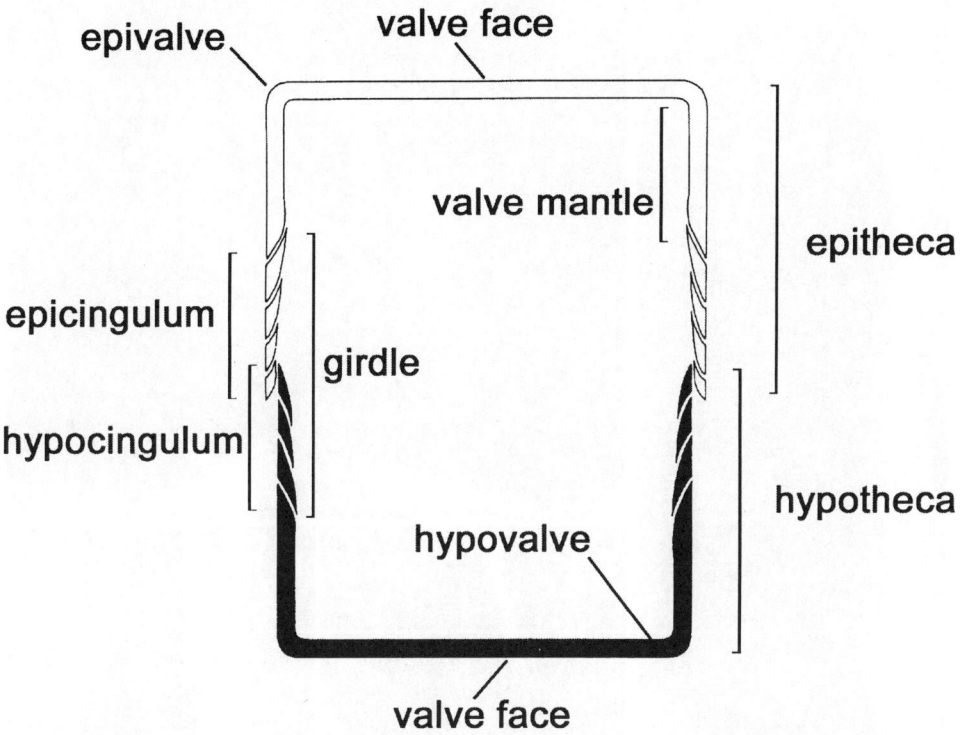

Figure 3. Diagrammatic section through a frustule, showing the relationship between the different elements. The epitheca is slightly larger and overlaps the hypotheca. The hypotheca was formed at the immediately preceding cell division, whereas the epitheca was formed earlier—perhaps many cell divisions previously. Note that the hypotheca in this case contains fewer girdle bands than the epitheca. Its complement of bands will be completed before the next cell division, as the epitheca and hypotheca slide apart to accommodate growth.

within a few minutes to several hours, after which no further material is added to the valve, even though it may remain part of the wall of a living cell for days or months. At some stage after the completion of the new valves, the two daughter cells usually separate from each other. One daughter cell inherits the epitheca of the parent cell and the new theca it has produced becomes the new hypotheca (Fig. 4a, left daughter cell). The other daughter cell inherits the parental hypotheca, which becomes its own epitheca and overlaps the new hypotheca. Thus, in any diatom cell, the two thecae are of different ages. The hypotheca will always have been formed just after the last cell division; the epitheca, on the other hand, may have been produced days or weeks earlier, when environmental conditions may have been quite different. Comparisons between the epitheca and hypotheca are thus a useful check on how reliable morphological characteristics of the frustule are for taxonomy. In a few cases, marked variation has been found between the two thecae of a single frustule. Such "Janus" cells are quite rare, however, and many characteristics of the

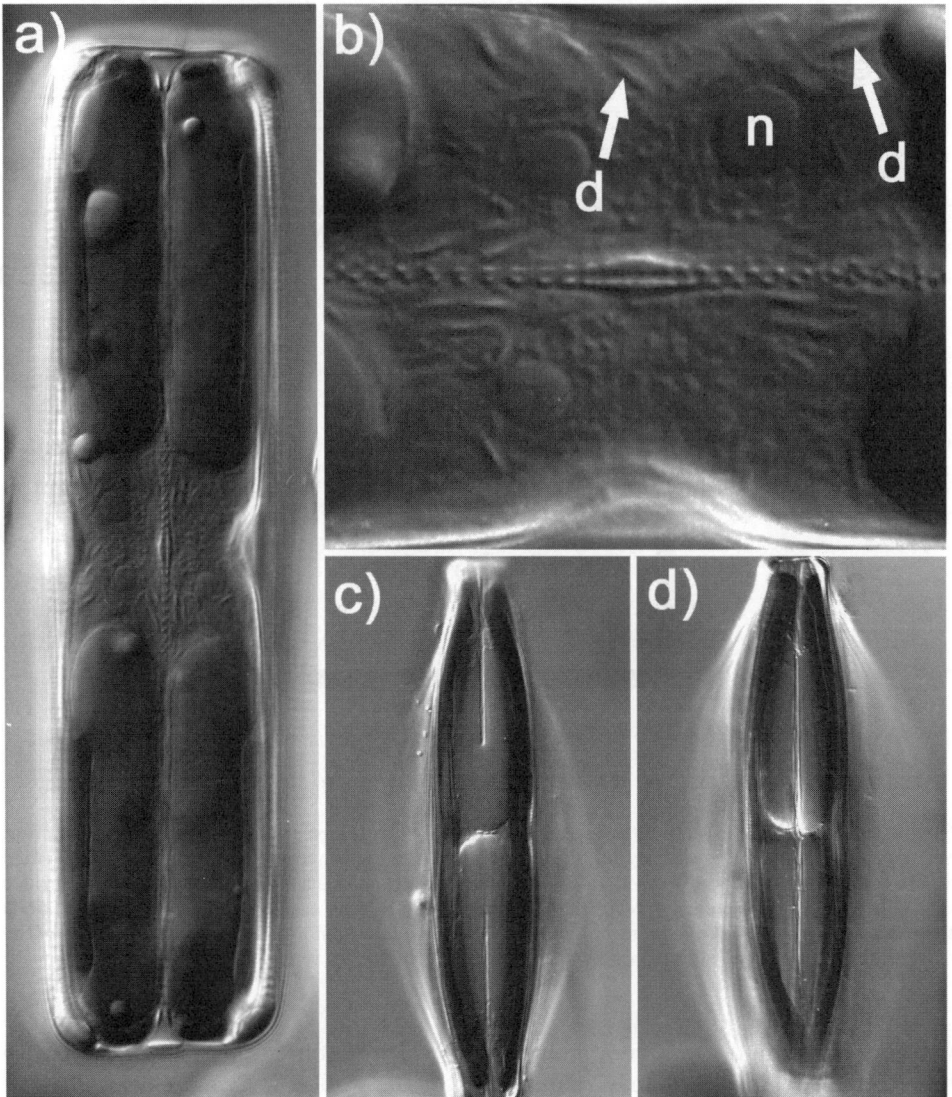

Figure 4. Dividing cells, differential interference contrast optics (DIC). a: *Pinnularia* cell in girdle view (cf. Fig. 2a) soon after division, during valve formation. The new cells remain enclosed within the parent frustule (epitheca to left). b: Detail of A: the thickened centers and the transverse ribs of the new valves are visible in section. On either side of the division plane are the daughter nuclei, which contain prominent nucleoli (e.g. n) and are surrounded by dictyosomes (e.g. d). c and d: Stages in the division of a *Craticula* cell: note the ingrowing cleavage furrow that cuts the cell in two.

valve are highly constant (Mann, 1999). Another check is to grow clonal lineages (derived from a single cell) in different culture conditions.

A very important consequence of the way diatom cells divide is that, except in cases where the girdle is constructed in such a way that tangential expansion is possible during the extension of the hypotheca (Crawford, 1981)—and such cases are apparently rare—the average cell size (measured at right angles to the long axis of the girdle cylinder) in a population will decrease with time.

2.4 The life cycle

Many plants and animals have complex life cycles, involving transformations between morphologically dissimilar phases. The metamorphoses of butterflies and dragonflies are well-known examples, and many red and brown algae exist as two discrete stages that differ in their size, appearance, structure and ecology. Diatoms are relatively simple. The vegetative cells are diploid (with two copies of each chromosome) and the only haploid cells (with one copy of each chromosome) are the sperm and eggs (isogametes in pennate diatoms), just as in higher animals and humans.

There is one striking difference between diatom life cycles and those of other organisms and this relates to cell size, which generally decreases with time within diatom populations (Section 2.3). One might expect, from the mode of cell division, that cell size distributions would conform to the binomial expansion (assuming that the two daughter cells have equal rates of progression through the cell cycle—in some cases they do not: see Mann, 1988), and that there would therefore always be some large cells left. However, even though this is theoretically possible, in practice the absolute upper limit within a population, as well as the mean, will drift downwards, as a result of losses to grazers and parasites, physiological damage, and quite possibly the "old age" of thecae (*ibid.*; Jewson, 1992a,b).

As cells diminish in size, their shape in valve view (the cross sectional shape of the cylinder) often changes also, unless the valves are already circular. Shape generally becomes simpler as size reduces (Geitler, 1932) and approximates more closely to a simple ellipse or circle (Fig. 5). Since the outline of the valve is created by being molded against the inside of the girdle cylinder during valve formation, it may at first seem surprising that any change in shape should occur. The reasons for change appear to be purely physical (Mann, 1994), reflecting the tendency of any body with a relatively homogeneous flexible envelope under tangential tension (often increased in diatoms by considerable internal hydrostatic pressure) to adopt a shape that yields the minimal area for the volume enclosed (e.g. Thompson, 1942). Thus, since the girdle is not inflexible, its deformation drives slow, continuous change in valve shape during size reduction. In some cases, however, shape can become more complex during size reduction (Mann, 1999 gives examples), presumably as a result of anisotropy or heterogeneity of the cell wall.

Size does not diminish indefinitely. Small diatoms can become larger again through one of two processes: auxosporulation or vegetative cell enlargement. The principal means of restoring size is via auxosporulation, which is usually associated with sexual reproduction. Once cells drop below a certain critical size limit, they gain the potential to become sexual and will transform into gametangia (gamete-producing cells) if environmental conditions are suitable. There is much variation in how the gametes are produced and in their behavior, and in how the zygote

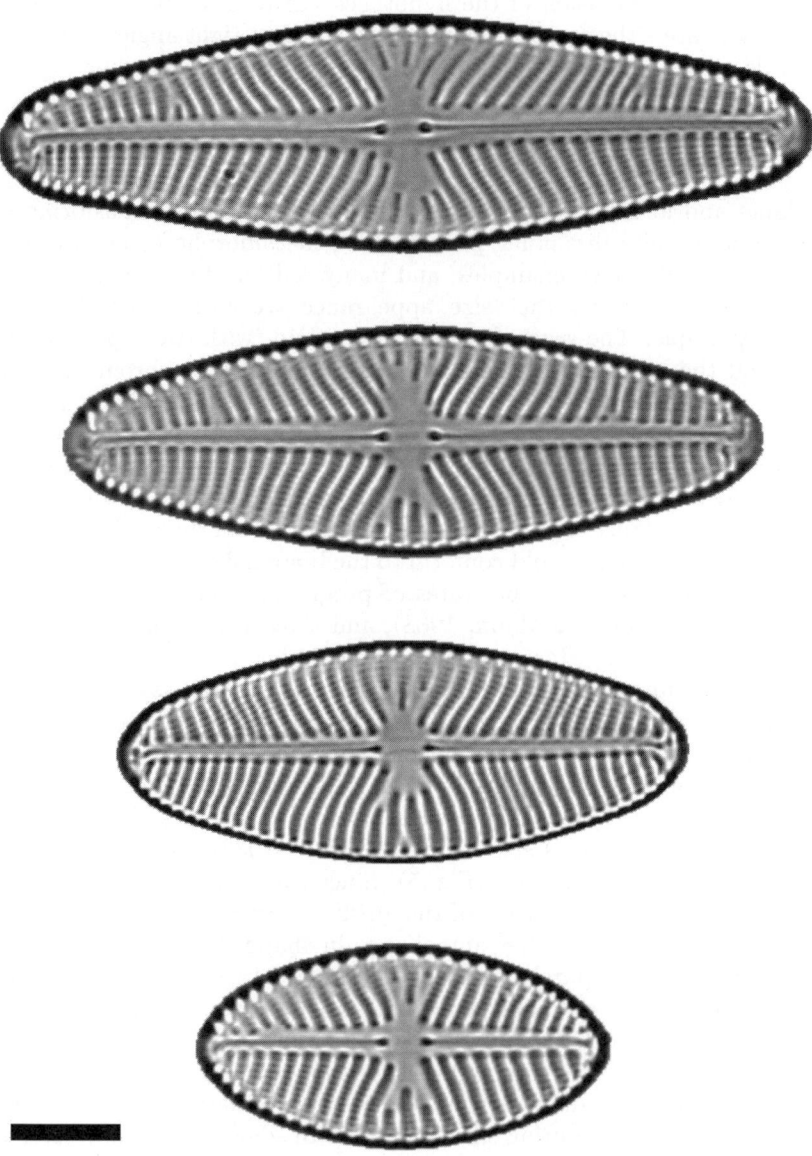

Figure 5. *Navicula reinhardtii*. Valves representing four stages in the life cycle. The largest valve is the oldest, i.e. it is one of the earliest to be formed in the life cycle, after expansion to maximum size during auxosporulation. Note that as the size decreases (length more than width), the shape simplifies. Light microscopy (brightfield). Scale bar 10 μm.

develops (reviewed briefly by Round et al., 1990), but in all cases the gametangial frustules are cast off during the formation of the gametes, or during fertilization (Fig. 6a). Then the zygotes expand considerably, often to around twice the length or diameter of the gametangia (Fig. 6b), and then each produces a new large frustule. The expanding zygote—which is called an auxospore—is surrounded by an organic wall, which often also contains silica scales or bands. Together, the organic wall and the silica elements constrain expansion, allowing auxospores of some species to develop anisometrically (Fig. 6b) and produce their characteristic shape (Mann, 1994). Vegetative cell enlargement has not been studied in great detail and it may perhaps be an artifact of culture. The frustule is cast off and the protoplast, probably surrounded by a diatotepic layer, expands before again forming a frustule.

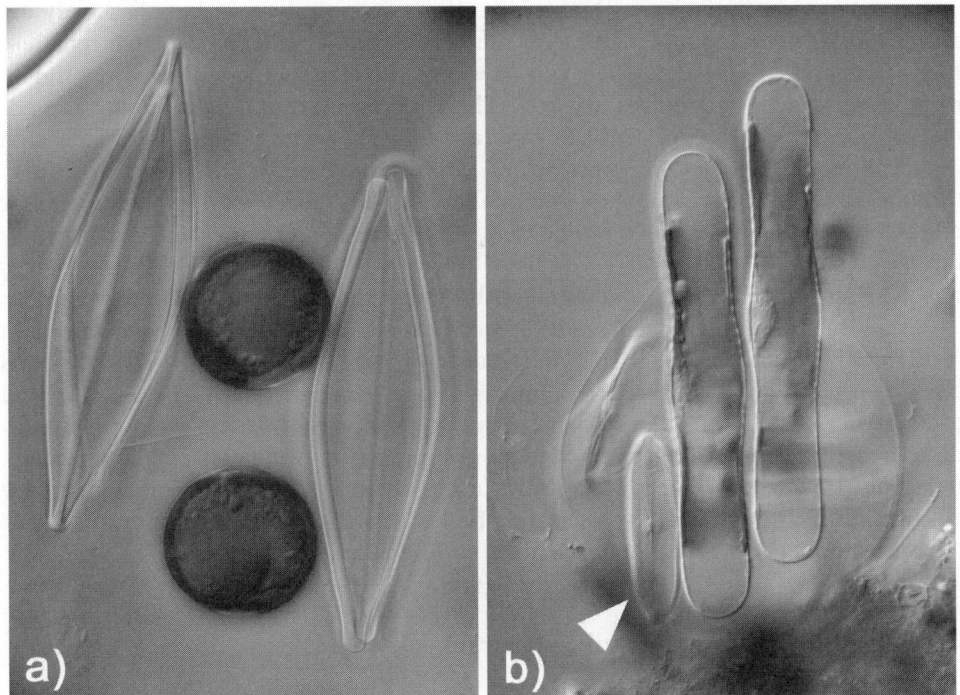

Figure 6. a: *Craticula*: spherical zygotes formed through the fusion of gametes produced by the two cells whose frustules are visible, empty, on either side. During expansion of the auxospore, it changes from spherical to elongate, through unequal growth. b: *Caloneis*: fully expanded auxospores. Compare the size of the cast-off theca (arrow). DIC.

Thus, the diatom life cycle consists of long periods of slow size reduction, which can last several years (Mann, 1988; Jewson, 1992a), alternating with enlargement via auxospores (and perhaps also vegetative cell enlargement), which is completed within a few days. Studies of natural populations indicate that auxosporulation (and therefore the sexual reproduction that precedes it in most cases) does not occur at a constant rate but is episodic (Mann, 1988; Jewson, 1992a,b), so that populations often consist of several size classes, which may or may not overlap, producing multimodal size spectra. The size achieved through the expansion of the auxospores is usually proportional to the sizes of the gametangia that gave rise to them (e.g. Davidovich, 1994) and gametangium sizes exhibit a more or less fixed range that is characteristic of the population, race or species. Thus, although the range of size exhibited by a species is wide, its upper and lower limits have taxonomic value (Mann, 1999). Unfortunately, for complex reasons (e.g. see Mann, 1988), it is very difficult to determine the full size range exhibited by a given species, unless the species is kept in culture, because natural populations usually exhibit size spectra that are strongly biased towards the lower end of the full size range.

Over time, therefore, the valve outline describes a trajectory in any hyperdimensional display of size–shape variation. The size–shape trajectory usually exhibits two singularities: it begins with the formation of new valves within the newly expanded auxospore and ends with the transformation of walled vegetative cells into naked gametes. A few cases are known where the size–shape trajectory is indefinite, because size reduction is avoided. The trajectories of different species are not necessarily parallel, in spite of the general tendency towards shape simplification noted above. Thus, diatom species that initially have quite different shapes may become almost identical as they get smaller, or they may begin by being similar and subsequently diverge: in either case, their size–shape trajectories apparently intersect. Large and small individuals of the same diatom species may sometimes appear less similar to each other than either does to individuals of other species.

Many mistakes have been made in taxonomy, and continue to be made, as a result of the changes in cell size and shape during the life cycle (Fig. 5), which are often accompanied by subtle changes also in the patterns of pores, ribs and other structures on the valves. Genera, species and varieties have sometimes been described that are nothing more than stages in the life cycle of a single organism. It is only through careful studies of variation within populations and laboratory cultures, and through effective recording and communication of information about the variation pattern, that such mistakes can be avoided.

3 Classifying and identifying diatoms

3.1 Remarks on the processes of classification and identification

Until some 30 years ago, classification and identification were essentially the same process in diatoms. Taxonomists studied samples of diatom populations and used particular characteristics to classify the diatoms into sets, they defined the sets on the basis of these characteristics, and they named them according to the rules laid down in the *International Code of Botanical Nomenclature* (Section 3.2). Taxonomists and others (ecologists, physiologists, biochemists) then used the same characteristics to categorize specimens that were not included in the original sets, with due allowance for the fact that the original definitions of the sets might not have been sufficiently inclusive (e.g. through incomplete knowledge of size–shape variation). Furthermore, until recently, most diatom species and genera were defined monothetically, i.e. the possession of a particular, single set of characteristics was both sufficient and necessary for inclusion in the group. These comfortable certainties have been lost.

One of the principal features of pre-1970s classifications of diatoms is that they were based on an extremely limited number of characteristics, almost all of them referring to the morphology of the valve as seen with the light microscope (e.g. Fig. 5). In part, this was inevitable. Electron microscopes, particularly the scanning electron microscope, were not routinely available to diatomists, and molecular genetic techniques were in their infancy; protoplast (e.g. chloroplast) characteristics could have been used for classification, and they were by some (e.g. Mereschkowsky, 1903, though his classifications were largely rejected), but they were less convenient to study than the silica frustule; and sexual reproduction was thought impractical to study. By contrast, samples of frustules could be preserved indefinitely after they had been cleaned with oxidizing agents (to remove the organic material, which would otherwise rot) and could be mounted in resins with a high refractive index (to provide contrast between the glassy silica of the frustule and the mountant) for study with the light microscope. However, during cleaning, the links between the valves and the girdle bands were weakened and the frustule usually fell apart, so that only the valves remained for study in many cases.

Since 1970 we have been able to use many more characteristics in taxonomy and this, coupled with better methods of analysis, has led to the development of more natural classifications, i.e. classifications that are more accurate reflections of evolutionary relationships. New information has been provided by electron microscopical investigations of the valves and girdle bands, and by studies of the protoplast, cell cycle, sexual reproduction, and auxosporulation (e.g. Round et al., 1990). Very few significant advances have been achieved using the older approach, of studying valves by light microscopy, and this is tribute to the quality of the work done by earlier diatomists: they left little to be done. However, although there has been a profound change in the kinds of data used for classification and in the classifications produced, identification methods have changed little. The reasons are:

- light microscopy of valves is relatively cheap and convenient.

- in the principal applications of diatoms, e.g. ecology, palaeoecology and stratigraphy, often the only characteristics available for identification are those of the valve.

- most species can be recognized from their valves alone, using light microscopy. The genus they belong to may have been defined on the basis of protoplast and reproductive characteristics (such is the case, for instance, with *Sellaphora*) and hence may be unrecognizable from valve morphology, especially as seen in the light microscope, but if the species can be identified, the genus can be determined easily, simply through a look-up table.

Thus, identification has become partially decoupled from classification. Furthermore, identification is a less rigidly hierarchical process than classification. We recognize elephants as such and do not need to work consciously through the characteristics of all the evolutionary groupings to which elephants belong—chordates, vertebrates, mammals, etc.—in order to do this.

It is therefore foolish to attempt to construct a single inflexible set of dichotomous identification keys, in which organisms are successively keyed out to lower and lower levels of the taxonomic hierarchy—first to divisions, then to classes, orders, families, genera, species and varieties. Such a key would be feasible only if each group was defined monothetically and if all the characteristics used in constructing the classification were routinely available for identification. They are not. Thus, the identification keys provided in existing, printed diatom floras (e.g. Hustedt, 1927-1966; Patrick and Reimer, 1966, 1975; Krammer and Lange-Bertalot, 1986-91) must be regarded as attempts to achieve the impossible. They are not worthless, since even an imperfect identification aid is better than nothing, but they must be regarded only as adjuncts to the descriptions and illustrations. Computer-aided sorting and identification methods such as those described in this book (see also Pankhurst, 1991), together with the very different approaches made possible by molecular genetics and flow cytometry, represent the first significant new approaches to the identification of plants and animals since the eighteenth century.

3.2 Nomenclature

The aim of a taxonomic investigation is to produce a hierarchical classification, usually one that its authors believe accurately reflects evolutionary relationships. This classification is often then converted into a hierarchy of names, corresponding to different levels in a hierarchy of taxonomic ranks; if the underlying classification is subsequently shown to be wrong, alterations have to be made to the names. A set of rules exists to regulate how names are to be established and altered; its latest edition is the St Louis Code (Greuter et al., 2000). Among the many taxonomic ranks allowed by the Code, only a few are used in diatoms, the main ones being (in descending order of inclusivity): order, family, genus, species, variety, and form. These categories, with the sole arguable exception of species, can only be defined relative to each other—e.g. that the family is more inclusive than the genus, but less so than the order. Many biologists maintain that the species category is special (Mann, 1999), corresponding approximately to the boundary between reticulate relationships (as a result of genetic interchange among individuals, through sexual reproduction) and hierarchical relationships (reflecting divergent evolution after populations have become reproductively isolated).

The correct application of all names from family downwards is determined by reference to "types"—preserved specimens or published images that are explicitly and permanently linked to the name (except in special circumstances)—and by the principle of priority. To illustrate how the Code works, imagine that two taxonomists, working in different places or at different times, each discover what appears to be a hitherto unknown species of the genus *Cognita*. They describe and name the new species (the Code specifies how this is to be done: among other things, it requires a description in Latin), one calling his species *Cognita prima*, the other calling his *Cognita secunda*. Each taxonomist specifies a type specimen to demonstrate the proper use of the name. The specimen does not have to be close to the population mean in its characteristics; it is simply a reference point to determine how names are used. Imagine now that a third taxonomist studies

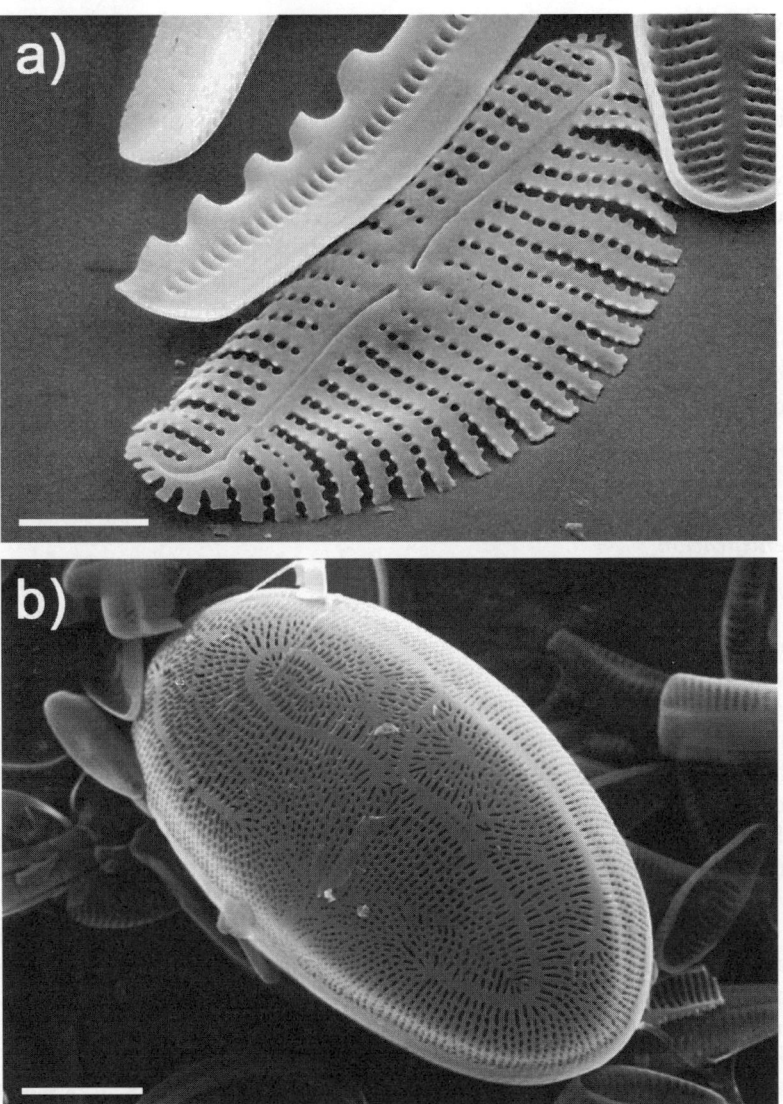

Figure 7. a: *Encyonema*, a partially complete valve, showing the rib–stria system extending out from the pattern center (here, the longitudinal raphe-sternum). SEM. Scale bar 2 μm. b: Initial epivalve of *Cocconeis*. Despite the extraordinary convolutions of the pattern center, the morphogenetic flexibility of the rib–stria system allows a complete valve to be formed. SEM. Scale bar 10 μm.

the genus *Cognita* and decides that *C. secunda*, as originally described, is actually not one species but two. So, there is one published species name, but two species: which inherits the name "*secunda*"? The Code says that it is the one that would include the type specimen of *C. secunda*; the other needs a new name, *tertia*, say, with its own type. A fourth taxonomist now studies the genus *Cognita*, using new evidence, and concludes that there is actually only one, very variable species and that the type specimens of *prima*, *secunda* and *tertia* all belong within it. What should the combined species be called? The Code states that the correct name is the first to have been published. *Cognita secunda* was published a day earlier than *C. prima* and much earlier than *C. tertia*, so *C. secunda* is the correct name and *C. prima* and *C. tertia* become synonyms of *C. secunda*. However, the names *prima* and *tertia* must not be forgotten, in case taxonomist 5 can prove that there is more than one species after all; in this case, the types become relevant again, to determine which name should be applied to which group.

One great weakness of the type system as it applies to diatoms is that it is currently permissible to designate a population of diatoms, or even a slide preparation containing a mixture of diatoms, rather than a single specimen. Since populations may prove heterogeneous, many diatom types are potentially ambiguous (Mann, 1998).

3.3 Valve morphology, with especial reference to light microscope based identification

3.3.1 The pattern center

In all but a few diatoms, the valves are essentially composed of ribs of silica, separated by rows of pores (Fig. 7a). The function of the pores is to allow the passage of water and dissolved solutes in and out of the cell, and to permit the secretion of polysaccharides and other organic material. The ribs often branch—it is rare for any to have two free ends—and can be traced in from the valve margin, which is composed of a strip of imperforate silica, to a "pattern center," which is usually either of a ring of silica, called the "annulus," or an elongate rib or strip, called the "sternum" (Mann, 1984). Two principal diatom types can be distinguished according to which of these types of pattern center is present. "Centric" diatoms have an annulus and bear radiating systems of ribs (Fig. 8a), whereas "pennate" diatoms have a sternum, with a series of ribs on either side (Figs 2b, 7a, 9). The annulus and sternum are also the ontogenetic centers of centric and pennate valves: during valve formation, the annulus and sternum are laid down first and then the valve ribs develop outwards from them (Li and Volcani, 1985; Chiappino and Volcani, 1977; see also Mann, 1984, and Pickett-Heaps et al., 1990). There are a few diatoms whose morphology cannot yet be reconciled with the two primary categories, such as *Nephroneis* (Amspoker, 1989), and there are also a few diatoms that have significantly different pattern centers, which may need to be recognized as further variants alongside the centric and pennate groups; these include *Toxarium*, *Synedrosphenia*, *Ardissonea* and *Climacosphenia* (Mann, 1984; Round et al., 1990), which have two parallel rib-like pattern centers, which could be interpreted as a bifacial annulus (i.e. an annulus in which ribs are initiated internally as well as externally), or as two independent sterna, or as a single reflexed sternum. Studies of valve development are needed to aid interpretation.

There is an important modification of the pennate pattern center, present in a very large group of diatoms, in which one or two longitudinal slits are incorporated into the sternum, forming a composite structure referred to as the "raphe-sternum"

Chapter 2: Organism and image 27

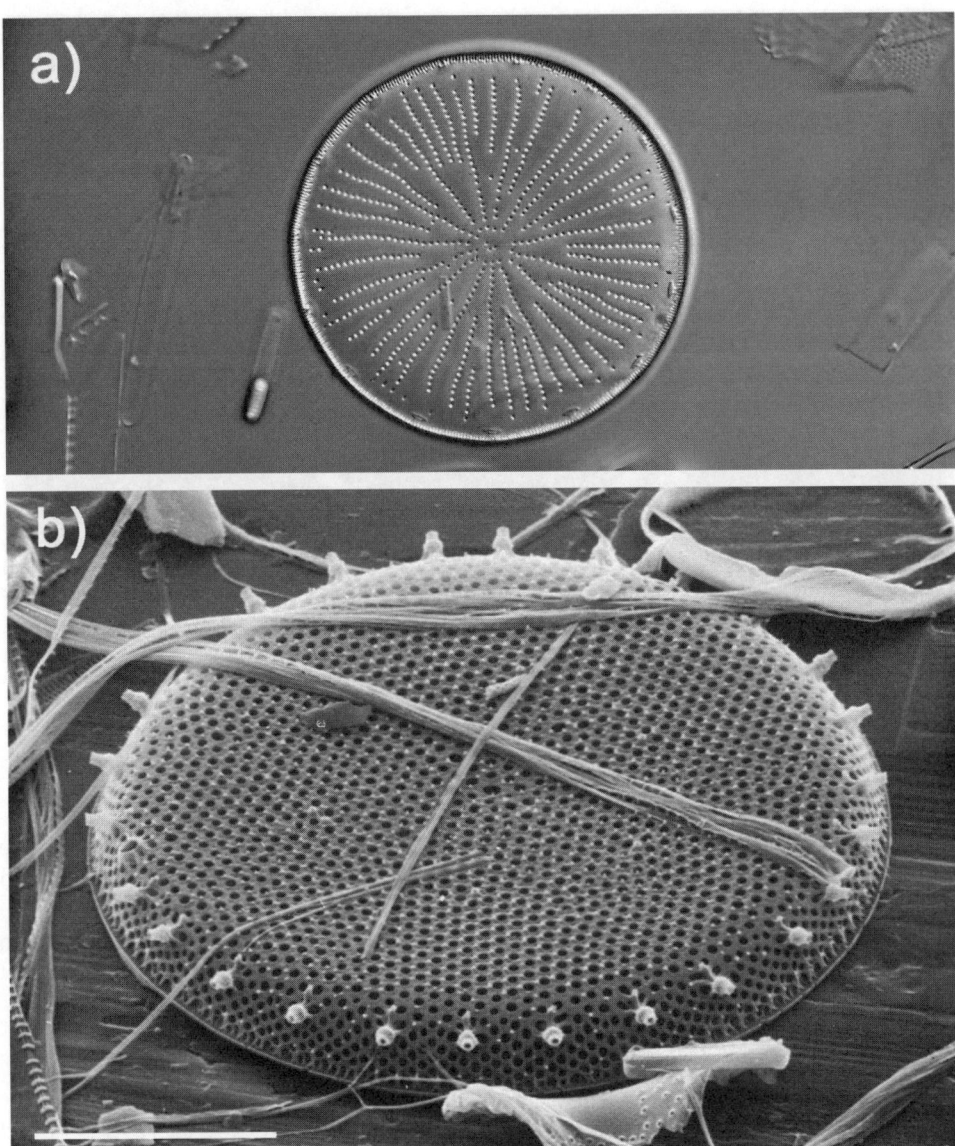

Figure 8. Centric diatoms. a: *Actinocyclus* valve, DIC. Note the radiating rows of pores, separated by wide flat ribs, which are subtended by a ring (annulus) at the center. b: *Thalassiosira* valve, SEM. Here, the radiating rib–stria system is obscured by development of a system of hexagonal chambers on top. Note also the ring of fultoportulae at the junction of the flat valve face and the sloping mantle. Scale bar 10 μm.

(Mann, 1984), see Figs 2b, 5 and 7a. These slits constitute part of the "raphe system," which is a complex of wall and protoplast components that enables diatom cells to move actively, at speeds of up to 25 μm/s or more (Edgar and Pickett-Heaps,

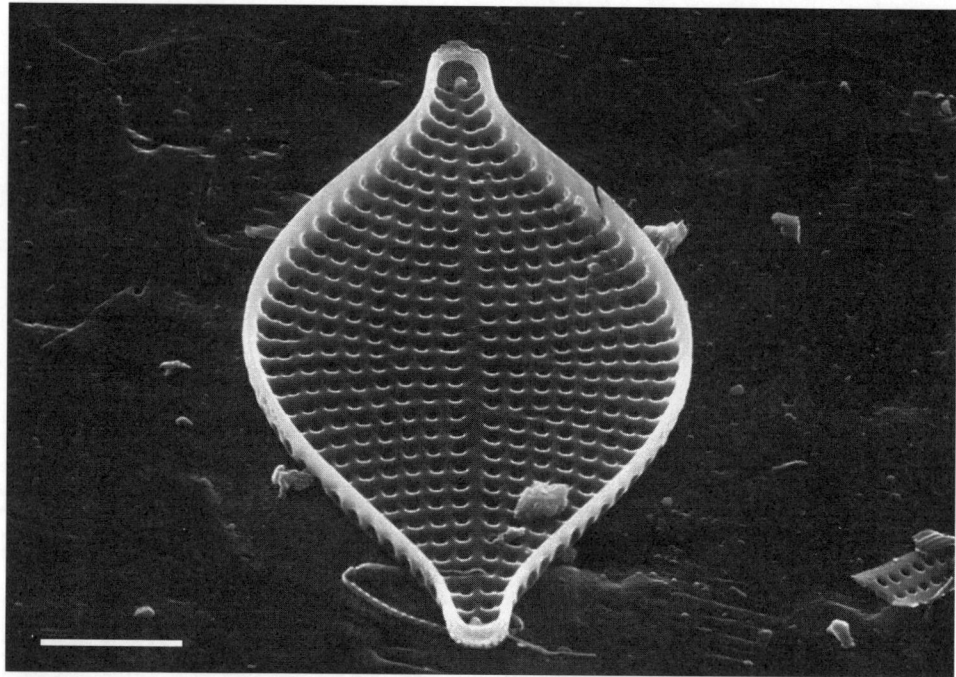

Figure 9. *Rhaphoneis*: an araphid pennate diatom. The rib–stria system is subtended by a simple longitudinal rib, the sternum. SEM. Scale bar 10 µm.

1984). The presence of the raphe and the shape of the raphe slits, especially at the center (Figs 10 and 11a) and poles, are important characteristics for classification and identification. In a relatively small but significant group, comprising *Eunotia* and its allies, the raphe slits are not integrated into the sternum but lie to one side of it.

3.3.2 Valve outline

The starting point for each size–shape trajectory is the initial cell, i.e. the first cell formed after auxosporulation. In the simplest case, the auxospore expands equally in all directions and the initial cells are molded by the spherical auxospore without modification. The valves of the initial cell are therefore circular and so are all the valves formed thereafter (Fig. 8). This occurs in many of the most primitive diatoms, which are all centric (Medlin et al., 1996).

Departures from circularity are achieved by constraining the expansion of the auxospore, usually through the incorporation into the auxospore wall of stiffer siliceous strips or bands (Mann, 1994). Bipolar (Fig. 6b), tripolar or multipolar outlines can thus be produced, and these shapes can be further modified through contractions of the protoplast within the expanded auxospore before the initial

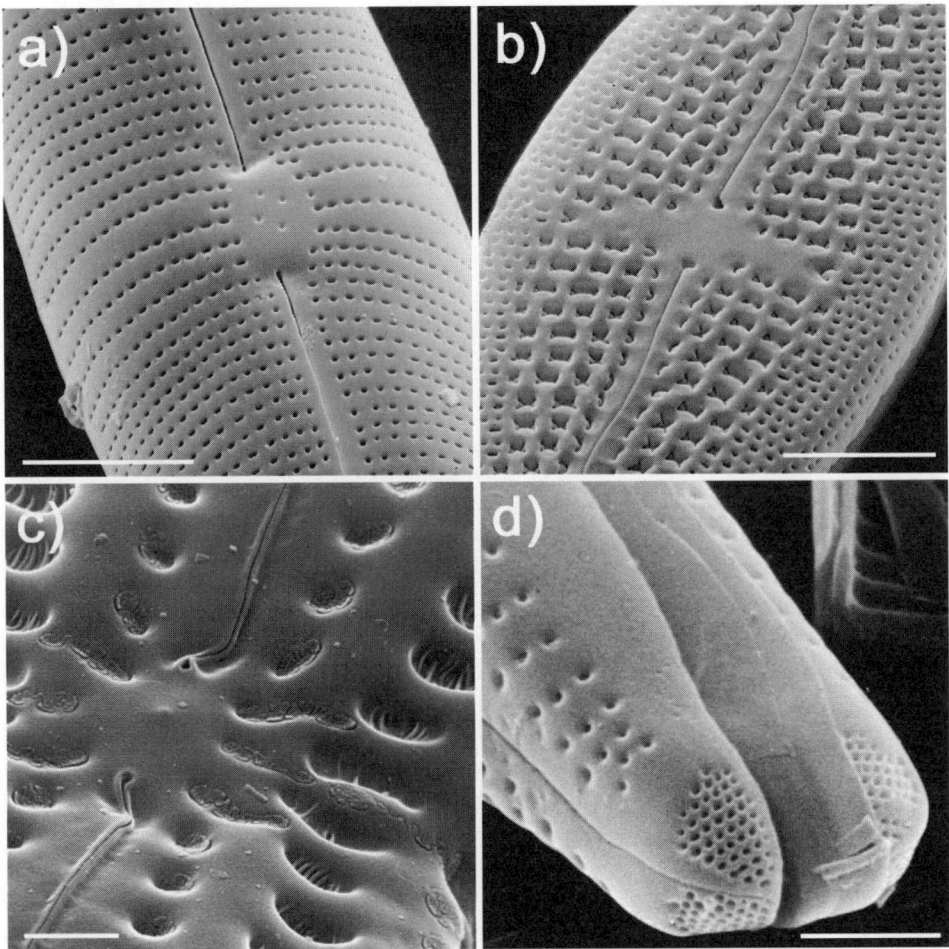

Figure 10. Raphid pennate diatoms. a: *Parlibellus*, with simple central raphe endings and a simple rib–stria system with round pores. b: *Aneumastus*, the central raphe endings are deflected and expanded, and the rib–stria system changes in linear density near the margin. c: *Diploneis crabro*, a highly complex valve structure, with different types of pore (overlying a chambered wall structure that is not shown) and twisted raphe endings. d: *Gomphonema*, with special areas of pores at the base of the valve, through which polysaccharide stalks are secreted. SEM. Scale bars: a/b 4 μm, c/d 2 μm.

valves are formed. Hence, within the centric diatoms, there are examples of elliptical, elongate, triangular, quadrilateral, or even more elaborate outlines. Pennate diatoms generally have elongate, bipolar valves and the majority are bilaterally symmetrical (Figs 2b and 5). However, some pennate diatoms are asymmetrical (Fig. 7a), as a result of anisometric expansion of the auxospore and/or variation in flexibility across the girdle; the asymmetry may be with respect to either the long axis of the valve or the short axis or both. These grosser aspects of shape tend to be constant at higher levels of the taxonomic hierarchy. Thus, for example, genera like

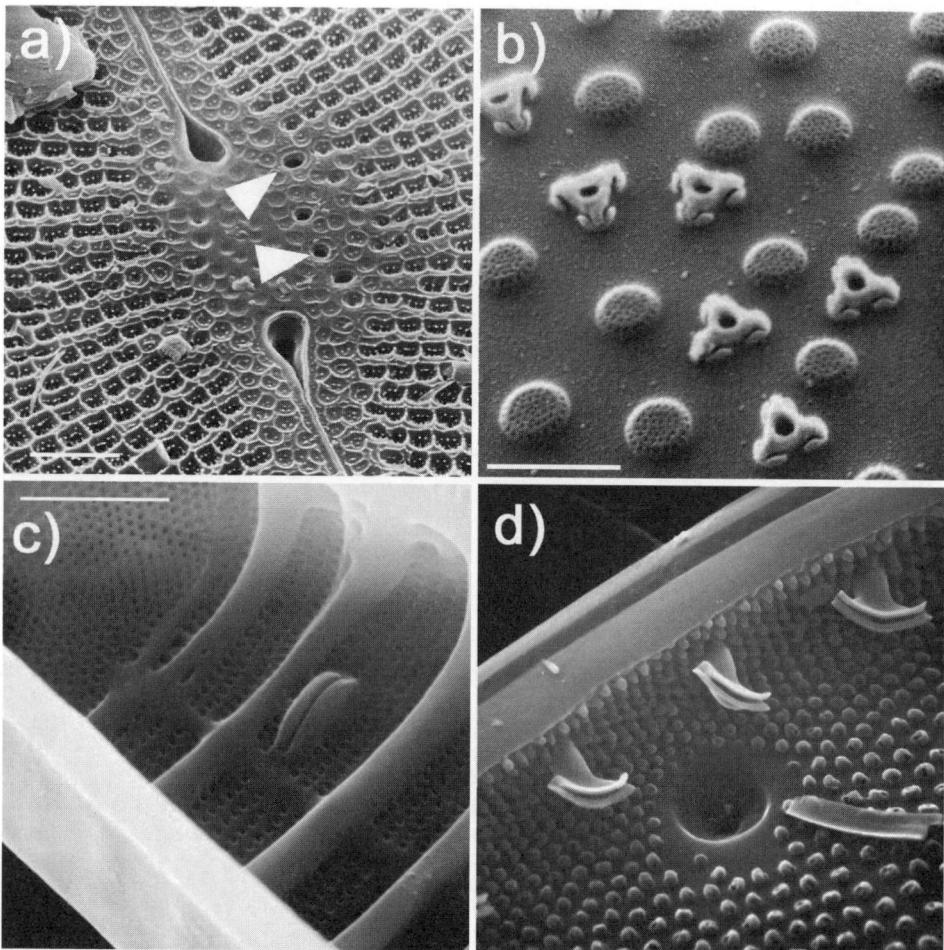

Figure 11. Special wall structures. a: Stigmata (e.g. arrows) near the central raphe endings of *Didymosphenia*. Scale bar 4 μm. b: Internal openings of fultoportulae, each with a central circular pore and a complex tripartite system of satellite pores and buttresses. Between the fultoportulae are the slightly raised, sieve-like openings of the areolae. Scale bar 1 μm. c: A simple sessile rimoportula in the araphid pennate *Diatoma*. Note also that some of the transverse ribs are thickened. Scale bar 2 μm. d: Elaborate stalked rimoportulae in *Actinocyclus*. The depression is a special structure, the pseudonodulus (one per valve) and the openings of the areolae are covered by domed porous membranes. c and d are internal views. SEM.

Cymbella and *Gomphonema* are characterized by being bilaterally asymmetrical and heteropolar, respectively. However, no character can be assumed *a priori* to be constant within a natural evolutionary grouping. Thus, although most *Lyrella* species have symmetrical, boat-like valves, at least one (*L. amphoroides*), does not (Mann and Stickle, 1997) and the genus *Biremis* also includes symmetrical and asymmetrical representatives (Round et al., 1990; Cox, 1990).

The subtlest features of the outline may be significant for identification and differences can be difficult or impossible to describe in words, even after standardization of terminology like that attempted by the Systematics Association Committee for Descriptive Biological Terminology (1962). Traditionally, this has been circumvented by recourse to illustrations. More recently, computer-aided morphometrics have been used to help analyze and describe variation, using methods based on Legendre polynomials (e.g. Stoermer and Ladewski, 1982), Fourier analysis (Pappas et al., 2001), or simpler descriptors, such as rectangularity (Droop, 1994; Droop et al., 2000). These by-pass the inadequacies of verbal communication, while still allowing the human observer to check visually whether differences in shape have been captured by the analysis.

3.3.3 The rib–stria system

The lines of pores lying between the ribs are called "striae," and the pores themselves, which can be circular (Fig. 10a), elliptical (Fig. 10c), rectangular (Fig. 11a) or polygonal (Fig. 10b), are called "areolae." Sometimes several rows of pores are formed between the ribs (bi- or multiseriate striae) and sometimes the basic stria–rib system is modified or hidden by the superimposition of further layers of silica (Fig. 8b), added later during the development of the valve, so that the areolae may become complex chambers (loculi, hence "loculate areolae"), whose apertures are constricted both internally and externally. Loculate areolae are frequently hexagonal (Fig. 8b) and can form honeycomb-like arrays. In the light microscope, the chambered nature of the areolae can scarcely ever be detected, but the presence of hexagonal arrays is a good guide to their presence. Even non-loculate areolae frequently have a complex structure, containing fine sieve-like membranes of silica, flaps, or meshwork (Mann, 1981; Round et al., 1990), but these too can rarely be detected with the light microscope. In some diatoms the structure and spacing of the ribs, striae and areolae change across the valve, from the pattern-center to the margin.

Within a species, the overall pattern formed by the ribs, striae and areolae is usually fairly constant (Fig. 5). In some pennate diatoms, for example, the striae are generally parallel (e.g. in *Neidium*), whereas in others they are strongly radiate, approaching the arrangement in centric diatoms. The rib–stria system is space-filling (Fig. 7). Diatoms are not like humans, where (except in rare cases) there are different sizes but constant numbers of limbs, fingers, ribs, vertebrae and vital organs. In diatoms, the pores and ribs vary little in size and spacing within a species, but the numbers change as the valves reduce in size during the life cycle. This correlates with the fact that the rib–stria system is formed from the center outwards: ribs are initiated from the annulus or sternum at a more or less fixed spacing (in some diatoms the control is tighter than in others) and grow out, branching and extending to fill the space available, while maintaining more or less the same distance from each other. Thus, in larger valves of a given species, more branching will usually occur than in smaller valves. Occasionally, as a result of disturbance to the cell during the formation of the valve (e.g. through osmotic effects or poisoning with chemicals that affect the cytoskeleton), the pattern center may be displaced from its usual position, but the rib–stria system can nevertheless develop as necessary to create a complete, functional valve (Fig. 7b).

Hence, for identification, it is more important to concentrate on the spacing of the striae and areolae than on absolute numbers. Descriptions of diatom species almost always express this information as a linear density—usually as the number per 10 μm. The stria density is generally between 5 and 50 per 10 μm. Although this range is almost wholly within the resolution limits of the light microscope, in practice stria densities above 35 per 10 μm are difficult to resolve and measure. The spacing of the areolae along the striae and the spatial relationship between the areolae of adjacent striae are rarely recorded in species descriptions (principally because, before the advent of computer-aided image analysis, they were so difficult to quantify—differences were recorded mostly via the drawn or photographic image), but they are undoubtedly important for species differentiation. In some diatoms, some of the valve ribs may be additionally thickened, as in *Denticula* or *Diatoma* (Fig. 11c); closely related species may differ in the linear density of these secondary ribs. Small silica bridges ("fibulae") are present below the raphe in some genera, e.g. *Nitzschia*, and these too have a characteristic spacing.

3.3.4 Special structures

In a few cases, valves consist of nothing more than the pattern center and the rib–stria system, e.g. *Chrysanthemodiscus* (Round et al., 1990). Usually, however, the valve possesses one or more extra structures that do not belong to the rib–stria system. Most centric diatoms, and also those pennate diatoms that lack a raphe system ("araphid" pennate diatoms), have one or more "rimoportulae" (labiate processes)—tubular structures of unknown function that pierce the wall and have a lip-like internal aperture (Figs 11c and d). The centric order Thalassiosirales possesses "fultoportulae" (Fig. 11b), which are structures involved in the secretion of chitin threads (see Round et al., 1990). Both types of portule generally appear in the light microscope only as dots or circles, but their presence and location can be important for identification. Obtaining this information automatically will provide a major challenge for computer-aided feature extraction and a special project on Thalassiosirales would be worthwhile (cf. Hasle and Syvertsen, 1996).

Other special features include special ribs unrelated to the rib–stria system, and areas of special pores involved in the secretion of pads or stalks that attach diatom cells to solid substrata (Fig. 10d). In pennate diatoms these secretory structures are located at the ends of the cell and hence may be almost invisible in the light microscope; in centric diatoms they often lie on special elevations of the valve and can appear like eyes (hence they are referred to as "ocelli"). Single special pores ("stigmata") may be present near the center of the raphe (Fig. 11a) and can be significant for species recognition. These, like rimo- and fultoportulae, pose a particular challenge for automated feature extraction. However, in many cases, it may be possible to achieve reliable identifications without them, despite their historical importance.

Chapter 2: Organism and image

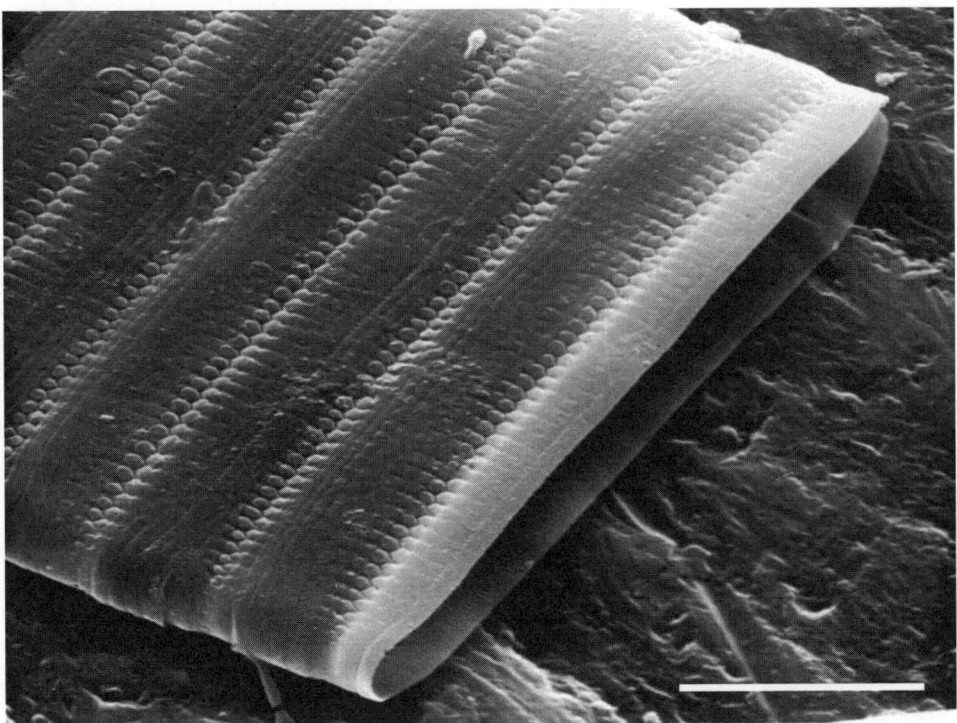

Figure 12. *Fragilaria*. In this diatom, sibling valves, i.e. valves formed back to back at a single cell division, are linked together by interdigitating spines. The chain-like colonies can only come apart through breakage of the spines or through death of a cell and disruption of its girdle, as has happened at the right. SEM. Scale bar 10 μm.

4 Complications

Section 3 mentions the main categories of morphological character that can be used for visual identification of valves: their size, the shape of the valve outline, the organization and density of the rib–stria system, areola spacing and structure, and the presence of special ribs, pores and spines. It remains to list some special difficulties in identification caused by the morphological characteristics of certain diatoms.

- If the depth of the valve is equal to or greater than its diameter or width, it will often lie in girdle view or obliquely, not in valve view. In some cases, valves always lie in girdle view. Identification systems must allow for this.

- A not inconsiderable number of diatoms form colonies. Often, the cells are connected together only by polysaccharide, which is easily removed during specimen preparation, so that the frustules and valves then separate from each other. However, in some centric and pennate diatoms, sibling valves (i.e. valves formed at the same time, following division of a single cell) have interlocking silica spines or processes, which cannot come apart unless broken (Fig. 12). In this case, the unit of identification is the pair of sibling valves.

- A few diatoms are twisted about the axis of the girdle cylinder. All diatoms are three-dimensional, but most can be represented adequately in a single two-dimensional projection; this is not possible for twisted forms.

- The valves of freshwater diatoms generally have a relatively simple cross-sectional shape, with a more or less flat top (valve face) and down-turned mantles (Fig. 10b). Many marine diatoms, by contrast, have a domed (Fig. 2b) or folded valve face. Sometimes, the morphology can be very complex indeed. Information on elaborate topologies has traditionally been recorded by drawing and is used in identification, largely via the drawn image itself. Digital imaging techniques need to be developed for diatoms that can detect, record and analyze three-dimensional shapes and which can also trace rib–stria systems across a complex topology.

- In most diatoms, the two valves of each frustule, though slightly different in size and varying stochastically with respect to aspects of the rib–stria pattern, portule number and position, etc, are essentially similar. However, in a significant minority of genera, the valves are constitutively dissimilar, in shape or pattern or both. Examples of "heterovalvy" are found in *Gephyria*, *Achnanthes*, *Achnanthidium*, *Cocconeis* and *Rhoicosphenia* (all with structural and morphological heterovalvy), and *Rhoikoneis* (morphological heterovalvy only) (Round et al., 1990). In terms of its effects on taxonomy and identification, this is analogous to the situation in many higher animals, where the two sexes are markedly different.

5 Final comments

I can think of no reason why the evolution of new species of diatoms needs to have been accompanied automatically by the evolution of new valve morphologies, i.e. some diatom species may be indistinguishable to the human eye and/or to computer-based classification–identification systems. We don't know of any such cases yet, however (Mann, 1999), and it is unlikely that morphological data will ever be superseded as the primary means of identifying the vast majority of diatom species. Unfortunately, this chapter cannot do more than touch on the basic features of diatom variation. A much more detailed account is given by Round et al. (1990).

References

Amspoker, M.C. (1989) *Nephroneis macintirei*, gen. et sp. nov., a marine araphid diatom from California, USA. Diatom Research, Vol. 4, pp. 171-177.

Anonymous (1703) Two letters from a gentleman in the country, relating to Mr Leeuwenhoek's letter in Transaction, no. 283. Communicated by Mr. C. Philosophical Trans. of the Royal Society of London, Vol. 23, pp. 1494-1501.

Björnland, T. and Liaaen-Jensen, S. (1989) Distribution patterns of carotenoids in relation to chromophyte phylogeny and systematics. In: The chromophyte algae: problems and perspectives, J.C. Green, B.S.C. Leadbeater and W.L. Diver (eds), Clarendon Press, Oxford, UK, pp. 37-61.

Borowitzka, M.A. and Volcani, B.E. (1978) The polymorphic diatom *Phaeodactylum tricornutum*: ultrastructure of its morphotypes. J. of Phycology, Vol. 14, pp. 10-21.

Borowitzka, M.A., Chiappino, M.L. and Volcani, B.E. (1977) Ultrastructure of a chain-forming diatom *Phaeodactylum tricornutum*. J. of Phycology, Vol. 13, pp. 162-170.

Brott, L.L., Naik, R.R., Pikas, D.J., Kirkpatrick, S.M., Tomlin, D.W., Whitlock, P.W., Clarson S.J. and Stone, M.O. (2001) Ultrafast holographic nanopatterning of biocatalytically formed silica. Nature, Vol. 413, pp. 291-293.

Cavalier-Smith, T. (1986) The kingdom Chromista: origin and systematics. Progress in Phycological Research, Vol. 4, pp. 309-347.

Chiappino, M.L. and Volcani, B.E. (1977) Studies on the biochemistry and fine structure of silica shell formation in diatoms. VII. Sequential cell wall development in the pennate diatom *Navicula pelliculosa*. Protoplasma, Vol. 93, pp. 205-221.

Cox, E.J. (1990) *Biremis ambigua* (Cleve) D.G. Mann, an unusual marine epipelic diatom in need of further investigation. In: Ouvrage dédié à la Mémoire du Professeur Henry Germain (1903-1989), M. Ricard and M. Coste (eds), Koeltz Scientific Books, Koenigstein, pp. 63-72.

Cox, E.J. (1996) Identification of freshwater diatoms from live material. Chapman and Hall, London.

Cox, E.J. (1999) Variation in patterns of valve morphogenesis between representatives of six biraphid diatom genera (Bacillariophyceae). J. of Phycology, Vol. 35, pp. 1297-1312.

Crawford, R.M. (1981) Some considerations of size reduction in diatom cell walls. In: Proc. 6th Symp. on Recent and Fossil Diatoms, R. Ross (ed.), Otto Koeltz, Koenigstein, pp. 253-265.

Davidovich, N.A. (1994) Factors controlling the size of initial cells in diatoms. Russian J. of Plant Physiology, Vol. 41, pp. 220-224.

Droop, S.J.M. (1994) Morphological variation in *Diploneis smithii* and *D. fusca* (Bacillariophyceae). Archiv für Protistenkunde, Vol. 144, pp. 249-270.

Droop, S.J.M., Mann, D.G. and Lokhorst, G.M. (2000) Spatial and temporal stability of demes in *Diploneis smithii/D. fusca* (Bacillariophyta) supports a narrow species concept. Phycologia, Vol. 39 , pp. 527-546.

Edgar, L.A. and Pickett-Heaps, J.D. (1984) Diatom locomotion. Progress in Phycological Research, Vol. 3, pp. 47-88.

Ehrenberg, C.G. (1838) Die Infusionsthierchen als vollkommene Organismen. Ein Blick in das tiefere organische Leben der Natur. Leopold Voss, Leipzig.

Field, C.B., Behrenfeld, M.J., Randerson, J.T. and Falkowski, P. (1998) Primary production of the biosphere: integrating terrestrial and oceanic components. Science, Vol. 281, pp. 237-240.

Ford, B.J. (1991) The Leeuwenhoek legacy. Biopress, Bristol, and Farrand Press, London.

Ford, C.W. and Percival, E. (1965) The carbohydrates of *Phaeodactylum tricornutum*. II. A sulphated glucuronomannan. J. of the Chemical Society (London), Vol. 1965, pp. 7042-7046.

Geitler, L. (1932) Der Formwechsel der pennaten Diatomeen (Kieselalgen). Archiv für Protistenkunde, Vol. 78, pp. 1-226.

Graham, L.E. and Wilcox, L.W. (2000) Algae. Prentice-Hall, New Jersey, USA.

Greuter, W., Mcneill, J., Barrie, F.R., Burdet, H.-M., Demoulin, V., Filgueiras, T.S., Nicolson, D.H., Silva, P.C., Skog, J.E., Trehane, P., Turland, N.J. and Hawksworth, D.L. (2000) International Code of Botanical Nomenclature (St Louis Code). Koeltz Scientific Books, Koenigstein. Also available online at: http://www.bgbm.fu-berlin.de/iapt/nomenclature/code/SaintLouis/0001ICSLContents.htm

Groombridge, B. and Jenkins, M.D. (2000) Global biodiversity: Earth's living resources in the 21st century. World Conservation Press, Cambridge, UK.

Hasle, G.R. and Syvertsen, E.E. (1996) Marine diatoms. In: Identifying marine diatoms and dinoflagellates, C.R. Tomas (ed.), Academic Press, London, pp. 5-385.

Hoek, C. van den, Mann, D.G. and Jahns, H.M. (1995) Algae. An introduction to phycology. Cambridge University Press, Cambridge, UK.

Holwill, M.E.J. and Taylor, H.C. (2000) Mechanisms of flagellar propulsion. In: The flagellates. Unity, diversity and evolution, B.S.C. Leadbeater and J.C. Green (eds), Taylor and Francis, London and New York, pp. 49-68.

Hustedt, F. (1927-66) Die Kieselalgen Deutschlands, Österreichs und der Schweiz, unter Berücksichtigung der übrigen Länder Europas sowie der angrenzenden Meeresgebiete. In: Dr L. Rabenhorsts Kryptogamen-Flora von Deutschland, Österreich und der Schweiz, Vol. 7 (3 parts). Akademische Verlagsgesellschaft, Leipzig.

Jeffrey, S.W. (1989) Chlorophyll c pigments and their distribution in the chromophyte algae. In: The chromophyte algae: problems and perspectives, J.C. Green, B.S.C. Leadbeater and W.L. Diver (eds), Clarendon Press, Oxford, UK, pp. 13-36.

Jewson, D.H. (1992a) Size reduction, reproductive strategy and the life cycle of a centric diatom. Philosophical Trans. of the Royal Society of London, series B, Vol. 336, pp. 191-213.

Jewson, D.H. (1992b) Life cycle of a *Stephanodiscus* sp. (Bacillariophyta). J. of Phycology, Vol. 28, pp. 856-866.

Krammer, K. and Lange-Bertalot, H. (1986–1991) Bacillariophyceae 1. Teil: Naviculaceae. In: Süsswasserflora von Mitteleuropa, H. Ettl, J. Gerloff, H. Heynig and D. Mollenhauer (eds), Vol. 2 (4 parts). G. Fischer, Stuttgart and New York.

Kröger, N., Bergsdorf, C. and Sumper, M. (1996) Frustulins: domain conservation in a protein family associated with diatom cell walls. European J. of Biochemistry, Vol. 239, pp. 259-264.

Kröger, N., Lehmann, G., Rachel, R. and Sumper, M. (1997) Characterization of a 200-kDa diatom protein that is specifically associated with a silica-based substructure of the cell wall. European J. of Biochemistry, Vol. 250, pp. 99-105.

Kröger, N., Deutzmann, R., Bergsdorf, C. and Sumper, M. (2000) Species-specific polyamines from diatoms control silica morphology. Proc. of the National Academy of Sciences of the United States of America, Vol. 97, pp. 14133-14138.

Kröger, N., Deutzmann, R. and Sumper, M. (2001) Silica-precipitating peptides from diatoms. The chemical structure of silaffin-A from *Cylindrotheca fusiformis*. J. of Biological Chemistry, Vol. 276, pp. 26066-26070.

Li, C.-W. and Volcani, B.E. (1985) Studies on the biochemistry and fine structure of silica shell formation in diatoms. VIII. Morphogenesis of the cell wall in a centric diatom, *Ditylum brightwellii*. Protoplasma, Vol. 124, pp. 10-29.

Liebisch, W. (1928) *Amphitetras antediluviana* Ehrbg., sowie einige Beiträge zum Bau und zur Entwicklung der Diatomeenzelle. Zeitschrift für Botanik, Vol. 20, pp. 225-271.

Liebisch, W. (1929) Experimentelle und kritische Untersuchungen über die Pektinmembran der Diatomeen unter besonderer Berücksichtigung der Auxosporenbildung und der Kratikularzustände. Zeitschrift für Botanik, Vol. 22, pp. 1-65.

Lundholm, N., Daugbjerg, N., Moestrup, Ø, Hoef-Emden, K. and Melkonian, M. (2001) Phylogeny of the Bacillariaceae with emphasis on *Pseudo-nitzschia* (Bacillariophyceae). Phycologia, Vol. 40(4), supplement, pp. 7-8.

Mann, D.G. (1981) Sieves and flaps: siliceous minutiae in the pores of raphid diatoms. In: Proc. 6th Symp. on Recent and Fossil Diatoms, R. Ross (ed.), O. Koeltz, Koenigstein, pp. 279-300.

Mann, D.G. (1984) An ontogenetic approach to diatom systematics. In: Proc. 7th Int. Diatom Symp., D.G. Mann (ed.), O. Koeltz, Koenigstein, pp. 113-144.

Mann, D.G. (1988) Why didn't Lund see sex in *Asterionella*? A discussion of the diatom life cycle in nature. In: Algae and the Aquatic Environment, F.E. Round (ed.), Biopress, Bristol, UK, pp. 383-412.

Mann, D.G. (1989) The diatom genus *Sellaphora*: separation from *Navicula*. British Phycological J., Vol. 24, pp. 1-20.

Mann, D.G. (1994) The origins of shape and form in diatoms: the interplay between morphogenetic studies and systematics. In: Shape and form in plants and fungi, D.S. Ingram and A.J. Hudson (eds), Academic Press, London, pp 17-38.

Mann, D.G. (1996) Chloroplast morphology, movements and inheritance in diatoms. In: Cytology, genetics and molecular biology of algae, B.R. Chaudhary and S.B. Agrawal (eds), SPB Academic Publishing, Amsterdam, pp. 249-274.

Mann, D.G. (1998) Ehrenbergiana: problems of elusive types and old collections, with special reference to diatoms. Linnean, Special Issue 1, pp. 63-88.

Mann, D.G. (1999) The species concept in diatoms. Phycologia, Vol. 38, pp. 437-495.

Mann, D.G. and Droop, S.J.M. (1996) Biodiversity, biogeography and conservation of diatoms. Hydrobiologia, Vol. 336, pp. 19-32.

Mann, D.G. and Stickle, A.J. (1988) Nuclear movements and frustule symmetry in raphid pennate diatoms. In: Proc. 9th Int. Diatom Symp., F.E. Round (ed.), Biopress, Bristol and O. Koeltz, Koenigstein, pp. 281-289.

Mann, D.G. and Stickle, A.J. (1995) Sexual reproduction and systematics of *Placoneis* (Bacillariophyta). Phycologia, Vol. 34, pp. 74-86.

Mann, D.G. and Stickle, A.J. (1997) Sporadic evolution of dorsiventrality in raphid diatoms, with special reference to *Lyrella amphoroides* sp. nov. Nova Hedwigia, Vol. 65, pp. 59-77.

Mann, D.G., Simpson, G.E., Sluiman, H.J. and Möller, M. (2001) *rbc*L gene tree of diatoms: a second large data-set for phylogenetic reconstruction. Phycologia, Vol. 40(4), supplement, pp. 1-2.

Medlin, L.K., Gersonde, R., Kooistra, W.H.C.F. and Wellbrock, U. (1996) Evolution of the diatoms (Bacillariophyta) II. Nuclear-encoded small-subunit rRNA sequence comparisons confirm a paraphyletic origin for the centric diatoms. Molecular Biology and Evolution, Vol. 13, pp. 67-75.

Medlin, L.K., Koositra, W.H.C.F., Potter, D., Saunders, G.A. and Andersen, R.A. (1997) Phylogenetic relationships of the "golden algae" (haptophytes, heterokont chromophytes) and their plastids. Plant Systematics and Evolution, supplement 11, pp. 187-219.

Mereschkowsky, C. (1903) Über *Placoneis*, ein neues Diatomeen-Genus. Beihefte zum Botanischen Centralblatt, Vol. 15, pp. 1-30.

Olson, R.J., Vaulot, D. and Chisholm, S.W. (1986a) Effects of environmental stresses on the cell cycles of two marine phytoplankton species. Plant Physiology, Vol. 80, pp. 918-925.

Olson, R.J., Watras, C. and Chisholm, S.W. (1986b) Patterns of individual cell growth in marine centric diatoms. J. of General Microbiology, Vol. 132, pp. 1197-1204.

Pankhurst, R. (1991) Practical taxonomic computing. Cambridge University Press, Cambridge, UK.

Pappas, J.L., Fowler, G.W. and Stoermer, E.F. (2001) Calculating shape descriptors from Fourier analysis: shape analysis of *Asterionella* (Heterokontophyta, Bacillariophyceae). Phycologia, Vol. 40, pp. 438-454.

Patrick, R.M. and Reimer, C.E. (1966) Diatoms of the United States, exclusive of Alaska and Hawaii. Volume 1. Fragilariaceae, Eunotiaceae, Achnanthaceae, Naviculaceae. Monographs of the Academy of Natural Sciences of Philadelphia, Vol. 13, pp. 1-688.

Patrick, R.M. and Reimer, C.E. (1975) Diatoms of the United States, exclusive of Alaska and Hawaii. Volume 2, Part 1. Entomoneidaceae, Cymbellaceae, Gomphonemaceae, Epithemiaceae. Monographs of the Academy of Natural Sciences of Philadelphia, Vol. 13, pp. 1-213.

Paterson, D.M. and Black, K.S. (1999) Water flow, sediment dynamics, and benthic biology. In: Estuaries, D.B. Nedwell and D. Raffaelli (eds). Advances in Ecological Research 29, Academic Press, London, pp. 155-193.

Paterson, D.M. and Hagerthey, S.E. (2001) Microphytobenthos in contrasting coastal ecosystems: biology and dynamics. In: Ecological comparisons of sedimentary shores, K. Reise (ed.). Ecological Studies 151, Springer Verlag, Heidelberg, pp. 105-125.

Paterson, D.M., Wiltshire, K.H., Miles, A., Blackburn, J., Davidson, I., Yates, M.G., McGrorty, S. and Eastwood, J.A. (1998) Microbiological mediation of spectral reflectance from intertidal cohesive sediments. Limnology and Oceanography, Vol. 43, pp. 1207-1221.

Patterson, D.J. (1989) Stramenopiles: chromophytes from a protistan perspective. In: The chromophyte algae: problems and perspectives, J.C. Green, B.S.C. Leadbeater and W.L. Diver (eds), Clarendon Press, Oxford, UK, pp. 357-379.

Pickett-Heaps, J.D., Schmid, A.-M.M. and Edgar, L.A. (1990) The cell biology of diatom valve formation. Progress in Phycological Research, Vol. 7, pp. 1-168.

Pritchard, A. (1861) A history of infusoria, including the Desmidiaceae and Diatomaceae, British and foreign. Whittaker, London (4th edition).

Reynolds, C.S. (1984) The ecology of freshwater phytoplankton. Cambridge University Press, Cambridge, UK.

Ross, R., Cox, E.J., Karayeva, N.I., Mann, D.G., Paddock, T.B.B., Simonsen, R. and Sims, P.A. (1979) An amended terminology for the siliceous components of the diatom cell. Nova Hedwigia, Beiheft 64, pp. 513-533.

Round, F.E. (1981) The ecology of algae. Cambridge University Press, Cambridge, UK.

Round, F.E., Crawford, R.M. and Mann, D.G. (1990) The diatoms. Biology and morphology of the genera. Cambridge University Press, Cambridge, UK.

Saunders, G.W., Potter, D., Paskind, M.P. and Andersen, R.A. (1995) Cladistic analyses of combined traditional and molecular data sets reveal an algal lineage. Proc. National Academy of Sciences of the United States of America, Vol. 92, pp. 244-248.

Schnepf, E., Deichgräber, G. and Drebes, G. (1980) Morphogenetic processes in *Attheya decora* (Bacillariophyceae, Biddulphiineae). Plant Systematics and Evolution, Vol. 135, pp. 265-277.

Stoermer, E.F. and Ladewski, T.B. (1982) Quantitative analysis of shape variation in type and modern populations of *Gomphoneis herculeana*. Nova Hedwigia, Beiheft 73, pp. 347-386.

Stoermer, E.F. and Smol, J.P. (1999) The diatoms: applications for the environmental and earth sciences. Cambridge University Press, Cambridge, UK.

Stosch, H.A. von (1975) An amended terminology of the diatom girdle. Nova Hedwigia, Vol. 53, pp. 1-35.

Stosch, H.A. von (1977) Observations on *Bellerochea* and *Streptotheca*, including descriptions of three new planktonic diatom species. Nova Hedwigia, Beiheft 54, pp. 113-166.

Stosch, H.A. von (1981) Structural and histochemical observations on the organic layers of the diatom cell wall. In: Proc. 6th Symp. on Recent and Fossil Diatoms, R. Ross (ed.), Otto Koeltz, Koenigstein, pp. 231-252.

Stosch, H.A. von and Reimann, B.E.F. (1970) *Subsilicea fragilarioides* gen. et spec. nov., eine Diatomee (Fragilariaceae) mit vorwiegend organischer Membran. Nova Hedwigia, Beiheft 31, pp. 1-36.

Systematics Association Committee for Descriptive Biological Terminology (1962) II. Terminology of simple symmetrical plane shapes (chart 1). Taxon, Vol. 11, pp. 145-156.

Thompson, d'A.W. (1942) On growth and form. Cambridge University Press, Cambridge, UK (2nd edition).

Volcani, B.E. (1978) Role of silicon in diatom metabolism and silicification. In: Biochemistry of silicon and related problems, G. Bendz and I. Lindqvist (eds), Plenum Press, New York and London, pp. 177-204.

CHAPTER 3

DIATOM APPLICATIONS

RICHARD J. TELFORD, STEVE JUGGINS, MARTYN KELLY
AND BERTRAND LUDES

> Diatoms are microscopic, single-celled algae which are found in almost all aquatic habitats. The often excellent preservation of their silica cell walls, together with their widespread distribution and sensitivity to a range of environmental factors, makes them excellent indicators of contemporary and historical environments. Diatom analysis has been used in a range of applications, including water quality assessment, environmental change (including climate change), forensic science, archaeology and oil exploration. This chapter presents a brief review of some of the most common applications of the technique.

1 Introduction

Diatoms are microscopic, single-celled algae of the division Bacillariophyta. They possess a number of attributes that make them excellent indicator organisms in a wide range of earth and life-science applications. Firstly, they are ubiquitous in the environment and form abundant and diverse communities in most aquatic and damp terrestrial habitats. Secondly, diatoms are characterized by a siliceous cell wall (frustule) that is preserved (fossilized) in freshwater and marine sediments after their death. Since diatom taxonomy and identification are largely based on the ornamentation of the silica frustule, both living and fossil material can be readily identified to species level using the same floras. Thirdly, diatoms are sensitive to a range of water quality and other habitat parameters, making them excellent environmental indicators. Fourthly, many diatom taxa are apparently cosmopolitan, so many floras and identification keys have a global, or at least continent-wide application. Finally, diatoms have a fossil record back to at least the lower Cretaceous, making them suitable for investigating environmental change over a range of human and geological timescales.

Table 1 lists a range of diatom applications in the earth, life and medical sciences and humanities, and highlights especially applications in environmental and global change covering a range of timescales from the recent to geological past. This chapter presents a brief review of some of the most important diatom applications in these fields. For a more complete review of diatom applications see Stoermer and Smol (1999) and Battarbee et al. (2001).

2 Forensic science: the diagnosis of drowning

The diagnosis of drowning is one of the most difficult tasks in forensic pathology, particularly when corpses are putrefied. In fresh bodies, the presence of fine froth at

Table 1. Diatom applications.

timescale (years)	1	10	100	1000–10000+	100000+
problem	environmental assessment/ surveillance	environmental impacts/ recovery	post-industrial pollution	natural variability	geological record
applications	biological water quality monitoring	monitoring freshwater quality (lakes and rivers)	lake acidification and eutrophication	estuarine and coastal change, sealevel change	oil and gas exploration
	toxic algal blooms	coastal monitoring	coastal eutrophication	past climate change (marine and continental records)	
	forensic science		UV impacts in Arctic	human impact on aquatic ecosystems	
				archaeology	

the mouth and nostrils may indicate drowning but histological investigations are of primary importance to highlight evidence of drowning-related pulmonary changes (Fornes et al., 1998). When bodies are badly decomposed these observations are not possible and the diagnosis of drowning is practically impossible without biological tests. The goal of these tests is to measure the concentration of a substance whose concentration in blood is affected by the penetration of water or salts into the lungs, such as chloride and iron.

Testing for the presence of diatoms in the cadaver has also been proposed to provide supportive evidence of drowning. Udermann and Schuhmann (1975) and Ranner et al. (1982) considered that an accurate diatom count could discriminate, in most cases, between drowning and non-drowning cases. However, in the absence of clearly accepted positive criteria, the reliability of this test has been disputed since diatoms can also be found in organs of non-drowned victims. Gylseth and Mowé (1979) and Schellmann and Sperl (1979) were of the opinion that this test could not be used as evidence of drowning.

Peabody (1977) showed that to support a positive diagnosis of drowning, rigorous criteria must be followed. Thus, the diatom investigation must be performed using proper techniques, avoiding contamination of glass containers and reagents, and must ensure the correct interpretation of the result, which involves a complete taxonomic analysis of the diatoms recovered from water samples and from the organs of the deceased.

To assess the diagnosis of drowning, Ludes et al. (1999) have established that the analysis may be considered positive if 20 diatoms are identified per 100 μl of a pellet obtained from a 10 g lung sample. For organs such as the brain, liver, kidney and bone marrow, more than 5 complete diatoms per 100 μl of a pellet obtained from a 10 g tissue sample are an indicator of water inhalation.

The qualitative analysis consists of identifying the diatoms and comparing them with those extracted from water samples taken at the site where the body was found. Following Buhtz and Burkhardt (1938), Ludes et al. (1999) have shown that a comparison of the diatom taxa from water samples and lung tissue may be a reliable indicator of the site of drowning, especially if the composition of the diatom flora is monitored throughout the year at sites where drownings occur most frequently. The degree of correlation between the relative abundances of taxa recovered from lung and water samples may be a reliable indicator of the site of drowning.

3 Environmental monitoring

An understanding of the spatial variation of the diatom flora is also a valuable means of inferring environmental quality, whereas repeated sampling at the same locality enables to follow temporal trends, from a few weeks to several years or even decades. In depositional environments such as lakes, it is possible for contemporary and palaeoecological monitoring to be complement each other (Montieth, 2000).

The main applications of such contemporary environmental monitoring are the same as for palaeoecological reconstructions, with techniques available for monitoring acidification, eutrophication and salinity (see below). In addition, there are several methods available for monitoring organic pollution. Some of these methods have been available for 40 years (e.g. Zelinka and Marvan, 1961) although it was only in the past two decades that diatom-based methods have started to take a prominent role in environmental monitoring programs.

Part of the attraction of benthic diatoms in monitoring programs is the ease with which they can be sampled. In rivers with stony bottoms at least five cobbles or small boulders are removed and their top surfaces scrubbed with a hard brush in order to remove the brown surface film. If such substrates are not available, then it is possible to use artificial substrates (such as a length of frayed polypropylene rope) or to collect samples from submerged or emergent macrophytes (Kelly and Whitton, 1998). Once back in the laboratory, the samples are subjected to preparation methods as described in the next chapter and standard slides are made. If stored in a properly curated herbarium, these provide a permanent record of conditions at a site that can be reanalyzed in the future if necessary.

The literature concerning the use of diatoms for pollution monitoring in running waters was reviewed recently by Prygiel et al. (1998) and Kelly (2002); the discussion in this chapter will be limited to three particular case studies which reveal the potential value of diatoms to aid environmental decision-making.

3.1 Use of the Indice Biologique Diatomique in France

French diatomists have been at the forefront of methods development for general water quality monitoring, developing a number of indices which give a single, integrated value for a suite of variables including organic pollution, eutrophication and salinity (see Prygiel and Coste, 2000). Initial success of these indices in the Artois-Picardie region of north-east France (Prygiel and Coste, 1993) led to a wider adoption of the technique throughout France, and prompted the development, in 1995, of the Indice Biologique Diatomique (IBD, Lenoir and Coste, 1996).

Whereas most diatom-based pollution indices use equations similar to the weighted average functions described below in the section on palaeolimnology, the

Table 2. The percentage of sites assigned to water quality classes based on invertebrate, diatom and fish-based indices of water quality measured throughout France in 1998. Source: Réseau National des Données sur l'Eau (2001).

quality status	invertebrates Indice Biologique Global Normalisé	diatoms Indice Diatomique Biologique	fish Indice Poissons
very good	10	3	31
good	42	28.3	31
fair	32	55.4	19
bad	12	13	12
very bad	4	0.3	7

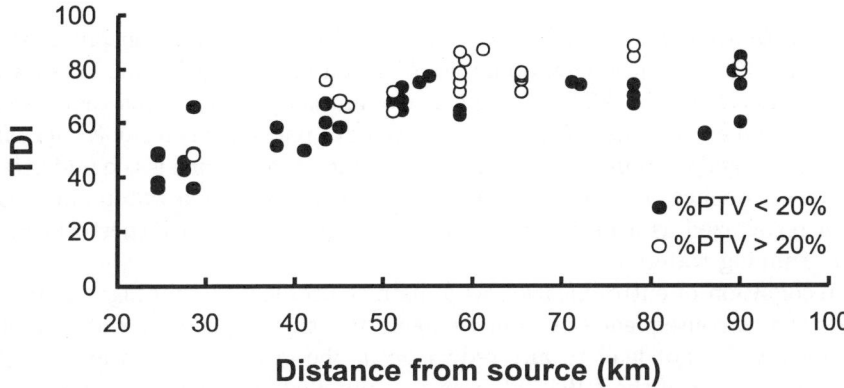

Figure 1. Plot of Trophic Diatom Index (TDI) vs. distance from source for the River Wear (UK); after Kelly (2001).

IBD uses presence-probability profiles for 209 taxa (Lenoir and Coste, 1996) and provides a value between 1 (very poor water quality) and 20 (very good quality) at each site. Associated methodological guidelines (Prygiel and Coste, 2000) enabled Agences d'Eau throughout France to use the technique in order to provide an overview of water quality throughout the country. This showed only 3% of all examined sites to have IBD values >16 (i.e. in the highest—least-polluted—water quality class), whereas 69% of all sites had IBD values indicating fair, bad or even very bad quality (Table 2). These figures are quite different to those for water quality indices based on benthic invertebrates and fish, both of which placed higher proportions of the same sites into the highest categories. Rather than simply replicating each other, indices based on different groups of organisms appear to be highlighting different aspects of water quality, with diatoms presumably reflecting the impact of nutrient enrichment at levels below which manifestations (e.g. on dissolved oxygen concentrations) are not reflected in the invertebrate or fish indices.

3.2 Monitoring acid streams in the UK

A consequence of the success of diatom-based methods for the reconstruction of historical changes in pH in lakes (see below) was new legislation in Western Europe to control the emissions of sulphur dioxide and other gaseous pollutants responsible for acidification. Follow-up monitoring continued at a number of standing and running water sites in order to examine the benefits of the reductions in "acid rain" on affected water bodies.

Contemporary diatom samples, along with chemical and other biological variables, were collected from a range of water bodies over the course of 12 years by the UK acid water monitoring network (Montieth, 2000). At some sites changes in diatom community composition indicated a reduction of acidity and a return towards pre-acidification communities.

3.3 Monitoring eutrophication in rivers

The Trophic Diatom Index (TDI, Kelly and Whitton, 1995) was conceived specifically to help the UK meet new obligations presented by the Urban Wastewater Treatment Directive (UWWTD, European Community, 1991). This required the installation of an extra phase of treatment to remove phosphorus from sewage effluents that discharged into eutrophic rivers. Eutrophication—which is the undesirable biological consequence of too many inorganic nutrients entering a watercourse—has long been recognized as a problem in standing waters, but had hitherto been neglected in running waters.

The recognition of eutrophication as a distinct problem in running waters was, in some ways, a consequence of improvements in other aspects of water quality, which typically did not lead to any reduction in the growth of "nuisance" algae. An important caveat of the UWWTD was that nutrient removal was required only if a benefit could be established. The TDI was based on a weighted-average equation and produced a value ranging from 1 (very low nutrients) to 100 (very high nutrients). It differed from many other indices mentioned in this chapter in that the interpretation of the TDI requires calculation of a second value, originally termed the "Percent Pollution Tolerant Valves" (%PTV). This was necessary to ensure that the ecological manifestations of high concentrations of inorganic nutrients were not confused with the effects of organic pollution. %PTV was, in effect, a measure of the reliability of the TDI as an estimator of eutrophication. A further point to note is that a consequence of high phosphorus concentrations is often that phosphorus is no longer the limiting nutrient (Kelly, 1998).

A feature of diatom-based indices and transfer functions (e.g. Pan et al., 1996; Coring et al., 1999; Rott et al., 1999) is that their predictive power is relatively low, with typically only 60–70% of total variation being explained by regressing the index against ambient phosphorus concentrations. This may reflect the spatial and temporal variability of phosphorus (Pan et al., 1999) as well as the influence of other factors such as grazing and hydrology on the diatom assemblage. Despite this variability, however, it is still possible to detect significant changes in the TDI between sites based on repeated measurements over a period of time (Kelly, 2001), see Fig. 1. In this example from the River Wear in north-east England, diatom indices provided one of a number of strands of evidence to decision-makers, eventually leading to an investment programme to raise the quality of sewage treatment that will cost approximately 11 million euros.

3.4 Future developments

The emphasis for evaluating environmental quality of running waters is slowly changing to one that emphasizes ecological, rather than chemical, characteristics of the water body and which measures these in relation to a "pristine" (or at least pre-impact) state. The main legislative focus for this in Europe has been the Water Framework Directive (WFD) which bases a designation of "good ecological status" on evidence from a suite of organisms, including (for freshwaters) invertebrates, fish, macrophytes, benthic microflora and phytoplankton. In practice, as diatoms are the most numerous group of the benthic microflora, they will almost certainly

Chapter 3: Diatom applications

play a key role in national implementations of the WFD.

Methods for assessing "good ecological status" (also known as "ecological integrity" and "ecosystem health") using diatoms are still in their infancy, with important questions about both the definition of the baseline conditions and the measurement of deviations yet to be answered. In the case of standing waters, it may be possible to define baseline conditions using palaeolimnological techniques (see below). In running waters, the "baseline" may involve either the "best available" contemporary sites or models based on the taxa expected to occur at a site in the absence of any anthropogenic influences. No systems have yet been developed to a workable state: Chessman et al. (1999) in Australia and Hill et al. (2000) in the USA have each tried different techniques, but there is still a long way to go. The only certainty is that diatoms are likely to become a more important part of the freshwater manager's monitoring toolkit over the next decade or so.

4 Palaeoecological applications

Palaeoecology is the attempt to reconstruct past biological communities and their relationships with the abiotic environment. Many biological, chemical and physical environmental indicators have been used (see Lowe and Walker, 1997 for a review), but diatoms are amongst the most useful biological indicators. The examples illustrated below show some of the range of habitats, timescales, places and reconstructed variables studied using diatom palaeoecology.

4.1 Recent pollution histories

Diatoms have proved useful in reconstructing pollution histories in aquatic environments, increasing the understanding of the causes of these changes. The two cases where diatom evidence has been most useful are eutrophication, caused by increased plant nutrient availability (Wetzel, 2001), and lake acidification. Lake acidification became an important environmental issue in the late 1960s when declining fish populations in Scandinavian (Jensen and Snekvik, 1972) and Canadian lakes (Beamish and Harvey, 1972) were blamed on "acid rain" resulting from the emission of sulphur and nitrous oxides from power stations burning fossil-fuels. Other scientists (Rosenqvist, 1978; Krug and Frink, 1983; Pennington, 1984) and industry rejected these claims, instead suggesting that acidification followed land-use changes (such as forestry) or was the result of long-term progressive leaching of catchment soils.

The timing and extent of acidification needed to be understood before these hypotheses could be evaluated. However, since no sites had chemical monitoring spanning the period of interest (since the start of the industrial revolution), alternative, proxy records were required. As diatoms are sensitive to pH and acid lakes contain a diverse diatom flora, they have been one of the key indicators used to study acidification.

The Round Loch of Glenhead is an intensively studied (Jones et al., 1989; Allott et al., 1992) acid upland waterbody in south-west Scotland. The bedrock is granite and the catchment is not forested, although surrounding areas are. A short (45 cm) sediment core spanning over 150 years revealed substantial changes in the diatom flora during this period (Fig. 2). Before 1900, *Brachysira vitrea* dominated the assemblage, *Eunotia incisa* was rare and *Tabellaria quadriseptata*, an acid-loving taxon was very infrequent. After about 1900 there is a progressive decline in the frequency of *B. vitrea* and an increase in *E. incisa* and *T. quadriseptata*. These changes can be interpreted as a drop in pH, as *B. vitrea* is less tolerant to low pH conditions than the other two species. A more quantitative reconstruction can be made using a transfer function, based on the distribution of diatom taxa in many lakes with various pH values. This shows pH declining from about 5.5 to less than 5.

A longer sediment core (Jones et al., 1989), spanning most of the Holocene, from the same lake shows no significant long-term acidification or response to catchment vegetation changes reconstructed from the pollen record. Instead acidification began about 1900. Increasing industrialization at this time is shown by an increase in spherical carbonaceous particles (from the incomplete combustion of fossil fuels) and heavy metal concentrations, both indicators of atmospheric pollution. The coincidence of acidification and atmospheric industrial pollution supports the acid rain hypothesis.

4.2 Holocene/Quaternary climate change

The Quaternary, spanning the last ca. 2 million years, was a period of repeated intense climatic change: diatoms have been used successfully to reconstruct these climatic fluctuations. The direct relationship between climate and diatoms is weak; lake chemistry is normally a more important constraint (but see Joynt and Wolfe, 2001). Indirect links, however, especially in closed basin lakes, as mediated by lake depth and salinity, can be strong (e.g. Barker et al., in press), but need to be interpreted carefully (Gasse et al., 1997).

Lake Rukwa (Barker et al., in press) is a large, shallow, saline, closed-basin lake in the rift valley in southern Tanzania. Evidence from the nearby Lake Malawi (Finney and Johnson, 1991) and climate models (Cohmap, 1988) suggested that the late Quaternary climate of this region was out of phase with the better-known climate signal in northern Africa (e.g. Gasse, 1977). A 12.8 m core spanning the last 22,000 years was analyzed for diatoms. The resulting diatom stratigraphy can be interpreted both in terms of lake depth and salinity. The basal section of the core, 22.0–19.5 ka BP, contains benthic and facultative planktonic freshwater species. From 16.1–15.0 ka BP the assemblage is dominated by saline diatom species including *Nitzschia frustulum* and *Thalassiosira* spp. Paradoxically, this is interpreted as an increase in lake level, possibly caused by a more humid climate, as the central saline lake floods marginal freshwater swamps. After a possible hiatus, synchronous with the Younger Dryas, a series of planktonic freshwater diatoms follow between 13.5 and 6.6 ka BP, indicating deep nutrient-rich freshwater conditions. The late Holocene sees a return to saline planktonic *Thalassiosira* spp. A transfer function derived from diatom distributions in African lakes of various chemistries (Gasse et al., 1995) was used to generate a quantitative reconstruction of lake chemistry. These data show that the late glacial was arid, the early Holocene was humid, followed by an arid late Holocene. This duplicates the northern Africa climate signal; either the climatic hinge is located further south or previous work

Chapter 3: Diatom applications

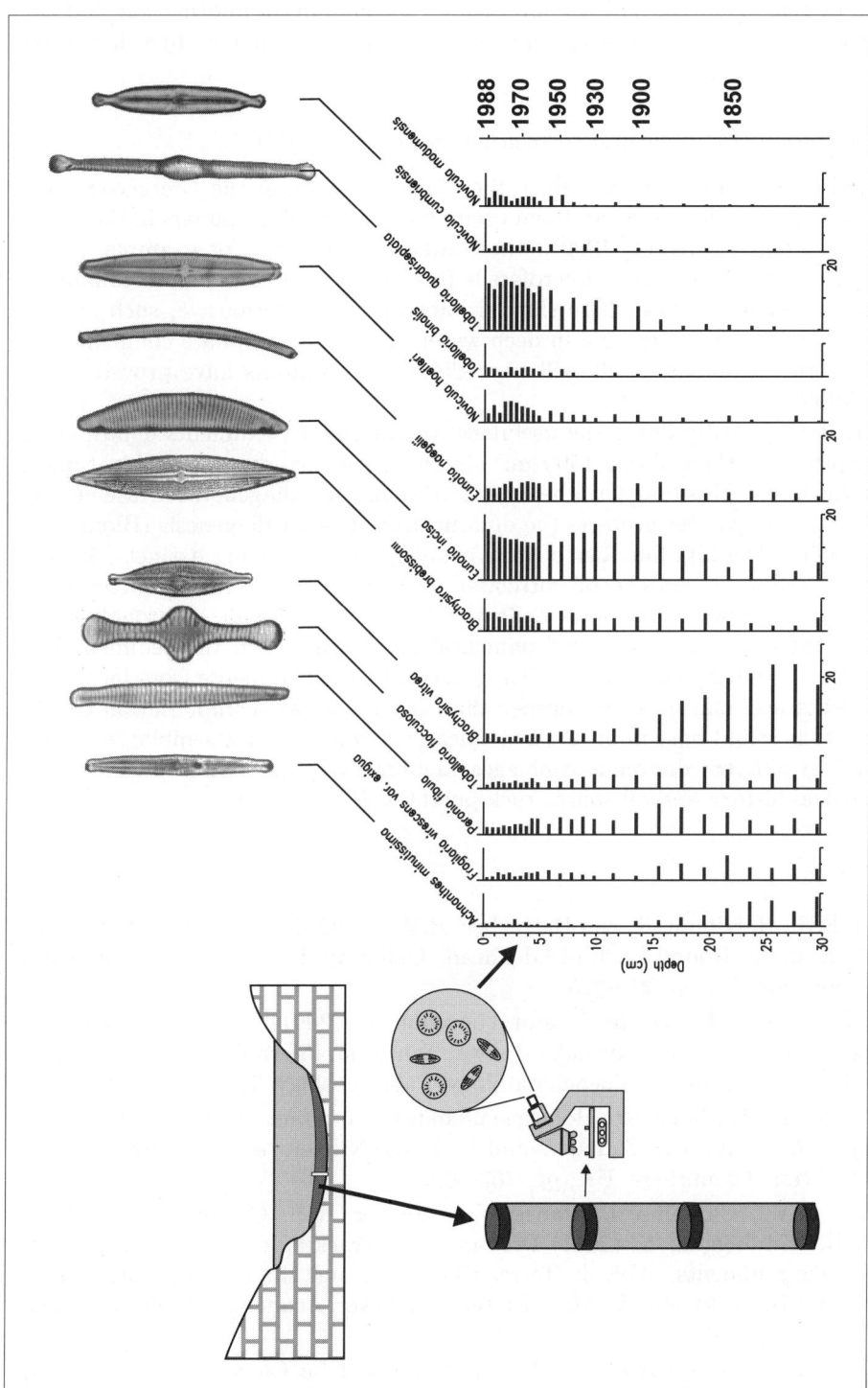

Figure 2. Diatom stratigraphy of the Round Loch of Glenhead (data from Allott et al., 1992).

needs to be reinterpreted. This example shows the amount of information that can be inferred from a diatom stratigraphy, especially when supported by other proxy data.

4.3 Biostratigraphy, geological correlation and oil exploration

The rapid evolution of diatoms, since they first appeared in the Cretaceous, and their diverse morphologies makes them useful biostratigraphic markers in the Cenozoic (last 65 ma, see Krebs, 1999), where they can be used, for example, in the exploration for oil. Calcareous microfossils (e.g. foraminifera) are more commonly used, but in environments where they are absent or uninformative, such as cold water, brackish/fresh water, and in deep water below the carbonate compensation depth (where calcareous fossils will have dissolved), diatoms have proved useful (Krebs, 1999).

Diagenesis typically limits the usefulness of diatoms to sediments deposited in water depths less than about 1500 m: dissolution makes diatoms unidentifiable beyond this limit. This range can be extended when early diagenetic encasement in calcite nodules or pyrites protects the diatoms from further diagenesis (Blome and Albert, 1985). The high number of taxa evolving and becoming extinct (Barron, 1983) allow sediment cores to be correlated with the first and last occurrences of taxa, and if these events are independently dated, they provide a chronology for the core. Information on the environmental conditions when the sediment was deposited can be deduced from the diatom assemblage. Sediments from lacustrine environments and marine water deeper than the carbonate compensation depth, both of which would host impoverished calcareous microfossil assemblages, would generate very different diatom assemblages, and have very different implications for the geological history and oil source rock potential (Krebs, 1999).

References

Allott, T.E.H., Harriman, R. and Battarbee, R.W. (1992) Reversibility of lake acidification at the Round Loch of Glenhead, Galloway, Scotland. Environmental Pollution, Vol. 77, pp. 219-225.

Barker, P., Telford, R., Gasse, F. and Thévenon, F. (2001) Late Pleistocene and Holocene paleohydrology of Lake Rukwa, Tanzania, inferred from diatom analysis. Palaeogeography, Palaeoclimatology, Palaeoecology (in press).

Barron, J.A. (1983) Miocene to Holocene planktonic diatoms. In: Plankton stratigraphy, H.M. Bolli, J.B. Saunders and K. Perch-Nielsen (eds), Cambridge University Press, Cambridge, UK, pp. 763-809.

Battarbee, R.W., Carvalho, L., Jones, V.J., Flower, R.J., Cameron, N.G., Bennion, H. and Juggins, S. (2001) Diatoms. In: Tracking environmental change using lake sediments. Vol. 3: Terrestrial, algal, and siliceous indicators, J.P. Smol, H.J.B. Birks and W.M. Last (eds), Kluwer Academic Publishers, The Netherlands (in press).

Beamish, J. and Harvey, H.H. (1972) Acidification of La Cloche Mountain lakes, Ontario, and resulting fish mortalities. Journal of the Fisheries Research Board of Canada, Vol. 29, pp. 1131-1143.

Blome, C.D. and Albert, N.R. (1985) Carbonate concretions – an ideal sedimentary host for microfossils. Geology, Vol. 13, pp. 212-215.

Buhtz, M. and Burkhardt, W. (1938) Die Festellung des Ertränkungortes an dem Diatomeenbefund der Lungen. Gerichtliche Medizin, Vol. 29, pp. 469-484.

Chessman, B., Growns, I., Currey, J. and Plunkett-Cole, N. (1999) Predicting diatom communities at the genus level for the rapid biological assessment of rivers. Freshwater Biology, Vol. 41, pp. 317–331.

Cohmap, M. (1988) Climatic changes of the last 18,000 years. Observations and model simulations. Science, Vol. 241, pp. 1043-1052.

Coring, E. Schneider, S. Hamm, A. and Hofmann, G. (1999) Durchgehendes Trophiesystem auf der Grundlage der Trophieindikation mit Kieselalgen. Deutscher Verband für Wasserwirtschaft und Kulturbau e. V., Koblenz, Germany.

European Community (1991) Council directive of 21 May 1991 concerning urban waste water treatment (91/271/EEC). Official J. of the European Community, Series L, Vol. 135, pp. 40-52.

Finney, B.P. and Johnson, T.C. (1991) Sedimentation in Lake Malawi (East-Africa) during the past 10,000 years — a continuous paleoclimatic record from the Southern tropics. Palaeogeography, Palaeoclimatology, Palaeoecology, Vol. 85, pp. 351-366.

Fornes, P., Pépin, G., Heudes D. and Lecomte D. (1998) Diagnosis of drowning by combined computer-assisted histomorphometry of lungs with blood strontium determination. J. of Forensic Science, Vol. 43, pp. 772-776.

Gasse, F. (1977) Evolution of Lake Abhé (Ethiopia and TFAI) from 70,000 BP. Nature, Vol. 265, pp. 42-45.

Gasse, F., Juggins, S. and Ben Khelifa, L. (1995) Diatom-based transfer functions for inferring past hydrochemical characteristics of African lakes. Palaeogeography, Palaeoclimatology, Palaeoecology, Vol. 117, pp. 31-54.

Gasse, F., Barker, P., Gell, P.A., Fritz, S.C. and Chalie, F. (1997) Diatom-inferred salinity in palaeolakes: an indirect tracer of climate change. Quaternary Science Reviews, Vol. 16, pp. 547-563.

Gylseth, B. and Mowé, G. (1979) Diatoms in the lung tissues. Lancet, Vol. 29, p. 1375.

Hill, B.H., Herlihy, A.T., Kaufmann, P.R., Stevenson, R.J., McCormick, F.H. and Johnson, C.B. (2000) Use of periphyton assemblage data as an index of biotic integrity. J. of the North American Benthological Society, Vol. 19, pp. 50-67.

Jensen, K.W. and Snekvik, E. (1972) Low pH levels wipe out salmon and trout populations in southernmost Norway. Ambio, Vol. 1, pp. 223-225.

Jones, V.J., Stevenson, A.C. and Battarbee, R.W. (1989) Acidification of lakes in Galloway, south west Scotland – a diatom and pollen study of the post-glacial history of the Round Loch of Glenhead. J. of Ecology, Vol. 77, pp. 1-23.

Joynt, E.H. and Wolfe, A.P. (2001) Paleoenvironmental inference models from sediment diatom assemblages in Baffin Island lakes (Nunavut, Canada) and reconstruction of summer water temperature. Canadian J. of Fisheries and Aquatic Sciences, Vol. 58, pp. 1222-1243.

Kelly, M.G. (1998) Use of the trophic diatom index to monitor eutrophication in rivers. Water Research, Vol. 32, pp. 236-242.

Kelly, M.G. (2001) Role of benthic diatoms in the implementation of the Urban Wastewater Treatment Directive in the River Wear, NE England. J. of Applied Phycology (in press).

Kelly, M.G. (2002) Water quality assessment by algal monitoring. In: The handbook of environmental monitoring, F. Burden, I. McKelvie, A. Guenther and U. Förstner (eds), McGraw-Hill, New York (in press).

Kelly, M.G. and Whitton, B.A. (1995) The Trophic Diatom Index: a new index for monitoring eutrophication in rivers. J. of Applied Phycology, Vol. 7, pp. 433-444.

Kelly, M.G. and Whitton B.A. (1998) Biological monitoring of eutrophication in rivers. Hydrobiologia, Vol. 384, pp. 55-67.

Krebs, W.N. (1999) Diatoms in oil and gas exploration. In: The diatoms: applications for the environmental and earth sciences, E.F. Stoermer and J.P. Smol (eds), Cambridge University Press, Cambridge, UK, pp. 402-412.

Krug, E.C. and Frink, C.R. (1983) Acid rain on acid soil: a new perspective. Science, Vol. 221, pp. 520-525.

Lenoir, A. and Coste, M. (1996) Development of a practical diatom index of overall water quality applicable to the French National Water Board Network. In: Use of algae for monitoring rivers II, B.A. Whitton and E. Rott (eds), Universität Innsbruck, Austria, pp. 29-43.

Lowe, J.J. and Walker, M.J.C. (1997) Reconstructing quaternary environments. Addison Wesley Longman, UK (2nd edition).

Ludes, B., Coste, M., North, N., Doray, S., Tracqui, A. and Kintz P. (1999) Diatom analysis in victims' tissues as an indicator of the site of drowning. Int. J. of Legal Medecine, Vol. 112, pp. 163-166.

Montieth, D. (2000) Biology discussion. In: United Kingdom acid waters monitoring network 10 year report, D.T. Monteith and C.D. Evans (eds), ENSIS, London.

Pan, Y.D., Stevenson, R.J., Hill, B.H., Herlihy, A.T. and Collins, G.B. (1996) Using diatoms as indicators of ecological conditions in lotic systems: a regional assessment. J. of the North American Benthological Soc., Vol. 15, pp. 481-495.

Pan, Y.D., Stevenson, R.J. and Hill, B.H. (1999) Spatial patterns and ecological determinants of benthic algal assemblages in mid-Atlantic streams, USA. J. of Phycology, Vol. 35, pp. 460-468.

Peabody, A.J. (1977) Diatoms in forensic science. J. of Forensic Science, Vol. 17, pp. 81-87.

Pennington, W. (1984) Long-term natural acidification of upland sites in Cumbria: evidence from post-glacial lake sediments. Freshwater Biological Association Annual Report, Vol. 52, pp. 28-46.

Prygiel, J. and Coste, M. (1993) The assessment of water quality in the Artois-Picardie water basin (France) by the use of diatom indices. Hydrobiologia, Vol. 269/270, pp. 343-349.

Prygiel, J. and Coste, M. (2000) Guide méthodologique pour la mise en oeuvre de l'Indice Biologique Diatomées. NFT 90-354. Étude Agences de l'Eau-Cemagref Bordeaux, March 2000, Agences de l'Eau, France.

Prygiel, J., Coste, M. and Bukowska, J. (1998) Review of the major diatom-based techniques for the quality assessment of continental surface waters. In: Use of algae for monitoring rivers, J. Prygiel, B.A. Whitton and J. Bukowska (eds), Agence de l'Eau Artois-Picardie, Douai, France, pp. 224-238.

Ranner, G., Juan, H. and Udermann M. (1982) Zum Beweiswert von Diatomeen im Knochenmark beim Ertrinkungstod. Zeitschrift für Rechtsmedizin, Vol. 88, pp. 57-65.

Reseau National des Données sur L'Eau (2001) La qualité biologique des cours d'eau en France. Reseau National des Données sur L'Eau, Toulouse, France.

Rosenqvist, I.T. (1978) Alternative sources for acidification of river water in Norway. The Science of the Total Environment, Vol. 10, pp. 39-49.

Rott, E., Pipp, E., Pfister, P., van Dam, H., Ortler, K. Binder, N. and Pall, K. (1999) Indikationslisten für Aufwuchsalgen in österreichischen Fliessgewässern. Teil 2: Trophieindikation. Bundesministerium für Land- und Forstwirtschaft, Wien, Austria.

Schellmann, B. and Sperl, W. (1979) Nachweis im Knochenmark (Femur) Nichtertrunkener. Zeitschrift für Rechtsmedizin, Vol. 83, pp. 319-324.

Stoermer, E.F. and J.P. Smol (1999) The diatoms: applications for the environmental and earth sciences. Cambridge University Press, Cambridge, UK.

Udermann, M. and Schuhmann, G. (1975) Eine verbesserte Methode zum Diatomeen-Nachweis. Zeitschrift für Rechtsmedizin, Vol. 76, pp. 119-122.

Wetzel, R. G. (2001) Limnology: lake and river ecosystems. Academic Press, USA (3rd edition).

Zelinka, M. and P. Marvan (1961) Zur Präzisierung der biologischen Klassifikation der Reinheit fliessender Gewässer. Archiv für Hydrobiologie, Vol. 57, pp. 389-407.

CHAPTER 4

ADIAC IMAGING TECHNIQUES AND DATABASES

MICHA M. BAYER AND STEVE JUGGINS

This chapter outlines the process of image acquisition and data management involved in ADIAC. To make ADIAC possible, large databases of digital images of diatoms were required which had to be compiled specifically for the project because pre-existing material was unavailable. A total of approximately 6000 images were acquired over the duration of the project, including large test sets which were used for the testing of prototype software packages. The databases cover approximately 500 diatom taxa, and represent a wide range of different habitats and morphologies.

1 Microscopy and image acquisition

1.1 Preparation of diatoms for microscopy

Diatom collection from the natural habitat usually involves filtering water *in situ* by means of a plankton net, or collecting some of the natural substratum in the case of benthic or terrestrial diatoms (e.g. mud, aquatic plants etc). Diatoms are ubiquitous organisms found wherever there is sufficient light for photosynthesis and at least a minimal amount of water: samples from almost all aquatic and damp terrestrial habitats will usually contain diatoms.

The preparation of diatom samples involves cleaning with strong oxidizing agents, e.g. concentrated acids, to remove all organic matter (for a detailed protocol see Round et al., 1990). Following the oxidization step, one is left with only the siliceous parts of the diatom frustules that were present in the sample, which at this point have usually disintegrated into their components, i.e. valves and girdle bands, along with any inorganic particles such as sand or silt, if these were present in the original sample. Some of the latter may be removed by differential sedimentation, which is based on different settling rates for diatoms (slow) and large mineral particles (fast). This method, however, is limited by the fact that settling rates may be similar for diatoms and particles of similar sizes and densities, and consequently a complete separation of diatoms and inorganic particles is usually impossible.

It is, however, possible to separate live raphid diatoms from their substratum by making use of their phototactic behavior and their ability to move through sediments. For this, samples of mud or sand are placed in a flat dish, and the water is removed after allowing the suspended particles and diatoms to settle. The surface of the sample is then covered in lens tissue which provides a barrier for particulate matter but can be penetrated by diatoms moving to the sediment surface for photosynthesis. This usually takes several hours, after which the lens tissue may be removed and dissolved in concentrated acids as described above, resulting in a suspension of diatom valves and girdle elements without contaminating particles.

After cleaning, diatoms are dried onto coverslips and mounted onto glass slides using a high refractive index mountant such as Naphrax (RI greater than 1.65). Preparations of diatoms in Naphrax are effectively permanent, enabling them to be kept in a reference collection for future use.

1.2 Microscopic technique

Slides were examined on one of two photomicroscopes: a Zeiss Axiophot with 63x and 100x oil immersion lenses (both with a numerical aperture of 1.4), and a Reichert Polyvar 2 with a 100x oil immersion lens (NA 1.32). To maximize resolution, the following measures were implemented:

- Immersion oil was used both between the objective lens and the slide, and between the slide and the condenser, which maximizes the numerical aperture and thus the optical resolution (Bradbury, 1984).

- A monochromatic green filter was used to reduce the range of wavelengths and the possibility of chromatic aberration (Bradbury, 1984).

- Most valves were photographed with the condenser diaphragm almost fully open to ensure maximum resolution. Resulting images tend to be "flat" in terms of contrast, but contrast and brightness can be adjusted digitally afterwards if necessary. Thus, full resolution of morphological detail is possible without the loss of contrast. However, for the purpose of feature extraction the contrast was left unadjusted.

All diatoms were photographed using brightfield optics only. There are illumination systems that offer a better contrast, such as phase and differential interference contrast (DIC), but it was decided not to use these since they introduce "artefacts" into the image that are very difficult to standardize. Both scanning electron microscopy (SEM) and DIC introduce artificial shading giving the impression of directional illumination, which makes automatic feature extraction irregular and irreproducible.

1.3 Image acquisition

For the ADIAC project we used two digital cameras at the Royal Botanic Garden Edinburgh: a Kodak MegaPlus ES1.0 attached to the Zeiss Axiophot photomicroscope, and a Kodak MegaPlus 1.4 attached to the Reichert Polyvar photomicroscope. Both cameras were interfaced with personal computers running Optimas imaging software, versions 5.2 and 6.2 (MediaCybernetics, Silver Spring, MD 20910, USA). The MegaPlus ES1.0 has a resolution of 1008×1018 pixels, and allows image resolutions between 7 and 18 pixels/μm using the 63x and 100x objective lenses (both with an NA of 1.4), with magnification changers giving an extra 1.25x and 1.6x magnification where necessary. The MegaPlus 1.4 has a resolution of 1312×1024 pixels, and allows image resolutions of between 13 and 32 pixels/μm using a 100x objective (NA 1.32), with the magnification changer giving additional magnifications of 0.8x, 1.25x and 2.0x. All images were acquired directly into Optimas by means of custom written macros that control the image acquisition process.

Chapter 4: Imaging techniques and databases

Table 1. Resolutions of digital cameras used for ADIAC, in relation to microscope magnification, based on (a) a Zeiss Axiophot light microscope with apochromatic lenses as specified below and a Kodak MegaPlus ES1.0 digital camera (field of view 1008 pixel wide), and (b) a Reichert Polyvar 2 light microscope with an apochromatic 100x lens and a Kodak MegaPlus 1.4 digital camera (field of view 1312 pixel wide). Camera resolutions were calculated by dividing the camera's pixel number (horizontally) by the width of its field of view (FoV). FoV width is given in μm.

microscope	lens	magnif. changer setting	eyepiece FoV width	camera FoV width	camera resolution (pixels/μm)
Axiophot	oil 63x	1	388	141	7.15
Axiophot	oil 63x	1.25x	309	112	9.00
Axiophot	oil 63x	1.6x	241	88	11.45
Axiophot	oil 100x	1x	250	92	10.96
Axiophot	oil 100x	1.25x	200	73	13.81
Axiophot	oil 100x	1.6x	157	57	17.68
Polyvar	oil 100x	0.8x	294	101	12.99
Polyvar	oil 100x	1.0x	236	83	15.81
Polyvar	oil 100x	1.25x	189	66	19.88
Polyvar	oil 100x	2.0x	118	41	32.00

Irrespective of their size, valves were captured using the maximum magnification that allowed for the entire specimen to be captured in the field of view. The resulting resolutions in pixels per μm of object are given in Table 1. Ideally, capture should result in a resolution of 8 pixels/μm or more if all optically resolved detail is to be visualized in the resulting digital image. This digital resolution is more or less the minimum that can capture all the optical resolution of which the microscope is capable. The theoretical limit of microscope resolution is about 0.2 μm (Bradbury, 1984), but nearer 0.25 μm in practice; this equates to a stria density of about 40 per 10 μm (4/μm), where the striae appear as alternating dark and light lines. To resolve this detail would require at least 2 pixels per stria (one dark and one light) or 8 pixels/μm.

Since digital cameras have a fixed number of pixels, it follows that the larger the camera's field of view in microns, the lower its resolution in pixels per μm of image. It also follows that the lower the microscope magnification (to fit a larger diatom into the field of view), the lower the resolution possible with the camera. The overwhelming majority of images in ADIAC, however, were taken at magnifications that result in digital resolutions of more than 8 pixels/μm, i.e. with the 63x and 100x lenses (Table 1).

1.4 Image preprocessing

Digital images usually contain artefacts or imperfections that are part of the magnification, illumination and imaging systems, and nothing to do with the specimens themselves. They include unevenness of illumination, dust particles on optical surfaces between the specimen and the CCD chip, and texture in coatings on glass surfaces. Each of these affects all images similarly, and can be removed relatively

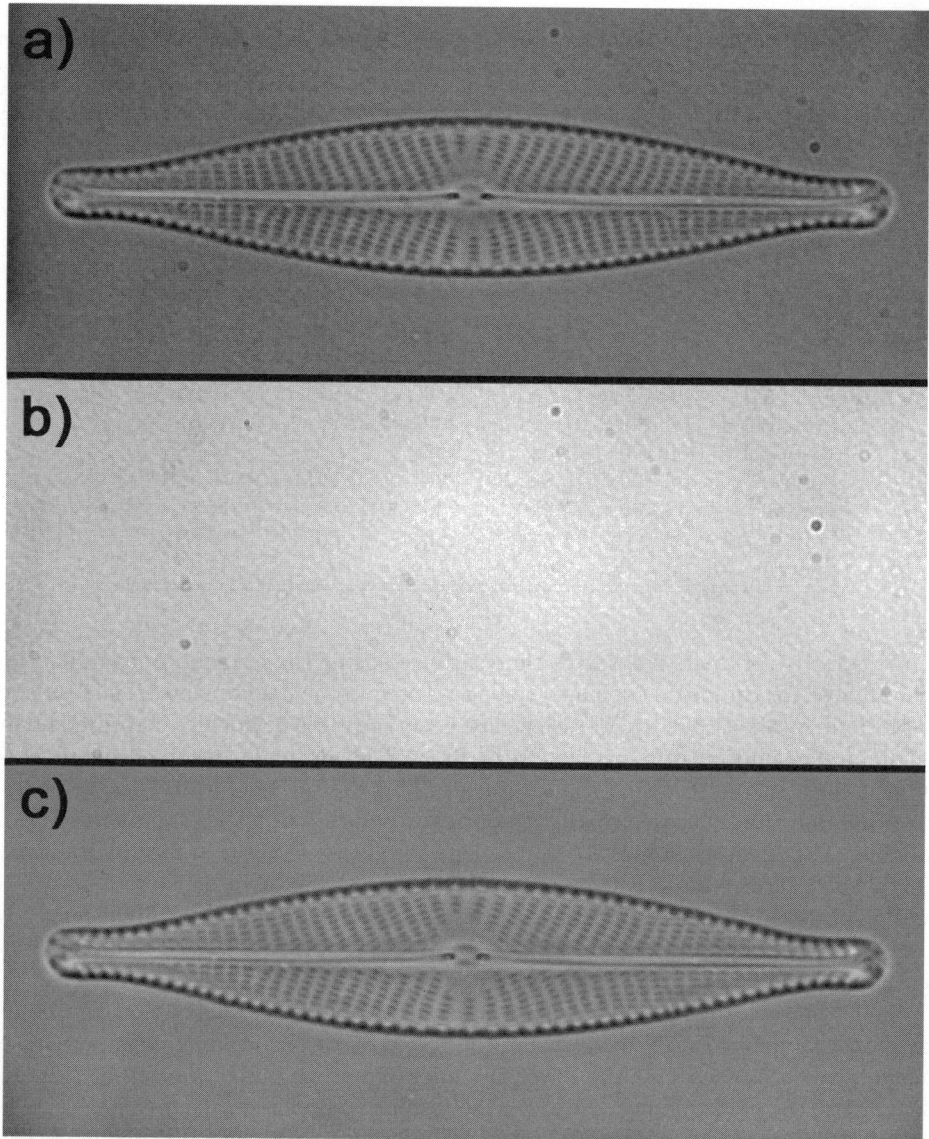

Figure 1. Removal of background texture and camera dust specks by image arithmetic. For the removal of background texture, resulting from the protective glass screen in the camera and dust specks adhering to it, a slightly darkened image of the specimen and the unwanted background objects that are to be removed (a) is divided by a brighter image without the specimen, which is simply removed by defocusing (b). The resultant image (c) contains the specimen, but not the dust particles and background texture which are common to both input images. Taxon illustrated: *Navicula rhynchocephala* Kützing 1844. The approximate length of the diatom is 50 μm.

easily using an image arithmetic function in the Optimas software. To achieve this, a darkened foreground image containing the specimen and the noise (Fig. 1a) is combined with a brightened background image of the noise only (Fig. 1b, where the specimen is simply defocused) using an arithmetical *divide* function, and the resultant image is of the specimen, but free of all imperfections common to both images (Fig. 1c). In the arithmetical operation underlying this, the foreground (numerator) pixel greyvalue is first augmented by 1 to eliminate the zero values resulting from black pixels, and then multiplied by a scaling factor, usually 255, to achieve final quotients in the range from 0 to 255 (Russ, 1992). The denominator is the greyvalue of the corresponding pixel from the background noise image. Thus, dust specks, which are dark in the foreground image (low greyvalue) and only slightly lighter in the background image, become disproportionately lighter than the object itself, which is dark in the foreground image and absent from the background image (high greyvalue). This procedure eliminates camera-related noise by blending it into the background.

A special source of image defects (but not corrected by the above procedure) are malfunctioning pixels on the CCD chip. These have to be corrected individually, but since they are always in the same place, the corrections can be done using an Optimas macro that changes the value of a faulty pixel to the value of an adjacent pixel.

An Optimas macro has also been developed that allows the placement of a scale bar as an image overlay. This places a 10 μm scale bar with a 1 μm square at one end, the latter to check for changes in pixel shape between different image viewers. Near the scale bar, the macro automatically writes the name of the calibration used, dependent on the magnification of the microscope. However, scale bars are undesirable for image analysis purposes and hence were used only with images set aside for viewing in the searchable WWW databases.

2 Image Databases

In ADIAC, digital images were required for several purposes:

- as sample material for the development of feature extraction algorithms
- as reference material against which unknown specimens can be compared, both by the software and by end users who want to verify a result given by the system
- as test material for the quantification of identification success

In the following section we describe the individual categories of image in detail.

2.1 General images

Images of this type were intended for algorithm development and as reference material for users wishing to verify automatically generated identifications. Diatoms for the general image category were selected according to strict quality criteria: valves had to be level (to avoid shading and focusing artefacts), undamaged and clean, with the valve facing up and no debris touching or over-/underlying the valve (Fig. 2). Most taxa covered by this type of image were represented by few specimens (approx. 1-3), and, in effect, this set of images represents taxonomic

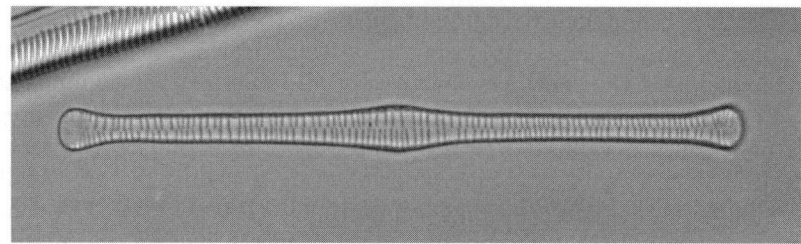

Figure 2. Example image from the general image database. All images in this category had to be of clean, level diatom valves, without over- or underlying material or debris touching the sides of the valve. Taxon illustrated: *Tabellaria quadriseptata* Knudson. Approximate length of diatom 70 μm.

breadth, rather than depth. Two key objectives guided the choice of taxa to be included in this image set. The first was to include taxa that covered a broad range of different morphologies, to ensure that the feature extraction algorithms covered most of the taxonomic characters present in diatoms. The second objective was to include diatoms from representative environments, that could be used as case studies to exemplify a real-world application of the identification system. To this end we scanned diatoms from a range of Northwest European coastal environments and UK freshwater rivers and lakes. The first set of material provides taxa useful for monitoring water quality in coastal environments and for studies of past sea-level change. The second set covers taxa commonly used in Europe to monitor the quality of flowing and standing freshwaters. Together these datasets include both benthic and planktonic taxa and provide a diverse range of centric and pennate morphologies. All of the images in this category were also made available to the public via the WWW image database.

2.2 Filled outlines data set

This image set was developed early in the project and was used for the development of outline analysis algorithms. It consisted of 100 manually thresholded, filled outlines of binarized diatoms (see Fig. 3 for an example) and consisted of a mixed set of 100 images, representing 69 taxa from a range of genera, each taxon counting between one and three specimens.

2.3 Mixed genera data set

This set of images served the primary purpose of training and testing prototype software with a view to obtaining robust predictions of identification rates. This required the inverse of the approach taken with the general image databases, i.e. taxonomic depth rather than breadth. With a specimen-based system such as that developed in ADIAC, taxonomic information is derived by the system from a training image set that has to convey a taxonomic concept of each taxon to the system. This clearly places critical importance on the process of specimen selection for the image set. For this data set, a minimum of 20 specimens per taxon was

Figure 3. Example image from filled outlines test set. Images were manually thresholded in Optimas image analysis software to include the exact outline of the diatom, then binarized and filled automatically. Taxon illustrated: *Hannaea arcus* var. *arcus* (Ehrenberg) Patrick in Patrick and Reimer. Approximate length of diatom 70 µm.

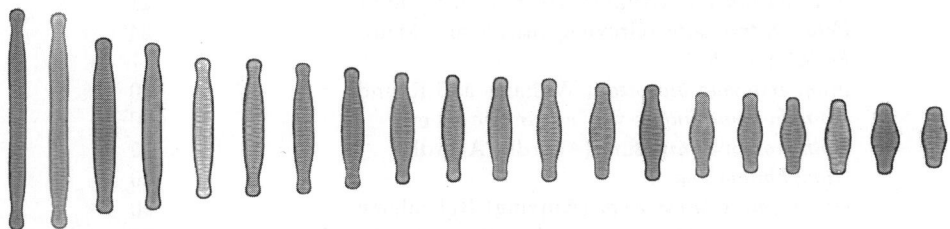

Figure 4. Example set of specimens used for taxon representation in the mixed genera test set (*Fragilariforma bicapitata* Williams and Round). Specimens were sampled to cover the entire range of sizes present on the slide under examination, and a minimum of 20 specimens were used per taxon. Length of longest specimen approx. 45 µm.

collected where possible, and the sampling of specimens was stratified such that the full size range was covered with specimens evenly spread out along the valve length gradient (Fig. 4). An ideal data set should also take into account other sources of morphological variation, such as among-populations variation, but this was beyond the scope of the current pilot project and will hopefully be covered in a future project.

Specimens for the mixed genera data set had to conform to the same quality criteria as images in the general image category (see Section 2.1 above), i.e. only slides containing small amounts of debris were used. The taxa selected for this set were mostly common European freshwater species, but several brackish water and marine taxa were also included (Table 2). Where possible, at least two taxa per genus were included to test for the discrimination of similar morphologies. In total, this data set included 48 taxa, represented by a total of 1009 images, but due to image acquisition continuing to the end of the project, and asynchrony in the testing, the actual size of data sets described in this book varies, although most

Table 2. The mixed genera data set counts 48 taxa and a total of 1009 images.

taxon name	images
Caloneis amphisbaena (Bory) Cleve	20
Cocconeis neothumensis Krammer	20
Cocconeis placentula var. *placentula* Ehrenberg	20
Cocconeis stauroneiformis (W. Smith) Okuno	22
Cymbella helvetica Kützing	26
Cymbella hybrida var. *hybrida* Grunow in Cleve and Möller	20
Cymbella subequalis Grunow in Van Heurck	21
Denticula tenuis Kützing	22
Diatoma mesodon (Ehrenberg) Kützing	26
Diatoma moniliformis Kützing	20
Encyonema silesiacum (Bleisch in Rabenhorst) D.G. Mann	25
Epithemia sorex var. *sorex* Kützing	20
Eunotia denticulata (Brébisson) Rabenhorst	22
Eunotia incisa Gregory	20
Eunotia tenella (Grunow) Hustedt in Schmidt	21
Fallacia forcipata (Greville) Stickle and Mann	24
Fallacia sp. 5	17
Fragilariforma bicapitata Williams and Round	20
Gomphonema augur var. *augur* Ehrenberg	20
Gomphonema minutum (Agardh) Agardh	25
Gomphonema sp. 1	20
Gyrosigma acuminatum (Kützing) Rabenhorst	20
Meridion circulare var. *circulare* (Greville) Agardh	20
Navicula capitata var. *capitata* Ehrenberg	20
Navicula constans var. *symmetrica* Hustedt	23
Navicula gregaria Donkin	11
Navicula lanceolata (Agardh) Ehrenberg	21
Navicula menisculus Schumann	20
Navicula radiosa Kützing	21
Navicula reinhardtii var. *reinhardtii* Grunow in Van Heurck	29
Navicula rhynchocephala Kützing	20
Navicula viridula var. *linearis* Hustedt	20
Nitzschia dissipata (Kützing) Grunow	20
Nitzschia hantzschiana Rabenhorst	20
Nitzschia sinuata var. *sinuata* (Thwaites) Grunow	20
Nitzschia sp. 2	27
Opephora olsenii Möller	20
Parlibellus delognei (Van Heurck) Cox	20
Petroneis humerosa (Brébisson ex Smith) Stickle and Mann	20
Pinnularia kuetzingii Krammer	21
Sellaphora bacillum (Ehrenberg) D.G. Mann	20
Stauroneis smithii Grunow	20
Staurosirella pinnata (Ehrenberg) Williams & Round	20
Surirella brebissonii Krammer and Lange-Bertalot	22
Tabellaria flocculosa (Roth) Kützing	20
Tabellaria quadriseptata Knudson	23
Tabularia investiens (Smith) Williams and Round	20
Tabularia sp. 1	20

groups worked with an earlier version consisting of 781 images of 37 taxa. Example images of the taxa in this data set are shown in the Appendix.

2.4 Sellaphora pupula *data set*

This data set consisted of taxa that are morphologically very similar but display subtle yet consistent differences. *Sellaphora pupula* Mereschkowsky is a complex of taxa that differ from one another in morphology (e.g. Mann, 1984), susceptibility to parasitism (Mann, 1999), reproductive strategy (Mann, 1984, 1989), sexual compatibility (Mann, 1989), and at the molecular level (Mann, 1999, and unpublished data). The evidence clearly suggests that these taxa deserve species status, but until formal descriptions have been published they are referred to as "demes." The term "deme" was first coined by Gilmour and Gregor (1939) and simply denotes a group of organisms that share a common feature. Prefixes may be added to define the feature in question (Gilmour and Heslop-Harrison, 1954), e.g. "geno-", which makes a genodeme a group of organisms that share the same genotype.

There are numerous demes of *Sellaphora pupula*, and their number may be as high as 100 or more worldwide (D.G. Mann, pers. comm.). Surveys of previously unexplored lakes regularly show up new demes, and for the purpose of this data set six new demes were used which occur in a reservoir near Edinburgh (Threipmuir).

The initial analysis and description of these demes was carried out by means of a visual classification and a multivariate morphometric analysis. Characters used for the analysis included valve length, valve width, stria frequency, stria genicularity, rectangularity, and a total of eight Fourier descriptors. The result of a principal components analysis of these variables is shown in Fig. 5 for the six most frequent demes, which subsequently were used for testing. Although some of the demes separated completely in morphospace, without any overlap between groups, others showed considerable overlap, but with clearly differing group centroids.

From the initial set of 200 specimens sampled, a stratified subset of 120 specimens was chosen, representing these six most frequent demes with exactly 20 randomly chosen specimens each. The demes were given the provisional names "large," "tidy," "elliptical," "pseudoblunt," "pseudocapitate" and "cf. rectangular" (Fig. 6). Morphological measurements of the demes are given in Table 3.

2.5 *Problem images*

Images for this category were taken with the intention of testing the behavior of feature extraction algorithms in more challenging situations. Most diatom slides contain at least some inorganic debris and fragments of diatom frustules, which can pose difficulties for automated feature extraction. The amount of debris varies with the type of sample, but slides from paleoecological studies of sediment cores in particular tend to be plagued by debris-related problems. The lens tissue extraction method (see Section 1.1) cannot be applied to these samples because it relies on the motility of live diatoms, whereas sediment cores contain valves from dead diatoms only, with the exception of the top few millimeters of the core. There are also a number of other potential complications which may affect the extraction of features from diatoms in images. These include:

Table 3. Descriptive statistics of *Sellaphora pupula* demes used for the test set. SD = standard deviation. All measurements are in μm.

deme		length	width	stria frequency	rectangularity	shape
elliptical	range	17.02–28.30	6.83–7.65	19.94–22.7	0.801–0.822	elliptical
	mean	22.25	7.24	21.10	0.810	
	SD	3.56	0.23	0.70	0.007	
large	range	23.01–39.76	8.63–9.86	18.41–21.4	0.842–0.880	rectangular-capitate
	mean	33.30	9.31	19.56	0.863	
	SD	3.92	0.31	0.69	0.010	
pseudoblunt	range	15.87–23.25	6.60–7.2	20.28–23.57	0.808–0.840	slightly capitate, blunt-ended
	mean	19.63	6.99	21.88	0.826	
	SD	1.59	0.17	0.86	0.009	
pseudocapitate	range	23.44–31.9	7.84–8.58	20.25–25.05	0.818–0.84	elliptical-capitate
	mean	27.44	8.15	22.60	0.831	
	SD	2.68	0.188	1.24	0.009	
cf. rectangular	range	20.18–26.67	6.75–7.33	22.02–24.62	0.825–0.854	rectangular-capitate
	mean	23.71	7.02	23.04	0.839	
	SD	2.10	0.19	0.69	0.007	
tidy	range	18.93–28.41	7.04–8.32	15.79–24.40	0.773–0.790	elliptical, ends rostrate
	mean	22.13	7.46	22.32	0.782	
	SD	2.21	0.28	1.76	0.005	

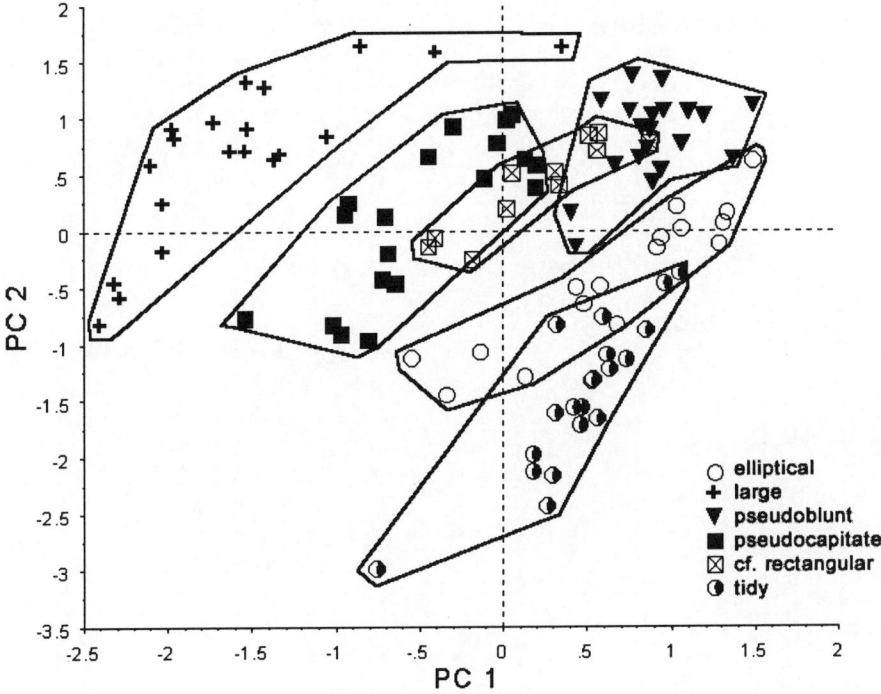

Figure 5. Principal component analysis of morphometric measurements from the six *Sellaphora pupula* demes used for testing. Of the three principal components extracted only the first two are shown here. These accounted for 61.2% (PC1) and 14.9% (PC2) of the total variation. The six demes have been outlined by drawing around outlying data points. Characters used for the analysis included valve length, valve width, stria frequency, stria genicularity, rectangularity, and a total of eight Fourier descriptors.

- other diatoms over-/underlying the diatom valve and creating a continuous shape
- inverted valves, which have a light-on-dark outline, as opposed to valves in the correct (i.e. valve face up) orientation, which have a dark-on-light outline
- fragmented valves, which tend to be particularly common in sediment samples
- valves which do not lie exactly horizontally and thus do not provide a homogeneous outline of a narrow range of grey values

A special case of diatom samples with extraneous material are forensic samples, which are discussed in Chapter 3. As part of ADIAC, Partner IML has accumulated approx. 600 problem images from forensic samples (see Fig. 7 for an example). At present the algorithms developed during ADIAC are not adapted for dealing with contaminated diatom samples, and this will have to be addressed as part of a future project. There is clearly a great need to do so, as few diatom samples conform to

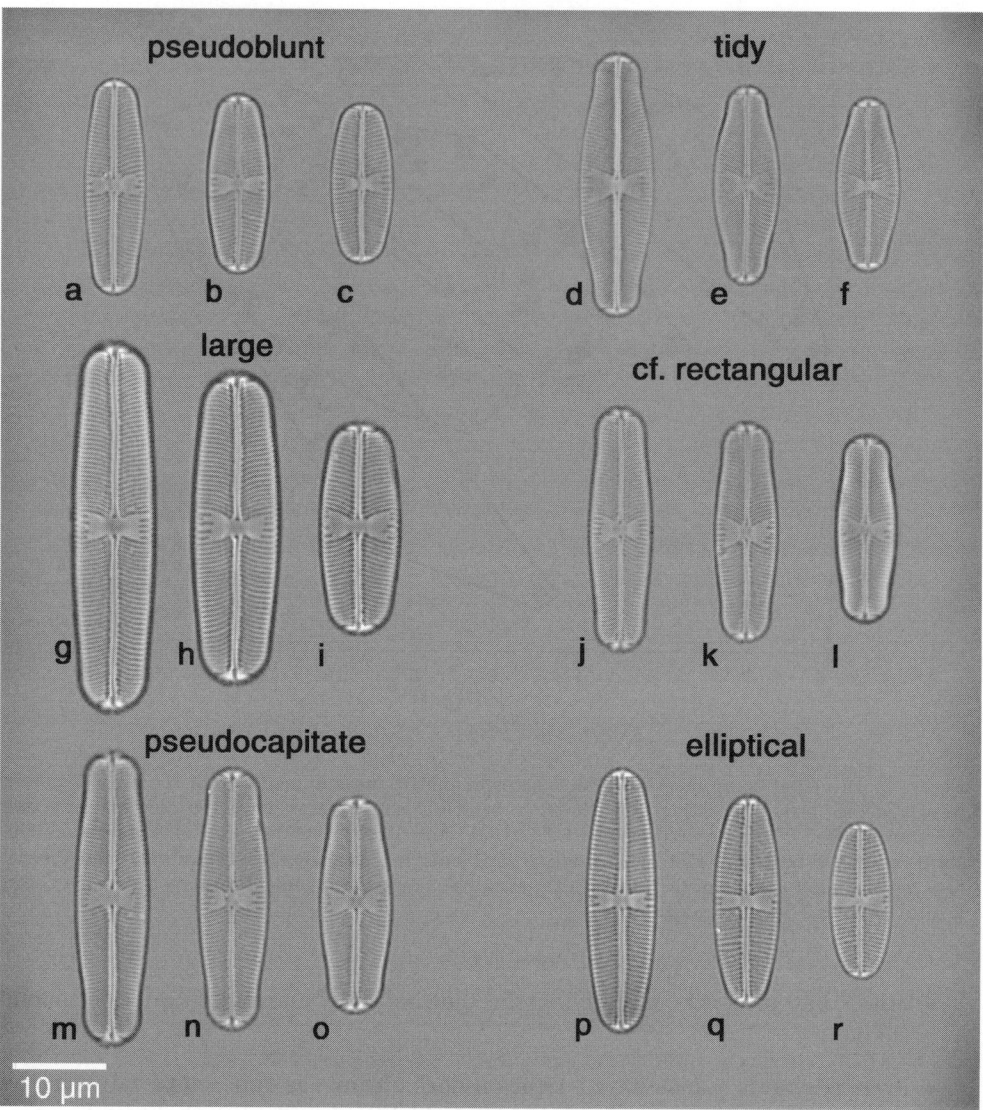

Figure 6. Example valves representing the six *Sellaphora pupula* demes used in the test set. For each deme, the longest, shortest and one valve of intermediate length are shown. All valves are to the same scale.

the strict criteria imposed on the collection of reference images during the project. However, it was clearly necessary to develop and test the algorithms with clean material before addressing the challenges of contaminated samples. Nevertheless, preliminary experiments have shown that striated regions can be extracted from valve images contaminated with debris (Chapter 9). Such regions could be used to steer contour extraction.

Chapter 4: Imaging techniques and databases

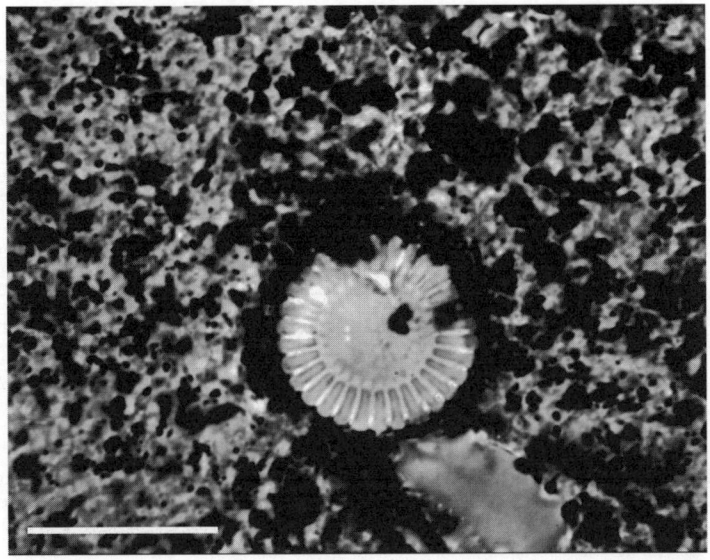

Figure 7. Example of test image used in the ADIAC image databases. This image shows a diatom extracted from forensic human tissue samples and exemplifies the difficulties that many diatom samples can cause for image analysis, in this case an obscured valve perimeter which impedes extraction of a clean outline (*Cyclotella meneghiniana* Kützing). Scale bar 10 μm.

3 Pandora taxonomic database

To provide a taxonomic context for the taxa shown on ADIAC images, a database had to be created that allowed the storage of taxonomic information about diatom specimens. The database system of choice for this was RBGE's Pandora database system.

Pandora is a specialist taxonomic database package which has been developed specifically for use with taxonomic data (Pankhurst, 1993), and has been used at the Royal Botanic Garden Edinburgh for many years. It is an implementation of the *Advanced Revelation* database system, which allows for indexed multi-value fields, an important feature for the storing of synonyms, where a one-to-many relationship is required.

The use of diatom data necessitated some changes to the existing Pandora version. Originally, Pandora was developed for use with higher plants, and data storage in the Records table is largely based on the assumption of individual plant specimens mounted on individual herbarium sheets. In the case of diatoms, however, specimens are usually stored on permanent microscope slides, which are in turn derived from field samples and, in some cases, subsamples, necessitating the implementation of a hierarchical data storage system. This was achieved by splitting the existing Records table into five subtables which are linked to each other hierarchically in a top-down fashion (Fig. 8):

- Collecting Events

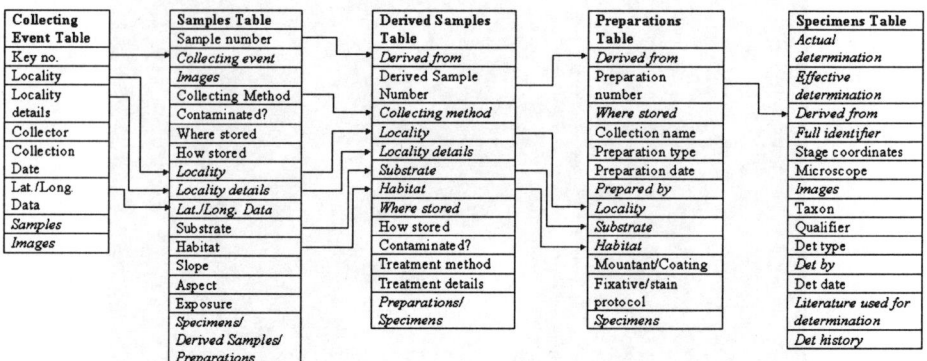

Figure 8. Diagram illustrating the revised specimen records section of the Pandora diatom database system; italicized field names indicate that the data are dependent on data in another field.

- Samples
- Derived Samples
- Preparations
- Specimens

Thus, a specimen record would be linked upwards to a preparation that the specimen is held on (e.g. a microscope slide), which would be linked upwards to a derived sample the preparation was prepared from (e.g. a lens tissue extraction), itself linked upwards to an original sample, and beyond this, to a collecting event (Fig. 8). Data from higher level subtables are displayed via "symbolic" fields in subtables of lower levels, e.g. the locality data from the Collecting Events table are also shown in the Samples, Derived Samples and Preparations tables, as access to this information may be required at any level.

Also, a new table had to be created to hold image data, and, like the revised Records table, this was divided into subtables ("Original Images" and "Derived Images") which store the technical detail of the images attached to a specimen record (Fig. 9). The remainder of the Pandora database system (Fig. 9) was used unchanged and includes:

- the Ptaxon table, which contains taxonomic data
- the Persons table, which holds the names and related data of researchers involved in the collecting, processing or naming of taxa
- the Author Index, which contains authority strings made up from records in the Persons table

Chapter 4: Imaging techniques and databases

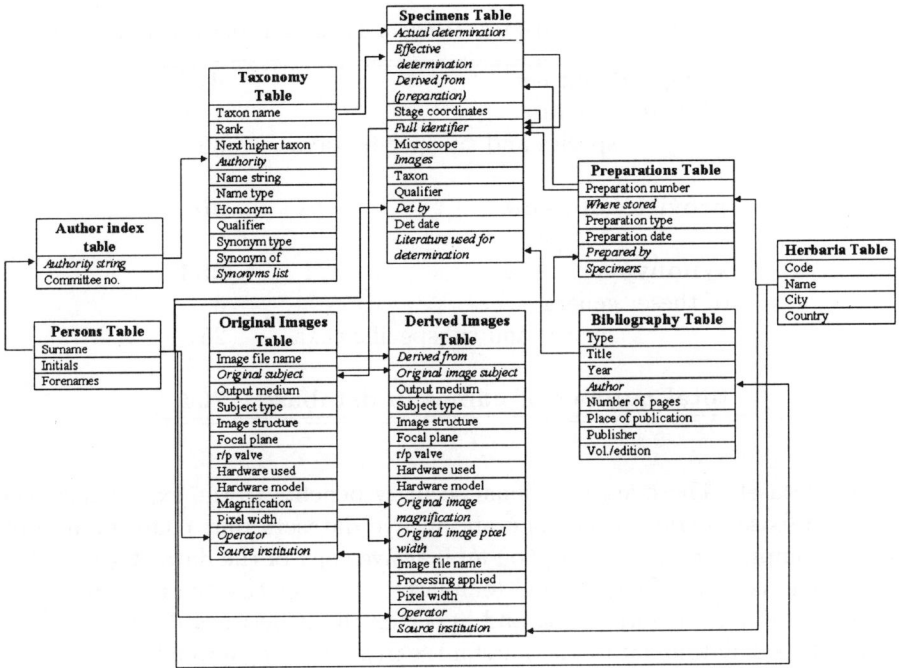

Figure 9. Diagram showing the most important relationships between data fields in the Pandora diatom database system; italicized field names indicate that the data are dependent on data in another field. Only selected tables and fields are shown.

- the Herbaria table, containing details about institutions involved in the provision of samples and images
- the Bibliography table, which stores full literature references

The complete Pandora diatom database comprises 247 data fields in 12 tables and subtables. The composition and number of taxa covered in the database at the end of the ADIAC project is summarized in Table 4.

4 WWW database

The Pandora taxonomic database system is an implementation of the *Advanced Revelation* database system, which is still DOS-based, and does not allow for direct interfacing with web browsers like other database systems. Therefore, Martin Pullan at RBGE has developed a data export facility, HTMLExport (Pullan, 1999), which allows to export HTML-formatted data from the Pandora database.

This module consists of a set of RBASIC routines which perform the data export process, and a PERL search engine which uses the standard Common Gateway Interface (CGI) on web servers. Data can be exported together with indexes, and the structure of the data in the Pandora database can effectively be mirrored in the

Table 4. Breakdown of taxa in ADIAC's Pandora database at the end of the project.

accepted names	**470**
of these: genera	109
species and subspecific taxa	361
provisional names	**40**
synonyms	**254**
of these: genera	3
species and subspecific taxa	251
total number of names in database	**764**

exported dataset. There are additional security benefits to the export mechanism in that database queries do not operate on live datasets that could be interfered with and compromized, but rather on an inactive copy of the data. Export of data is controlled via a set of customized templates which specify the data and indexes to be exported, output formatting and hypertext links between data fields.

The ADIAC diatom dataset is therefore searchable using a web browser, and has been made accessible for public use on the Internet: http://www.rbge.org.uk/ADIAC/db/adiacdb.htm. In addition to the standard form search facility, browse lists were also implemented, giving users an overview of the taxa currently held in the database. Searches are carried out using taxon names or part thereof, and can use either accepted names or synonyms. A search carried out using a synonym as the search term will result in the user being redirected to a search option for the accepted name.

Successful searches return all available specimens of the taxon searched for, and thumbnail images are displayed which are themselves links to larger size images of the specimens. Clicking on the taxon name directs the user to a page containing synonyms and ecological information about the taxon. The RBGE database has been linked to the ecological database at the University of Newcastle via a simple link in the exported data which activates a taxon-code based query into the ecological database and returns ecological information for a given taxon (see the next section). Images included in the database are restricted to clean valves only; problem and test images were excluded due their limited usefulness for the general public. All images are in either in GIF (thumbnails) or JPEG format (larger size images). There are approximately 2300 images available in the WWW database.

4.1 Ecological database

Diatoms are sensitive ecological indicators of water quality and environmental change. Within Europe they are routinely used to monitor the eutrophication and acidification of freshwaters, and current research is developing practical methods for biomonitoring of the coastal zone for a range of pollutants (see Chapter

3). The image database and automated identification procedures developed within ADIAC will thus be directly applicable to monitoring the quality of a wide range of European waters. Correct identifications and consistently applied taxonomy are primary requirements for any practical biomonitoring tool. However, the use of diatoms or any other organism as a bioindicator of environmental conditions also requires knowledge of the taxon's ecological response to the pollutant, or chemical parameter(s) being monitored. Thus, an important part of any biomonitoring system is the linking of taxonomic information derived from species identifications to ecological information describing the environmental conditions represented by a particular diatom assemblage, or association of different taxa.

Traditionally, diatomists have used an indexing system to encode information about the distribution of a taxon with respect to a chemical or other environmental parameter. The majority of these indices simply classify a taxon into one of a small number of categories representing its indicator value or optima, and/or range. To date, indices have been developed for a range of parameters including pH, salinity, trophic status, saprobity or organic pollution, and many common taxa have been classified according to these systems (e.g. Van Dam et al., 1994; Kelly and Whitton, 1995). In ADIAC, we have included a subset of this information for a limited number of key ecological parameters most relevant to current environmental monitoring activities: pH, salinity and trophic status. Information about the indicator values of each taxon in the ADIAC image database has been collated from literature searches and entered into a relational database running on a web server under Microsoft Access RDMS. The ecological database is accessed using Microsoft Active Server Pages (ASP) code that dynamically generates the appropriate SQL and queries the database via a web interface. Although the web interface could operate in "stand-alone" mode, it is currently used to provide information to the main ADIAC image database webpages at RBGE. Information from the ecological database in Newcastle is dynamically embedded into the RBGE webpages when users query the main ADIAC image database. This solution allows the two laboratories to enter data independently into their own databases, but ensures that the various types of information are seamlessly and transparently integrated in a single web interface.

One major problem encountered when attempting to search for diatom taxon names stored in different databases is that different names (synonyms) are often used to refer to the same taxonomic entity. Normally, the use of multiple synonyms is a result of frequent taxonomic revisions of various diatom groups, and the delay in updating older records. However, in some cases there is disagreement among diatomists as to the correct name for a taxon, with different laboratories adopting different naming conventions for what is the same taxon. To avoid the problem of failed searches because a user has asked for information using an old or disputed name, we have also included a complete list of synonyms for all taxa in the ecological database, and embedded appropriate ASP code in the web pages to search for ecological data by valid name or synonym. This means that the database will return the correct ecological information for a taxon, regardless of whether the database is searched using the currently valid name or a synonym. As already mentioned above, searches carried out with synonyms as the search term redirect the user to another search interface where the accepted name is automatically used for a new search. This is clearly not an optimal option, but rather a compromise which reflects the problem of multiple classifications in taxonomy. In an ideal world, only a single set of taxon names would be in use at any time, but the nature of the taxonomic process leads to new names being generated for taxa already described under another name (see Chapter 2). This needs to be taken into account by any new technology designed to help users of taxonomy, such as paleoecologists.

One potential solution to this problem is the use of complex database systems, such as the Prometheus system, currently under development at RBGE and Napier University Edinburgh (Pullan et al., 2000). One of the long-term aims for the development of this system is to give users control over which classification they use to view a given set of taxa, thereby avoiding the complications of multiple classifications. This will involve combining visualization tools, which are already available from phase 1 of the project, with datasets that allow for the definition of taxon boundaries on the basis of, for example, suites of extracted values of morphological characters. Taxa will then be defined by their member specimens, and the concepts behind a classification will become apparent, because differences between taxa in competing classifications can be visualized as differing combinations of member specimens. In the meantime, however, with this technology still under development, automatic identification software, as presented in other chapters of this book, could be adapted relatively easily to offer identification output in a choice of several existing classifications.

References

Bradbury, S. (1984) An introduction to the optical microscope. Oxford University Press, Oxford, UK.

Gilmour, J.S.L. and Gregor, J.W. (1939) Demes: a suggested new terminology. Nature (London), Vol. 144, pp. 333-334.

Gilmour, J.S.L. and Heslop-Harrison, J. (1954) The deme terminology and the units of micro-evolutionary change. Genetica, Vol. 27, pp. 147-161.

Kelly, M. and Whitton, B. (1995) The Trophic Diatom Index: a new index for monitoring eutrophication in rivers. J. Applied Phycology, Vol. 7, pp. 433-444.

Mann, D.G. (1984) Observations on copulation in *Navicula pupula* and *Amphora ovalis* in relation to the nature of diatom species. Annals of Botany, Vol. 54, pp. 429-438.

Mann, D.G. (1989) The species concept in diatoms: evidence for morphologically distinct, sympatric gamodemes in four epipelic species. Plant Systematics and Evolution, Vol. 164, pp. 215-237.

Mann, D.G. (1999) The species concept in diatoms. Phycologia, Vol. 38, pp. 437-495.

Pankhurst, R.J. (1993) Taxonomic databases: the Pandora system. In: Advances in computer methods for systematic biology: artificial intelligence, databases, computer vision, R. Fortuner (ed.), Baltimore, USA, pp. 229-240.

Pullan, M.R. (1999) HTMLEXPORT. See
http://rbge-sun1.rbge.org.uk/data/htmlexport/

Pullan, M.R., Watson, M.F., Kennedy, J.B., Raguenaud, C. and Hyam, R. (2000) The Prometheus Taxonomic Model: a practical approach to representing multiple classifications. Taxon, Vol. 49, pp. 55-75.

Round, F.E., Crawford, R.M. and Mann, D.G. (1990) The Diatoms. Biology and Morphology of the Genera. Cambridge University Press, Cambridge, UK.

Russ, J.C. (1992) The image processing handbook. CRC Press, Boca Raton (FL), USA.

Van Dam, H., Mertens, A. and Sinkeldam, J. (1994) A coded checklist and ecological indicator values of freshwater diatoms from the Netherlands. Netherlands J. of Aquatic Ecology, Vol. 28, pp. 117-133.

CHAPTER 5

HUMAN ERROR AND QUALITY ASSURANCE IN DIATOM ANALYSIS

MARTYN G. KELLY, MICHA M. BAYER, JOACHIM HÜRLIMANN AND RICHARD J. TELFORD

Diatom analysis is subject to a variety of errors, most of which are controllable to some extent. In this chapter, sources of errors in human diatom identification are described and the results of two exercises to quantify human error rates are presented. These exercises demonstrated considerable variation between participants that was not related to the length of individual experience, as well as variation in the consistency with which particular taxa and genera are identified. These results are put into context through an interlaboratory calibration exercise ("ring test") involving 21 analysts on an epilithon sample from a European stream. This demonstrated that variability in the number of taxa recorded and species diversity is greater than that in pollution indices computed on the same samples, presumably because much of the taxonomic variability relates to less common taxa that do not make a large contribution to weighted average calculations. These results emphasize the need for agreed taxonomic protocols for both human and computer-assisted applications of diatom analysis.

1 Introduction

In this chapter the automated identification of diatoms will be placed in a wider context. In particular, we present results from a unique experiment investigating human error rates in diatom identification, in order to provide an objective baseline against which automated identification systems should be evaluated. There are numerous practical applications of diatoms (see Chapter 3), of which environmental monitoring will be considered in this chapter. The use of diatoms in monitoring has increased over the past two decades or so, to the extent that diatom data are becoming an important part of the decision-making process for some environmental issues (e.g. acidification, eutrophication). Since the enforcement of such legislation can result in an investment of several million euros per waterbody, there are clearly serious implications if the quality of the data is prejudiced in any way.

The application of formal quality assurance (QA) mechanisms is one means by which data quality can be assured. Various international standards provide frameworks for QA, of which ISO 9000 (International Standards Organisation, 1994) is the best known, whereas ISO 17025 (International Standards Organisation, 1999) has specific applications for analytical laboratories. However, these existing standards lack a specifically ecological dimension and CEN, the European Standards Organisation, is presently developing a guidance standard for the QA of ecological fieldwork. One important element of this—of practical relevance to ADIAC—will be protocols for ensuring consistency in identifications. In assessing the success

Table 1. Comparison of the reported and the revised composition of planktonic diatoms in Lough Neagh, November 11, 1910. Original analysis by Dakin and Latarche (1913), re-analysis by Battarbee (1979). Table modified from Table 2 of Battarbee (1979). Note that several of the names used have been revised subsequently. The status of *Asterionella gracillima* is disputed, with most modern workers regarding it as a morphotype or variety of *A. formosa*. Dakin and Latarche (1913) recorded both *A. gracillima* and *A. formosa* in Lough Neagh. See text for more details.

Dakin and Latarche (1913)		Battarbee (1979)	
taxon	%	taxon	%
Tabellaria fenestrata var. asterionelloides	81	Tabellaria flocculosa	80
Asterionella gracillima	18	Asterionella formosa	13
Coscinodiscus lacustris	0.2	Melosira italica subsp. subarctica	3
Fragilaria crotonensis	0.2	Stephanodiscus astraea	1.4
Melosira sp.	0.1	Melosira ambigua	0.8
Synedra sp.	0.04	Melosira islandica subsp. helvetica	0.25
Stephanodiscus astraea	0.01	Stephanodiscus dubius	0.2
		Cyclotella comta	0.2
		Stephanodiscus astraea var. minutula	0.2

of ADIAC, there is a need to ensure that error rates associated with automated identification do not exceed those associated with human identification, and that results generated by identification software are consistent and repeatable.

The need for consistency in taxonomy and identification is well illustrated by a re-examination of plankton samples collected from Lough Neagh, Northern Ireland, in 1910 (Battarbee, 1979; Table 1). This table illustrates good agreement between the two most common genera (*Tabellaria* and *Asterionella*) but disagreement about the correct species in both cases. Battarbee (1979) also recognizes a number of discrepancies in the identification of less common taxa in the sample. A comparison with a recent checklist (Whitton et al., 1998) reveals a number of further nomenclatural changes since Battarbee (1979), with the *Melosira* spp. now all transferred to *Aulacoseira*. Such revisions are scientifically justified but are potentially confusing to non-specialists.

This study is a useful starting point for the present chapter because it highlights the types of problems that future workers may encounter when re-examining modern records. In some ways, the situation is less severe for diatoms than it is for some other groups of algae, as the ability to prepare permanent slides provides the scope for re-examination. However, this is only possible where slides can be re-located, and Battarbee (1979) noted that only two of Dakin and Latarche's original slides were available for re-examination.

Human errors in identification can arise from a number of causes, including:

- Practical limitations of traditional hierarchical keys (Tilling, 1984, 1987).

- Nomenclatural confusion, caused by a proliferation of names and overlapping boundaries between taxa adopted by different authorities (Kelly and Haworth, 2001).

- Phenotypic plasticity within taxa (Mann, 1999).

- Procedural issues such as the quality of the preparation, the presentation of the material on a slide and the optical quality of the microscope.

- The experience of the analyst, which is partially the result of "training," but also touches upon the psychological basis of perception which "...presupposes constant activity on our part in making guesses and modifying them in the light of our experience. Wherever this test meets with an obstacle, we abandon the guess and try again..." (Gombrich, 1977). Such a process draws on more than just formal taxonomic information, but also on context and on qualities that an organism possesses which are more than simply a sum of its parts. It is by such intuitive processes that a willow tree (*Salix* sp.), for example, can be identified from 100 m distance or more, without any reference to its botanical characteristics; however, repeated, "intuitive" identifications without reference to specialist literature must also be regarded as a potential source of systematic error.

- Fatigue, attributable to the repetitive and time-consuming process of diatom counting on slides.

Some of these issues are also relevant to ADIAC. Taxonomy and nomenclature influence the input data that are used to train the software and the procedural issues are relevant insofar as they provide a specification for the hardware and preparatory stages. On the other hand, issues such as the limitations of hierarchical keys and operator fatigue will not be relevant to computer-based identifications. The identification of a diatom is the output from the process and will integrate the errors associated with each of these stages. It is also worth noting that the relative importance of each of the issues listed above will vary depending upon the taxon in question. For many of the small species of *Navicula* (*sensu lato*), for example, optical quality may be of paramount importance, whereas for *Fragilaria* (*sensu lato*) nomenclatural confusion creates practical problems. Similarly, the reproducibility that is achievable in practice will depend to some extent upon the taxonomic level used. Kelly (1999) reports a ring-test in which genus-level identifications were consistent, but where there was considerable variation in species-level identifications. Similar results are reported by Charles et al. (1987) and Munro et al. (1990), but, overall, the subject of human error rates in diatom identification seems to have been neglected in the literature.

2 Material and methods

Experimental studies were carried out in two phases, the first examining human error rates in identification and the second placing these results into a broader context involving an analysis of a diatom sample from a European river site.

2.1 Human error rates in diatom identification

Two separate tests were carried out over the Internet as part of this study. The first test evaluated the identification performance of diatomists who were asked to identify common diatom taxa from a subset of images that is also being used to evaluate ADIAC software (the mixed genera test set, see Chapter 4). In the second test, a different group of diatomists were asked to identify specimens of the diatom *Sellaphora pupula* that had been provisionally classified into six demes (see Chapter 4) and that was also used in testing ADIAC software.

2.1.1 Test 1: mixed genera

A total of 74 images representing 34 taxa were used in this test (*n.b.* the list shown in Table 2 includes 38 taxa, as not all participants suggested the same name for a particular taxon). Each taxon was represented by two or three specimens, in order to test the consistency of identification of each participant.

2.1.2 Test 2: Sellaphora pupula

A subset of the original test set (Chapter 4) was used for this test. This consisted of 60 images (as opposed to the original set of 120 images) and included all of the six demes present in the original test set. Demes were represented by between 8 and 12 images each. Participants taking the *Sellaphora pupula* test were trained prior to taking the test by providing them with a series of images of valves representing the range of variability within each deme, along with a description of each deme in tables, complete with ranges of morphometric measurements for length, width and stria density. This information continued to be available while the participant took the test.

2.1.3 Data recording

In both tests, images were presented to all participants as greyscale JPEG images (maximum quality setting); the order of images was randomized using random number tables. A Java Applet, embedded in a standard HTML web page, conducted both tests. The applet read a list of image filenames from a text file, and then displayed the images one at a time. For each image participants could record a name: for the *Sellaphora pupula* test (test 2) this was limited to selecting a name from the list of the six demes, whereas for test 1 the name needed to be entered into a text box. Once all the images in the test had been shown, the participants were asked to fill in a form at the bottom of the page requesting their name, e-mail address and diatom experience in years. When this form was submitted, JavaScript code interrogated the Applet for the list of answers. Together with the participant's details, this was sent as a query to the server. An ASP (Active Server Pages) application on the server received the query and entered the participant's details into a Microsoft Access database table. The answers to each test were entered into separate tables, both linked to the participant table. The data were analysed in Microsoft Excel after being extracted from Access.

2.1.4 Data analysis

For the mixed genera test, participants were asked to use whatever nomenclature they normally use, and accordingly, homotypic synonyms of correct identifications were accepted as correct answers.

To avoid bias in the evaluation, the test results were compared against both the identifications made by the test organizers and against a "majority vote" of identifications from the participants themselves, i.e. the correct identification for each specimen was assumed to be the one that the majority of participants had put forward. The original identifications and those of the majority vote did differ, and produced differing results in 11 out of 74 cases. Accordingly, only the results from the majority vote analysis are presented here and should be interpreted as measures of data "precision" rather than "accuracy."

Identifications were scored as belonging to one of the following categories:

1. Correct, including identification to variety where appropriate.

2. Correct species, but lacking necessary identification of the variety.

3. Correct species, incorrect variety.

4. Correct, but with uncertainty shown by "cf." or "?".

5. Identification to genus level only.

6. Unknown or blank.

7. Incorrect.

In addition, an internal consistency analysis was carried out that tested for the identification consistency of replicate specimens from the same taxon. Consistencies were calculated for each taxon, to test how consistently replicate images of the same taxon had been identified as such, and for each participant, to test whether participants had different consistency levels in repeated identifications.

A total of 26 participants took the tests, with 16 participants taking test 1 and 10 participants taking test 2. Participants had been chosen from the wider diatomist community, with a wide range of levels of experience. For test 1, participants were selected with a bias towards freshwater diatom experts, since this test consisted almost entirely of taxa found in freshwater habitats. All of the taxa used are described in the "*Süsswasserflora von Mitteleuropa*" (Krammer and Lange-Bertalot, 1986-1991), a standard work used for diatom identification (*n.b.* one taxon, *Parlibellus delognei*, had been erroneously included in the test, but was later removed from the analysis).

2.2 Replicate analysis of a single epilithon sample

The error rates observed in the above exercises form only part of the total process error encountered in the analysis of a diatom slide. For this reason, a second stage of testing involved an interlaboratory calibration exercise (ring-test) in which 21 diatomists analyzed slides prepared from a single epilithon sample. This allowed identification error rates to be considered alongside errors associated with the distribution of valves on the coverslip and with other procedural issues such as the optical quality of the microscope. This sample was collected from the River Jonen downstream of a sewage works at Hausen, Switzerland, on 29th March 1999. The River Jonen at this point is a small stream, about 2 m wide, at an altitude of about 580 m. It is moderately eutrophic and shows only mild influence of organic pollution. Dominant taxa include *Navicula lanceolata* and *Nitzschia dissipata*. The instruction given to participants was simply to provide a species list based on a count of at least 400 valves, following their usual protocols.

Data analysis involved examining these species lists and computing species diversity and pollution indices. In addition, two similarity measures were used: Dominance Identity (DI = Bray-Curtis similarity, see Renkonen 1938) and Jaccard's (1901) Coefficient of floral community (Jaccard's Index, JI). Dominance Identity measures the overall similarity between two samples, taking the relative abundance of each taxon in the sample into account. It is calculated as

$$\text{DI}_{1,2} = \sum q_i, \qquad (1)$$

where $\text{DI}_{1,2}$ is the similarity between samples 1 and 2 and q_i is the smaller of the two relative abundances of species i. DI can vary between 0 and 100% with values greater than 60% generally regarded as indicating replicate samples (Gauch, 1982; Hürlimann, 1993; Kelly, 2001). Jaccard's Index (JI) simply compares the number of taxa common to both samples, in relation to the total of the two samples irrespective of their relative abundance, and is also expressed as a percentage. The formula for JI is

$$\text{JI} = [w/(A + B - w)] \cdot 100, \qquad (2)$$

where w is the number of taxa common to both samples, A is the number of taxa in sample 1 and B is the number of taxa in sample 2. Each sample was compared with every other sample in the set, giving a total of 210 comparisons.

Participants were also asked to state their level of experience with river diatoms: either "beginner," "advanced" or "expert." The interpretation of these categories overlapped to some extent, because "beginners" quoted 0–1 years of experience, "advanced" diatomists quoted 0–16 years and "expert" diatomists 4–19 years.

3 Results

3.1 Human error rates in diatom identification

3.1.1 Test 1: mixed genera

Agreement with the "correct" designation ranged from a complete (100%) agreement for *Navicula capitata* var. *capitata* to less than 20% for *Tabularia investiens* (Table 2). There was a positive correlation between the percentage of correct designations and internal consistency (Fig. 1), suggesting that at least part of the

Table 2. Identification rates and identification consistency scores (both in %) for the mixed genera test, by taxon, averaged across replicate specimens (images). Consistency data are only available for taxa which were represented by more than one specimen after the taxon names from the majority vote had been applied to the data. N/A = not available due to insufficient data.

taxon	ID rate	consistency
Navicula capitata var. *capitata* Ehrenberg	100.0	100.0
Meridion circulare (Greville) C. Agardh	97.8	93.3
Epithemia sorex var. *sorex* Kützing	96.9	93.8
Nitzschia sinuata var. *sinuata* (Thwaites) Grunow	93.8	N/A
Sellaphora bacillum (Ehrenberg) D.G. Mann	93.8	100.0
Stauroneis smithii Grunow	93.6	100.0
Caloneis amphisbaena (Bory) Cleve	93.3	100.0
Navicula radiosa Kützing	90.3	92.9
Diatoma mesodon (Ehrenberg) Kützing	80.7	93.3
Navicula reinhardtii var. *reinhardtii* Grunow in Van Heurck	75.0	85.7
Gomphonema augur var. *augur* Ehrenberg	74.4	93.3
Navicula rhynchocephala Kützing	73.8	66.7
Eunotia incisa Gregory	71.9	75.0
Tabellaria flocculosa (Roth) Kützing	70.9	58.3
Denticula tenuis Kützing	65.7	100.0
Cymbella subequalis Grunow in Van Heurck	64.6	64.3
Gyrosigma acuminatum (Kützing) Rabenhorst	59.6	92.9
Fragilariforma bicapitata D.M. Williams and Round	58.2	53.8
Nitzschia frustulum (Kützing) Grunow	56.3	N/A
Encyonema silesiacum (Bleisch in Rabenhorst) D.G. Mann	56.3	73.3
Navicula menisculus Schumann	53.2	92.9
Staurosirella pinnata (Ehrenberg) D.M. Williams and Round	51.8	38.1
Tabellaria quadriseptata Knudson	50.0	N/A
Navicula constans var. *symmetrica* Hustedt	46.9	68.8
Cocconeis placentula var. *placentula* Ehrenberg	46.7	N/A
Nitzschia hantzschiana Rabenhorst	46.7	86.7
Navicula viridula var. *linearis* Hustedt	43.8	40.0
Gomphonema minutum (C. Agardh) C. Agardh	43.8	73.3
Cymbella hybrida var. *hybrida* Grunow in Cleve and Möller	42.3	60.0
Opephora olsenii Möller	40.7	81.3
Diatoma vulgaris Bory	40.0	100.0
Cymbella helvetica Kützing	39.0	35.7
Cocconeis neodiminuta Krammer	37.5	N/A
Cocconeis placentula var. *euglypta* Ehrenberg	37.5	N/A
Nitzschia dissipata (Kützing) Grunow	25.0	N/A
Nitzschia dissipata var. *media* (Hantzsch) Grunow	25.0	91.7
Cocconeis neothumensis Krammer	18.8	N/A
Tabularia investiens (W. Smith) D.M. Williams and Round	18.8	N/A

problem encountered by the test participants was caused either by morphological variability in the taxa themselves or by deficiencies in their descriptions in popular floras. An alternative explanation is that participants gave consistent results for those taxa with which they were most familiar, but two or more names for the taxa

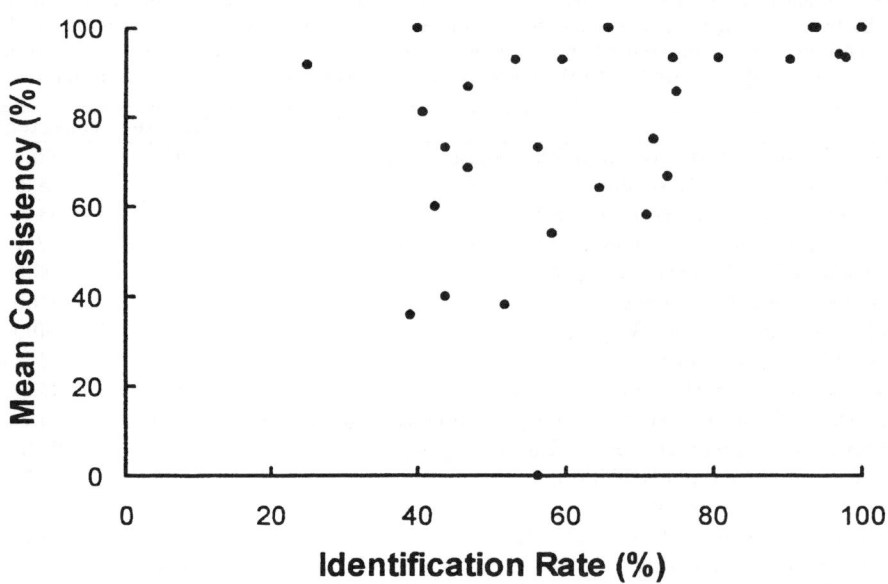

Figure 1. Relationship between percentage of "correct" designations in the mixed genera test and the mean level of identification consistency of each taxon by test participants.

that they knew less well. Such a situation will occur when an analyst encounters unfamiliar taxa during the course of analyzing "real" samples.

The ability of each participant to correctly identify each specimen ranged from 33.8 to 86.5%, with a mean of 63.3% (Table 3). This analysis, however, was based on the very strict criterion that an identification was "correct" only if it conformed to the "majority vote" at the lowest possible taxonomic level (i.e. category 1 in Section 2.1.4). If the evaluation criteria are relaxed to allow uncertainty (category 4) and to require correct identification only to species level (categories 2 and 3), the mean identification rate increases by approx. 10% to 73.7% (Table 3). Participants' individual identification rates were not significantly correlated with the length of diatom experience (which varied from 3 to over 25 years), although this may reflect the wide variety of habitats from which the test specimens were drawn, bearing in mind that few of the participants may have had the specialist skills across the whole range of these environments.

Table 4 summarizes results by genus, for those genera that were represented by more than one species. Both the consistency of identification per participant and identification rate (i.e. category 1) varied between genera, with *Cocconeis* scoring the lowest value for ID rate (35.1%) and *Cymbella* scoring the lowest value for consistency (53.3%), but in neither case was the variation statistically significant (ANOVA, $P = 0.621$ and 0.121, respectively; only cases with two or more species or varieties per genus considered).

Table 3. Results of the mixed genera test, summarized by participant. Maximum and minimum values for identification rate and consistency are in bold type. "Best case scenario" refers to identification rates of individual participants in which the categories 1–4 were scored as correct (see Section 2.1.4 for full definitions) and category 6 ("unknown") was omitted from the analysis, i.e. not scored as errors.

participant number	years of experience	ID rate (%)	"best case scenario"	consistency (%)
17	11–15	**33.8**	55.4	68.1
36	6–10	43.2	87.5	**100.0**
40	N/A	43.4	58.7	**46.3**
14	>25	48.6	59.5	67.6
25	>25	50.0	56.8	82.0
21	>25	56.8	68.1	62.8
33	3–5	58.1	68.1	82.4
19	11–15	63.5	68.9	74.5
26	16–25	68.9	75.7	83.1
39	16–25	71.6	75.7	87.9
27	11–15	72.5	81.6	69.6
24	11–15	74.3	78.4	86.6
16	6–10	78.4	82.4	89.5
13	16–25	79.7	83.8	86.5
18	3–5	83.8	**89.2**	98.3
23	6–10	**86.5**	**89.2**	84.5
mean:		63.3	73.7	79.4

Table 4. Identification rates and identification consistencies within taxa, averaged across participants and replicate species within those genera represented by more than one species in the majority vote analysis. SD = standard deviation, n = number of taxa.

	ID rate (%)			consistency (%)		
	mean	SD	n	mean	SD	n
Cocconeis	35.1	11.7	4	N/A	N/A	N/A
Cymbella	48.6	13.9	3	53.3	15.4	3
Diatoma	60.4	28.8	2	96.7	4.7	2
Gomphonema	59.1	21.6	2	83.3	14.1	2
Navicula	69.0	21.8	7	78.1	21.0	7
Nitzschia	49.4	28.4	5	89.2	3.5	2
Tabellaria	60.5	14.8	2	N/A	N/A	N/A

3.1.2 Test 2: Sellaphora pupula

This test extended the scope of Test 1 by looking at variation within a species complex but in a situation where the participants were trained using a set of on-

Table 5. Identification rates (%) of *Sellaphora pupula* demes.

correct designation: scored as:	elliptical	large	pseudo-blunt	pseudo-capitate	cf. rectangular	tidy
elliptical	98	1	0	0	0	0
large	0	78	1	1	0	0
pseudoblunt	2	9	85	6	8	0
pseudocapitate	0	8	3	66	23	2
cf. rectangular	0	4	9	24	69	1
tidy	0	0	3	3	0	97

line images and descriptions in advance of and during the test. The ability to distinguish demes of *Sellaphora pupula* ranged from 98% for "elliptical" specimens to 66% and 69% for "pseudocapitate" and "cf. rectangular" specimens, respectively (Table 5). These two demes were most likely to be mistaken for one another, rather than for other demes in the test. Identification rates of individual participants ranged from 60.0 to 98.3% correct, with a mean of 82.0% (Table 6). Participants' experience with diatoms ranged from less than one year to more than 25 years, but, as for Test 1, there was no significant correlation between the experience of the participants and their identification rate.

Table 6. Identification rates of individual participants in the *Sellaphora pupula* demes test.

participant no.	diatom experience in years	ID rate (%)
7	> 25	98.3
15	16–25	98.3
37	< 1	91.7
31	3–5	90.0
34	1–2	88.3
38	16–25	80.0
20	> 25	76.7
35	< 1	73.3
32	> 25	63.4
22	3–5	60.0
mean:		82.0

3.2 Replicate analysis of a single epilithon sample

Results of index calculations for the epilithon sample from the River Jonen are summarized in Table 7 and show relatively low levels of variation (coefficient of variation < 10%) between replicate analysts. Variability in the total number of species iden-

Table 7. Variability in replicate analyses of a single diatom sample collected from the River Jonen on 29 March 1999. Shannon-Weaver diversity: see Shannon and Weaver (1949). DI-CH (Swiss Diatom Index): see Bundesamt für Umwelt, Wald und Landschaft (2001). Trophic index: see Schmedtje et al. (1998). Saprobic index: see Rott et al. (1997).

	mean (all data) n=21	mean (max and min excluded) n=19	coefficient of variation (%)	min	max
number of valves counted	564	545	38.2	400	1103
number of taxa recorded	27	27	24.4	18	48
Shannon-Weaver diversity	3.28	3.29	8.54	2.75	3.72
DI-CH	4.3	4.35	8.4	3.3	4.9
trophic index	2.39	2.40	4.2	2.0	2.5
saprobic index	2.16	2.17	3.8	1.9	2.3

tified was greater (CV = 24.4%) and was significantly correlated with the number of valves counted (Spearman's Rank Correlation = 0.49, $P < 0.05$). This variability was composed of two components: variation in the number of "rare" taxa found in the sample, which is partially a consequence of searching time ("sampling effort"), and the taxonomic conventions adopted, with those analysts who "split" species into varieties ending up with longer taxon lists than those who "lumped" varieties under a single species. Neither type of variation has a great effect on the calculation of pollution indices (demonstrated by the lower variability observed in index calculations) and it may be controlled, if necessary, by the adoption of more rigorous analytical protocols. However, there was a significant positive relationship between the number of taxa recorded and the Shannon-Weaver diversity index.

DI and JI had a linear relationship (Fig. 2; $r = 0.59$, $P < 0.001$) with DI values of less than 60% typically having JI values of 20–40%. As DI > 60% is the criterion for "replicate" analyses adopted in most studies, these results suggest that much of the variability in JI is due to taxa in the sample that are relatively uncommon. It is particularly noteworthy that many comparisons with DI > 60% had a JI between 20 and 40%.

The results for one species, *Achnanthes minutissima*, put some of these observations into context (Table 8). This species was found in all samples, with many analysts also finding varieties. Eleven of the 21 participants found just one form, nine found two forms and one analyst found three. It is not clear whether those who only recorded the nominate variety consciously rejected the possibility that other varieties were present. There is, in any case, disagreement amongst specialists about the taxonomic limits of these varieties. Whereas most who found more than one form regarded the nominate variety as dominant, one participant attributed all *A. minutissima* in the sample to var. *jackii* and two others regarded var. *saprophila* to be more abundant than the nominate variety. At least part of this

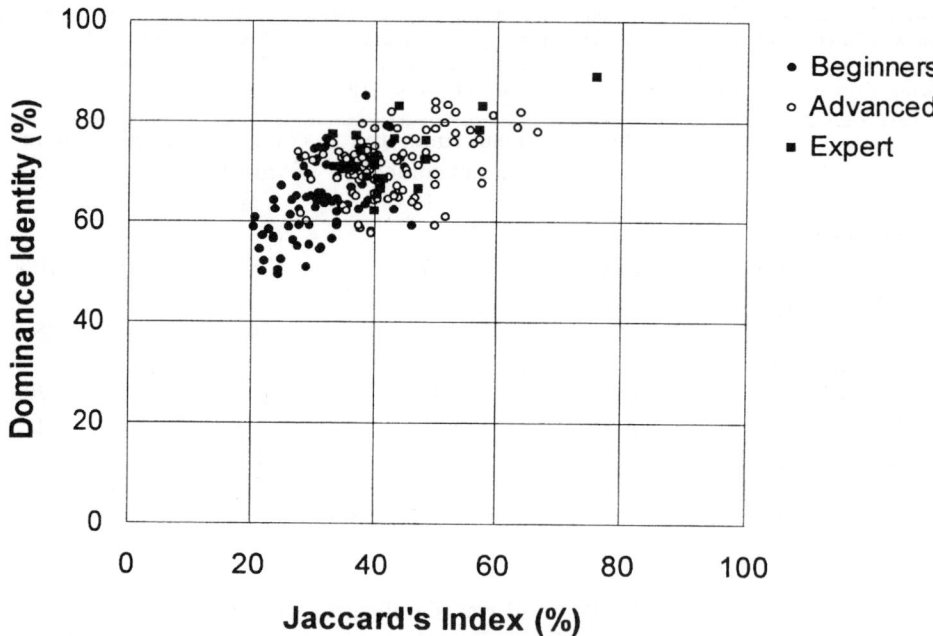

Figure 2. Relationship between Dominance Identity (DI) and Jaccard's Index (JI) in 21 replicate analyses of a single diatom sample collected from River Jonen on 29 March 1999. The categories "beginner", "advanced" and "expert" are defined in the text. In this graph, the lowest of the two experience categories in each comparison is plotted.

Table 8. Variability (%) of *Achnanthes minutissima* and varieties recorded in the River Jonen during the ring-test. Note that mean values are based only on those records where the taxon is present.

taxon	number of records	min.	max.	mean
Achnanthes minutissima Kützing	19	1.1	16.0	8.4
Achnanthes minutissima var. *affinis* (Grunow) Lange-Bertalot	3	0.1	1.5	0.7
Achnanthes minutissima var. *inconspicua* Oestrup	1	0.2	0.2	0.2
Achnanthes minutissima var. *jackii* (Rabenhorst) Lange-Bertalot	1	10.6	10.6	10.6
Achnanthes minutissima var. *saprophila* Kobayasi and Mayama	8	0.2	9.4	3.7
all *A. minutissima*	21	4.1	16.0	10.4

variation is due to the relative experience of the analysts involved. The opinion of one of us with particular experience of this species (JH), is that two of the five taxa were probably correct, with the nominate variety dominating in the sample. One of the other three taxa was definitely incorrect (var. *jackii*) and the other two were either incorrect or very rare. On the whole, results from "advanced" and "expert" diatomists involved in the study agreed with this opinion.

4 Discussion

The results of the mixed genera test have demonstrated a generally lower identification rate in human determinations of diatoms (mean = 63.3%), than in that associated with the computer-based identifications described in Chapter 12 (mean > 96.9%). However, these results do need to be viewed in context. The computer-based identification was "trained" on a dataset including all of the taxa included in the test datasets but no others, whereas few of the human participants would be expected to encounter all of the taxa used in the test on a regular basis, although they would be "aware" that taxa existed beyond their immediate cognizance. When analyzing "real" samples, diatomists will base their identifications on a larger "training set" than that used for training ADIAC software. It is reasonable to assume that the risk of misidentification by ADIAC software will increase with the number of taxa in the training set. It follows from this that the variability in the results of the ring-test is manifested more in the numbers of taxa and the JI rather than in the DI or the pollution index calculations. Whereas the index calculations are influenced more by the common (and, presumably, more readily recognized) taxa in the sample, the rarer taxa in a sample require participants to draw deeper on their mental reserves, with a greater scope for error as a result.

Experience was not a good predictor of identification rates in either the mixed genera (Table 3) or the *Sellaphora pupula* test (Table 6). It is interesting to note that success in distinguishing taxa varied considerably between genera (Table 4), although differences were not statistically significant. In informal conversations, many diatomists express the opinion that some genera are inherently more "difficult" to identify than others. The cause of such difficulty will lay in the lack of discrete morphological characteristics available to distinguish taxa and, in turn, affects the ease with which "traditional" hierarchical keys are both written and used. ADIAC software may, indeed, provide facilities to develop benchmarks against which new keys can be tested and improved, and become part of training programmes for new staff.

These issues are well illustrated by the studies on *Sellaphora pupula* (Tables 5, 6) and *Achnanthes minutissima* (Table 8), both of which highlight the problems encountered at finer levels of visual perception. Although the use of infraspecific categories may contribute additional ecological information to a study, it is clear from these data that this information may be lost owing to the size of the associated error component, through human identification. Results of earlier ring tests (Munro et al., 1991; Kelly, 1997) suggest that some of this taxonomic variability might be removed through workshops leading to the adoption of agreed taxonomic protocols. On the other hand, such discussions need to take place against a backdrop of the ecological information that such infraspecific forms contribute, especially where boundaries between varieties are often vague, as in *A. minutissima*.

A further issue that arises is that the computer-based identification is actually more reliable at distinguishing *Sellaphora pupula* demes than most experienced human analysts: identification rates obtained with ADIAC software ranged from 89 to 100%, depending on the methods employed (see the "identification" chapters), compared to the average of 82.0% achieved by the participants. In the case of *S. pupula*, these demes effectively deserve species status (e.g. Mann, 1999) and reflect a new paradigm in diatom research, in which valve morphology, as perceived by human observers, is only just sensitive enough to predict taxonomic status, and then only after considerable training (two highly experienced participants actually competed closely with the top rates achieved by the ADIAC software, see Table 6). Overall, these results suggest that the extra sensitivity of computer-based identification may be a more reliable basis for such separations than human judgement.[a] The utility of such a finding for practical monitoring will depend upon the ecology of each deme being sufficiently distinct to offer useful information to end-users. Conversely, unless the demes can be reliably identified, their ecology cannot be studied.

An implication of these observations is that the level of identification needs to be appropriate to the analysis in question. This has two practical benefits: it provides a verifiable target and it prevents misdirection of effort. The ability to verify identifications is important, as it opens the way for the application of formal quality control procedures to ecological analyses (Kelly, 2001) and provides an objective basis for scrutiny that can underpin the use of ecological data in legal and quasi-legal contexts. The issue of misdirection of effort suggests the "80:20 rule" familiar to computer programmers (80% of a problem is solved with the first 20% of effort) and will be familiar to anyone involved in the analysis of ecological samples, where the effort involved in identifying one or a few specimens of a rare organism is out of proportion to their representation in the sample as a whole. The practical question that leads on from this is whether an analyst with a finite amount of available time is better directed to spend this identifying these rare specimens or moving on to another sample. Such a question can only be answered by reference to the aims of the study.

By contrast, the effort required to accommodate more taxa in a computer-based identification system such as ADIAC lies not in the architecture of the system, which automatically builds (in this case) a decision tree on the basis of the data from the training set, but in the acquisition of the training data. For each new taxon that is added to these data, about 20–30 high quality images are required (although some testing will be required in the future to see whether these sample sizes may possibly be reduced). The 80:20 rule still applies, but as an initial investment of effort rather than as a "recurrent cost." The underlying issue, both for human and computer-based identification, is whether this additional effort can be justified in terms of the extra information that it contributes to the study. The benefits of exhaustive

---

[a] Note added by editor HdB: visual discrimination is indeed characterized by just-noticeable differences beyond which no difference can be detected. But accepting computer results without the possibility to verify them may be a philosophical issue...

studies of a few samples need to be weighed against what has been termed the "data-rich but information-poor syndrome" in environmental monitoring (Ward et al., 1986).

The diatoms are a large group, with over 2000 freshwater and brackish taxa recorded from within the British Isles alone (Whitton et al., 1998), a figure that is almost certainly an underestimate of the true situation (Mann and Droop, 1996). Bearing this in mind, training either human analysts or computer-based systems such as that developed by ADIAC, being able to identify all taxa is probably unrealistic, and well-designed Data Quality Objectives (DQO; Environmental Protection Agency, 1996) that balance information needs and data integrity, will be a necessary component of any practical implementations of ADIAC. In effect, a DQO provides a "chain of evidence" between the field sample and the type specimens and descriptions of each taxon recorded in that sample, comparable to the well-established principle of "traceability" in conventional QA terminology. Through the use of agreed taxonomic conventions, standardizing floras within a study and including verification procedures, it should be possible for any future data user to link a name such as "*Achnanthes minutissima*" back to Kützing's original publication and to know whether such a designation implies that infraspecific forms are included or excluded. For many groups of soft-bodied organisms that do not preserve as well as diatoms there may be a temptation to overspecify DQOs in anticipation of future data needs; however, the ease with which diatom samples can be stored in herbaria and be re-analyzed in the future reduces the value of such an overspecification.

References

Battarbee, R.W. (1979). Early algological records: help or hindrance to palaeolimnology? Nova Hedwigia, Vol. 64, pp. 379-393.

Bundesamt für Umwelt, Wald und Landschaft (2001) Methoden zur Untersuchung und Beurteilung der Fliessgewässer Kieselalgen Stufe F (flächendeckend). Vollzug Umwelt. Mitteilungen zum Gewässerschutz. CD-ROM Version of 11 May 2001.

Charles, D.F., Whitehead, D.R., Anderson, D.S., Bienert, R., Camburn, K.E., Cook, R.B., Crisman, T.L., Davis, R.B., Ford, J., Fry, B.D., Hites, R.A., Kahl, J.S., Kingston, J.C., Kreis, R.G., Mitchell, M.J., Norton, S.A., Roll, L.A., Smol, J.P., Sweets, P.R., Uutala, A.J., White, J.R., Whiting, M.C. and Wise, R.J. (1987) The PIRLA project (Paleoecological Investigation of Recent Lake Acidification): preliminary results from the Adirondacks, New England, N. Great Lakes and N. Florida. Water, Air and Soil Pollution, Vol. 31, pp. 355-366.

Dakin, W.J. and Latarche, M. (1913) The plankton of Lough Neagh. Proc. Royal Irish Academy, Series B, Vol. 30, pp. 20-96.

Environmental Protection Agency (1996) Performance-based methods system (PBMS). Revision to Rapid Bioassessment Protocols for Use in Streams and Rivers: Periphyton, Benthic Macroinvertebrates, and Fish. See http://www.epa.gov/OWOW/monitoring/AWPD/RBP/ch4main.html

Gauch, H.G. (1982) Multivariate Analysis in Community Ecology. Cambridge University Press, Cambridge, UK.

Gombrich, E.H. (1977) Art and illusion. A study in the psychology of pictorial representation. Phaidon, London (5th Edition).

Hürlimann, J. (1993) Kieselalgen als Bioindikatoren aquatischer Ökosysteme zur Beurteilung von Umweltbelastungen und Umweltveränderungen. Inaugural Dissertation, Universität Zürich, Switzerland.

International Standards Organisation (1994) ISO 9000-1:1994: Quality management and quality assurance standards – Part 1: Guidelines for selection and use. International Standards Organisation, Geneva, Switzerland.

International Standards Organisation (1999) ISO/IEC 17025: General requirements for the competence of testing and calibration laboratories. International Standards Organisation, Geneva, Switzerland.

Jaccard, P. (1901) Etude comparative de la distribution florale dans une portion des Alpes et du Jura. Bulletin de la Société Vaudoise des Sciences Naturelles, Vol. 37, pp. 547-579.

Kelly, M.G. (1997) Sources of counting error in estimations of the trophic diatom index. Diatom Research, Vol. 12, pp. 255-262.

Kelly, M.G. (1999) Progress towards quality assurance of benthic diatom and phytoplankton analyses in the UK. In: Use of Algae for Monitoring Rivers III, J. Prygiel, B.A. Whitton and J. Bukowska (eds), Agence de l'Eau Artois-Picardie, Douai, France, pp. 208-215.

Kelly, M.G. (2001) Use of similarity measures for quality control of benthic diatom samples. Water Research, Vol. 35, pp. 2784-2788.

Kelly, M.G. and Haworth, E.Y. (2001) *Phylum Bacillariophyta* (Diatoms). In: The Freshwater Algal Flora of the British Isles, D.M. John, A.J. Brook, and B.A. Whitton (eds), Cambridge University Press, Cambridge, UK, pp. 273-277.

Krammer, K. and Lange-Bertalot, H. (1986-1991) *Bacillariophyceae.* In: Süsswasserflora von Mitteleuropa, Vol. 2, H. Ettl, J. Gerloff, H. Heynig and D. Mollenhauer (eds), Gustav Fischer Verlag, Stuttgart, Germany.

Mann, D.G. (1999) The species concept in diatoms. Phycologia, Vol. 38, pp. 437-495.

Mann, D.G. and Droop, S.J.M. (1996) Biodiversity, biogeography and conservation of diatoms. Hydrobiologia, Vol. 336, pp. 19-32.

Munro, M.A.R., Kreiser, A.M., Battarbee, R.W., Juggins, S., Stevenson, A.C., Anderson D.S., Anderson, N.J., Berge, F., Birks, H.J.B., Davis, R.B., Flower, R.J., Fritz, S.C., Haworth, E.Y., Jones, V.J., Kingston, J.C. and Renberg, I. (1990) Diatom quality control and data handling. Philosophical Trans. of the Royal Society, Series B, Vol. 327, pp. 257-261.

Renkonen, O. (1938) Statistisch-ökologische Untersuchungen über die terrestrische Käferwelt der finnischen Bruchmoore. Annales Botanici Societatis Zoologicae Botanicae Fennicae "Vanamo," Vol. 6, pp. 1-231.

Rott, E., Hofmann, G., Pall, K., Pfister, P. and Pipp, E. (1997) Indikationslisten für Aufwuchsalgen. Teil 1: Saprobielle Indikation. Bundesministerium für Land- und Forstwirtschaft, Wien, Austria.

Schmedtje, U., Bauer, A., Gutowski, A., Hoffman, G., Leukart, P., Melzer, A., Mollenhauer, D., Schneider, S. and Tremp, H. (1998) Trophiekartierung von aufwuchs- und makrophytendominierten Fliessgewässern. Informationsberichte Bayerisches Landesamt Wasserwirtschaft, Vol. 4/98, pp. 1-516.

Shannon, C. and Weaver, W. (1949) The mathematical theory of communication. University of Illinois Press, Urbana, USA.

Tilling, S.M. (1984) Keys to biological identification: their role and construction. J. Biological Education, Vol. 18, pp. 293-304.

Tilling, S.M. (1987) Education and taxonomy: the role of the Field Studies Council and AIDGAP. Biological J. of the Linnean Society, Vol. 32, pp. 87-96.

Ward, R.C., Loftis, J.C. and McBride, G.B. (1986) The "data-rich but information-poor" syndrome in water quality monitoring. Environmental Management, Vol. 10, pp. 291-297.

Whitton, B.A., John, D.M., Johnson, L.R., Boulton, P.N.G., Kelly, M.G. and Haworth, E.Y. (1998). A coded list of freshwater algae of the British isles. LOIS publication number 222. Institute of Hydrology, Wallingford, UK.

CHAPTER 6

CONTOUR EXTRACTION

STEFAN FISCHER, HAMID R. SHAHBAZKIA AND HORST BUNKE

This chapter deals with the automatic extraction of diatom contours from microscopic brightfield images. This is a difficult task and not trivial, because diatoms may lie on top of each other, or be surrounded by debris. In a first step, the unstructured background is suppressed by employing a texture-based region extraction. Then, for each of the remaining object regions, the precise position of the outline is detected. Using a set of 808 diatom images, our method yielded 776 good contours (i.e. 96% of all images). Remaining contours currently need to be extracted semi-automatically or even manually.

1 Introduction

In the later stages of the ADIAC project, two test image sets were created: the *Sellaphora pupula* set and the mixed genera set (Chapter 4). The first one is small enough—only 120 images—to extract the contours manually, i.e. supervised. At the end of the project the second set counted over 1000 images, hence it is preferable to extract the contours automatically or unsupervised. But this is not a trivial problem, because two diatom valves may touch one another or one may occlude part of the other. Often, debris or a broken valve touches the contour of a potentially usable valve, or a valve may not be level such that part of its contour is not well focused. As an example, a typical low resolution image is shown in Fig. 1. This image counts six diatoms labelled 1 to 6. As can be observed, contours can touch or intersect (2 and 3), they can have a low contrast and even be partially blurred (1). In addition, debris and background texture may be visible in the image; in many cases also dust specks caused by particles in another focal plane and therefore blurred. In view of these problems, a very robust method for contour extraction is necessary.

Many different approaches for segmenting gray level images have been described (Haralick and Shapiro, 1992; Pal and Pal, 1993). These include thresholding, edge detection and region extraction, see also Fu and Mui (1981). Thresholding requires one or more gray levels to separate objects from each other and from the background. Edge detection locates object outlines (contours) by evaluating local gray level profiles. In region extraction, also called segmentation, similar image regions are grouped together until all pixels in the image have been assigned to a class. This requires a similarity measure, for example the local gray level difference, but there are many other measures.

All approaches mentioned above require assumptions about the objects that are to be segmented. For example, in thresholding it is assumed that objects have

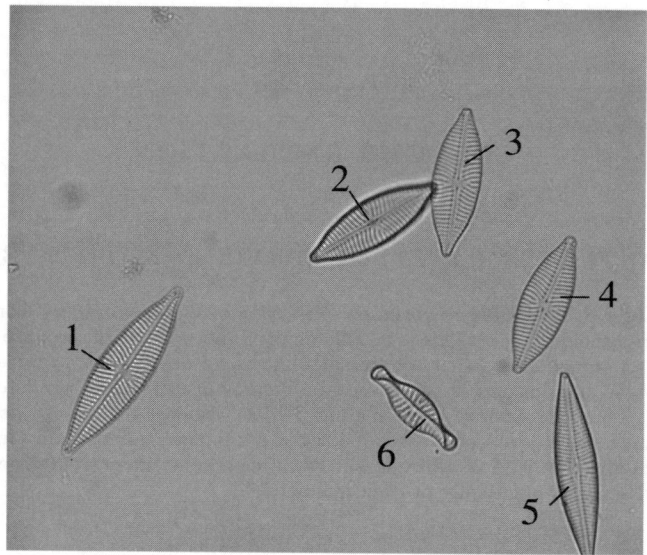

Figure 1. Example image showing different diatoms as well as minor debris and dust specks.

different gray level values, whereas in edge detection a large contrast over contours is required. Region extraction is a more flexible method, because the similarity criterion can be adapted to the specific segmentation task. Apart from simple gray level differences, differences in texture can be used to define a similarity criterion (Reed and du Buf, 1993). One problem with such criteria is that regions between different objects often cannot be uniquely assigned to one of the neighboring objects. Thus, a post-processing step is necessary to find the exact location of object outlines.

In the approach presented here, two steps are used to obtain objects and background. First, a region extraction approach separates the relatively smooth background region from the more structured—striated—object regions. Then, each of the extracted object regions is post-processed by a thresholding or edge detection procedure, in order to detect the exact location of the object's outline.

Most diatoms in brightfield images are characterized by a prominent outline caused by the silicified cell wall. For most diatoms, the outline can be detected by thresholding as well as edge detection methods. But if the illumination around a diatom is not uniform, thresholding using a single threshold value will be problematic. In such cases, sometimes only parts of the outline can be detected, or large areas of the surrounding background will also be detected. Nevertheless, edges forming the diatom contour are often preserved, such that edge detectors can be applied to obtain the complete outline. However, if the diatom is not properly focused, or if it is curved in pervalvar direction, i.e. through the valve, edges are often blurred and can only be detected partly. In such cases, the gray levels on the outline are often distinct enough to allow for thresholding, using a carefully chosen threshold value.

Most gray level threshold selection methods are based on statistics of the gray

Chapter 6: Contour extraction

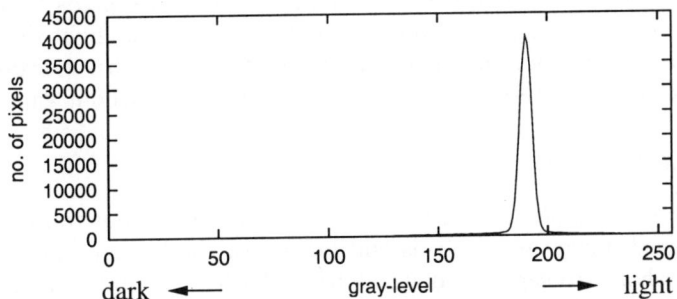

Figure 2. Gray level histogram of the image shown in Fig. 1.

level distribution—the histogram—of the image. Such methods show good results in cases where the histogram has separate peaks for gray levels belonging to different objects. For example, in optical character recognition there is one pronounced peak for the written text (black) and a second one for the background (white). For images with such bi-modal histograms, good thresholds can often be determined completely automatically by methods like the one proposed by Otsu (1979).

Unfortunately, diatoms in microscopic images can exhibit the same gray levels as the background. This can be seen from the histogram of the image shown in Fig. 1, which is shown in Fig. 2. There is only one peak and no separate peak for e.g. the dark contours. Only small ranges to the left and right of the peak characterize the dark contour and the typical, light diffraction halo around the diatoms. Mostly, these gray levels do not occur in the background, except for very dark gray levels which are related to debris like ash or sand grains. This property of the gray level distribution will be used below in order to separate diatoms from the background, and also to find the exact location of the outlines.

The overall procedure for the segmentation of diatoms is as follows. In a pre-segmentation step, multiple objects in an image are detected by suppressing the background. The remaining regions are analyzed in detail to find the exact location of object contours. The pre-segmentation step is described in Section 2. The procedure for the extraction of outlines is explained in Section 3. Finally, results obtained with ADIAC's mixed genera data set are presented in Section 4.

2 Pre-segmentation

In the pre-segmentation step, the image is segmented into regions of possible objects and background, using an approach based on texture. In contrast to areas occupied by objects like diatoms, the background of microscopic images is mostly unstructured, showing only very smooth gray level transitions and a small variance. Based on this property, images are separated into structured and unstructured regions using a region extraction approach. In order to obtain regions of possible objects, nested regions are post-processed by analyzing their local neighborhood. The procedure will be visualized using the example image shown in Fig. 1. Below we describe the individual steps: (a) how to select thresholds to separate the gray level range belonging to diatoms from those of dark or light debris, (b) the texture-based approach for the identification of structured object regions, and (c) how nested regions are merged to obtain uniform regions without gaps.

2.1 Threshold selection

If the microscopic sample is sparse, that is, when there are not too many diatoms in each captured image, most pixels will belong to the background. A first idea is to find thresholds to separate diatoms from the background. Unfortunately, as was shown above, in brightfield microscopy the gray levels of the background often correspond to those of the diatoms. But there are exceptions. The gray levels of dark or light areas near the outlines do mostly not appear in the background. This observation is used to select meaningful thresholds. Similar heuristics yielded good results in the segmentation of cells from microscopic images (Anoraganingrum et al., 1999).

To find an appropriate threshold, the histogram of the entire image is computed. Starting from the peak, first a threshold is searched for going towards the left tail of the histogram. In this heuristic procedure the first value is selected which has a frequency that is smaller than 15% of the mean frequency. This threshold separates dark image parts from the rest. Then, the same procedure is applied to the right tail of the histogram, in order to separate light image parts.

As can be seen in Fig. 2, the histogram of the image shown in Fig. 1 has a strong peak at gray level 191. Most image pixels fall into a small range around this peak value. The heuristic search procedure selects levels 185 and 195 as thresholds τ_{dark} and τ_{light} for the suppression of dark and light image regions.

2.2 Local texture analysis

For all remaining image pixels between thresholds τ_{dark} and τ_{light} the texture is analyzed using the local variance. The variance inside or near the diatoms' outlines is large if compared to that of the background. In other words, the local variance allows us to distinguish structured (striated) regions from the unstructured background. At each image position, the gray level variance is computed within a 3×3 neighborhood. The local variance is thresholded using $\tau_{var} = (2/3) \cdot (\tau_{light} - \tau_{dark})$. Because this processing is applied *after* the first thresholding, some very dark or light pixels may cause gaps in the variance result. In order to eliminate the gaps, a merging step groups together adjacent pixels with similar properties.

Figure 3 shows the result of this procedure. The background is white whereas object regions are shown with gray levels from the original input image. As can be seen from the long diatom on the left, white background pixels (gaps) can occur within a contour. These must still be eliminated by post-processing.

2.3 Neighborhood graph-based merging

To obtain the entire object regions, possibly diatoms, gaps need to be filled and regions smaller than the smallest possible diatom must be rejected. This is ac-

Chapter 6: Contour extraction 97

Figure 3. Image regions (gray) after background suppression.

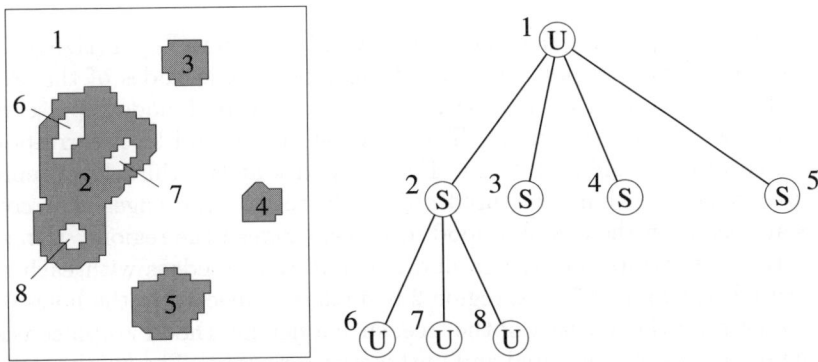

Figure 4. Illustration of a neighborhood graph of connected regions. Left: artificial image with four objects (gray) on an unstructured background. Right: corresponding neighborhood graph with structured (S) and unstructured (U) nodes.

complished by a neighborhood graph-based merging of connected regions. Each region in the image is labelled as either structured or unstructured. An unstructured region has no gaps; only structured regions can have gaps. Based on the detected image regions, a neighborhood graph of connected regions is built. This procedure will be described using the example shown in Fig. 4 (left). This example consists of four structured object regions on an unstructured background, and the structured region on the left (2) has three unstructured gaps (6, 7 and 8). The structured regions are gray and the unstructured background as well as the holes

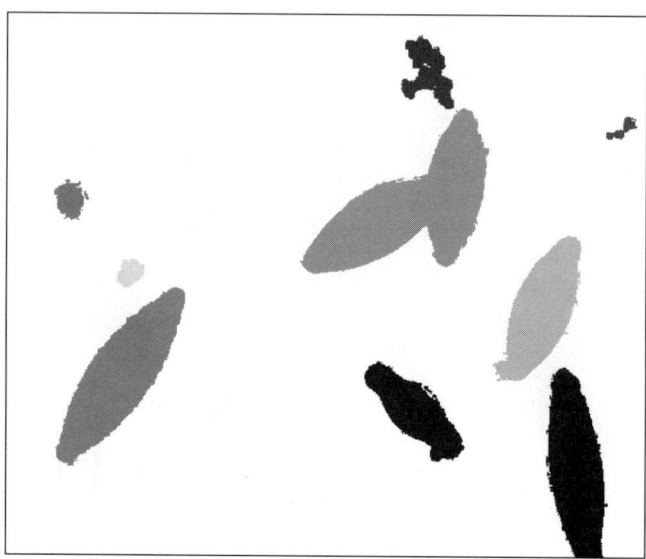

Figure 5. Detected object regions marked by different gray levels.

are white. The corresponding neighborhood graph is shown in Fig. 4 (right). There is a node for each image region numbered from 1 to 8. The nodes of the tree are labelled either structured (S) or unstructured (U). The root node represents the unstructured background region (1). The nodes at the second level correspond to the structured object regions (2 to 5). Finally, nodes at the third level represent unstructured regions within structured objects (6 to 8). The edges that connect the nodes are based on the neighborhood relations between the regions. Hence, the root node that represents the background is connected by edges with each of the four structured regions 2 to 5, and region 2 is further connected to the holes 6 to 8. As can be seen from the left part of the tree, such a neighborhood graph consists of alternating sequences of structured and unstructured regions. The number of nodes can be reduced by merging subsequent nodes in a sequence. As a merging criterion we use the size of individual regions—simply the number of pixels. In addition, if a certain region has a size smaller than a minimum diatom size, it will be merged with its parent node.

This procedure can be directly applied to the images obtained by the texture analysis described above. All regions of the image are used as nodes of the neighborhood graph. Figure 5 shows the final object regions with different gray levels.

For each object region, the smallest rectangular bounding box is determined and the corresponding part of the original input image is used for further processing. Figure 6 shows these individual images extracted from Fig. 5. As can be seen, an object image can still include two (or more) connected or overlapping objects. Debris, too, can be attached to the objects. Hence, further post-processing is necessary to find the exact location of the diatoms' outlines.

Chapter 6: Contour extraction　　　　　　　　　　　　　　　　　　　　　　　　99

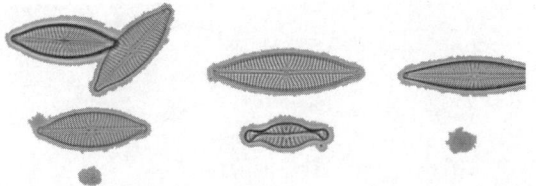

Figure 6. Individual region images, without black borders, extracted from the image shown in Fig. 1.

3 Contour extraction

After the pre-segmentation of input images into initial object regions, the location of the boundary between objects and background is determined using two different contour extraction methods. These methods differ in how the contours, to be used in a subsequent contour-following step, are detected in a binary image. The first method automatically selects global gray level thresholds in order to label the dark diatom border. The second method employs an edge detector in order to extract all edges. The best method has to be chosen depending on the quality of the input image, such that binary images with closed diatom contours are always obtained. These closed contours are then traced using a conventional contour-following algorithm.

In Sections 3.1 and 3.2 the two different approaches for the contour detection are presented. The contour following is described in Section 3.3.

3.1 Global thresholding

Of all global thresholding methods which determine a single threshold value from the gray level histogram, the method proposed by Otsu (1979) is known as one of the best (Sahoo et al., 1988). This method is based on a discriminant criterion, which is equivalent to the selection of a threshold such that the interclass variance between dark and bright regions is maximized (Papamarkos and Atsalakis, 2000). There are other statistical methods which use different criteria, such as the average pixel classification error rate. For a survey of thresholding methods see Sahoo et al. (1988).

Most methods rely on bi- or multi-modal histograms with distinct minima between separate peaks. As shown in the previous section, this condition does not occur in the case of brightfield diatom images. Hence, a heuristic procedure, similar to the one explained in Section 2, is used to select a threshold for the detection of the exact contour location. In contrast to the threshold selection method used during the pre-segmentation, a *single* threshold will be selected, since only the location of the dark diatom border is required and not the location of the light diffraction halo (the latter is the result of light diffracted by the vertical walls of the valve—the mantle—and tiny dust particles that stick to a valve's surface). Starting from the peak in the histogram, the search process goes towards the left (dark) tail of the histogram. Again, the first bin to the left of the peak with a frequency below a

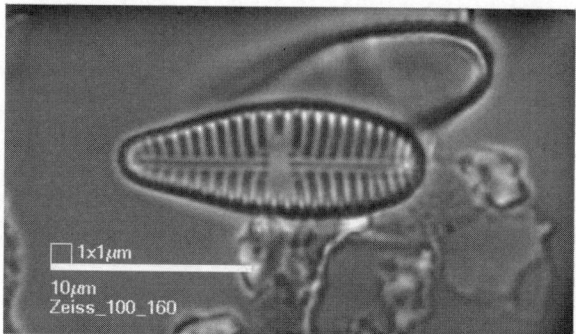

Figure 7. Example image for illustrating the contour extraction.

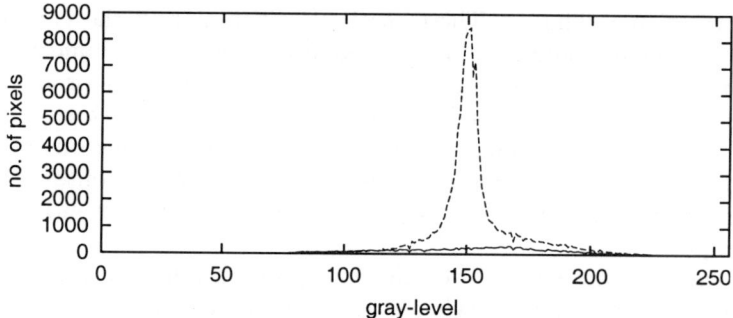

Figure 8. Histograms of the central diatom (solid) and its surround (dashed) shown in Fig. 7.

certain threshold is located. A threshold of 15% of the mean frequency yielded stable results for most images in ADIAC's database (we note that 15% was also used in Section 2).

We will compare the results obtained with our heuristic threshold selection and Otsu's method, using the image shown in Fig. 7. This image contains two diatoms, one of which is usable, and some debris. The distributions of the gray levels of the central diatom and its background (or surround) are shown by the histogram in Fig. 8. As can be seen, both curves overlap almost completely. While the histogram of the background has a strong peak around its mean value, the histogram of the central diatom is very flat.

The Otsu method selects a global threshold of 175, which is to the right of the main peak of the histogram. Using our heuristic procedure as described above, a threshold value of 140 is obtained, which is to the left of the peak. Figure 9 shows the thresholded images obtained with these two values plus one even lower value. It is obvious that the result obtained with our method is clearly superior. Although the human eye has absolutely no problem in locating the position of the contour, this is very difficult for an automatic procedure. The middle image shows that the black contour of the central diatom must be close to the real contour, that at

Chapter 6: Contour extraction 101

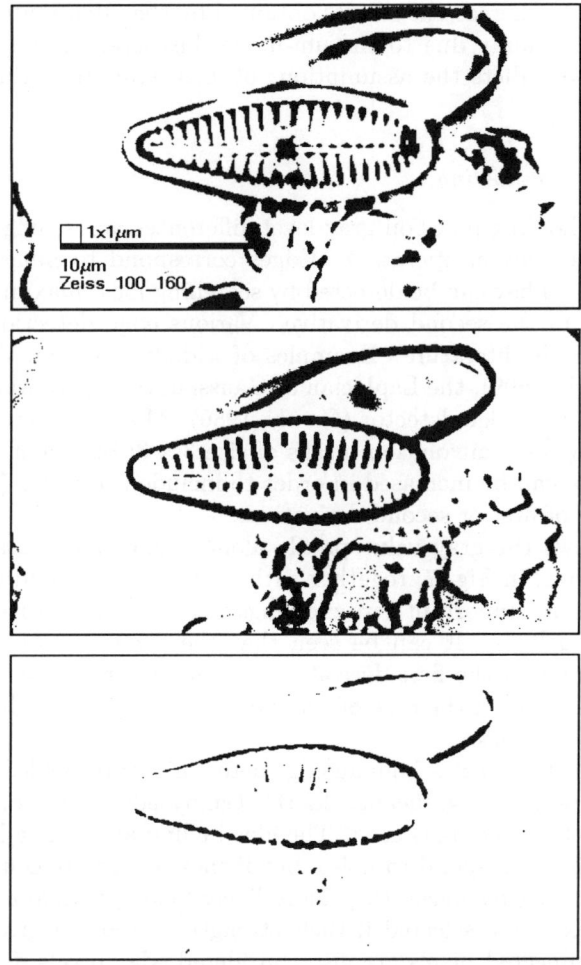

Figure 9. Binary images obtained for the image shown in Fig. 7 using different threshold values: 175 (top), 140 (middle), 102 (bottom).

least it is completely closed, but also that it is connected to part of the unfocused diatom. Further processing is required in order to select the right contour and to separate it from the other.

If the histogram is examined in detail, it can be seen that there is a small local minimum near the threshold obtained by the Otsu method. To verify the relevance of this minimum, the gray level image was smoothed by a 3×3 median filter and the threshold selection procedure was repeated. Now, a threshold of 102 is obtained by the Otsu method, whereas a value of 141 is found by the heuristic search. As can be seen in the bottom image in Fig. 9, only some small parts of the dark border are selected if the threshold value of 102 is used, and the contour is not closed. Hence, in this case the heuristic search provides the most stable threshold selection for the dark image border, whereas the results obtained by the Otsu method are unstable. This instability is mainly due to the uni-modal histogram of microscopic diatom images, which contradicts the assumptions of most statistical threshold selection methods.

3.2 Edge-based thresholding

Edge-based thresholding relies on gray level differences across edges, which can be located by edge detection operators. Edges correspond to image locations with strong transitions. They can be detected by searching local maxima in the gradient or zero-crossings in the second derivative. Various edge detection methods have been described in the literature. Examples of widely used edge detectors are the Sobel (Sonka et al., 1999), the Laplacian of Gaussian (LoG, see Marr and Hildreth, 1980), and the Canny edge detector (Canny, 1986). Most operators, like the Sobel, are small, i.e. they use convolution masks of size 3×3; but when there is noise in the image the size can be increased in order to include averaging (noise reduction) in the estimation of first or second derivatives.

Figure 10 shows the gray value profile along a line that crosses the left edge of the diatom shown in Fig. 7, together with its first and second derivatives. The locations of the most important zero crossings of the second derivative are marked by vertical dashed lines. It can be seen that these zero crossings coincide with minima and maxima of the first derivative, depending on the polarity of the gray level transition. Hence, in the case of the dark border around a diatom, edges will be detected on both sides.

The edge map that results from an edge detector is thresholded in order to keep the edges that are most significant. In the Canny edge detector (Canny, 1986), hysteresis-based thresholding is used. The idea behind applying a hysteresis is that weak edges usually correspond to noise, but if these edges are connected to any of the pixels with strong response, they more likely belong to real edges. Thus, in a first step edge pixels are selected if their strength is above a given threshold. In a second step, connected pixels are also considered edge pixels if their response is above an additional, second threshold. The resulting binary edge map is visualized in Fig. 11. Detected edges are marked in black while all other pixels are white. As can be seen, even the Canny operator, which is considered a benchmark in this field, can create gaps in a diatom contour. Small gaps can be closed by a subsequent dilatation of the edge image, which is a standard procedure from mathematical morphology (Sonka et al., 1999), but bigger gaps must be closed by hand, using cursor and mouse.

In general, the method proposed by Canny (1986) showed good results in the case of our microscopic diatom images, especially when the threshold-based method was unable to select a good threshold value and its resulting contour was disappointing.

3.3 Contour following

In a final step of the contour extraction, region borders are traced in the binary image. The edge image is searched, starting at the top-left corner, in a left-to-right

Chapter 6: Contour extraction

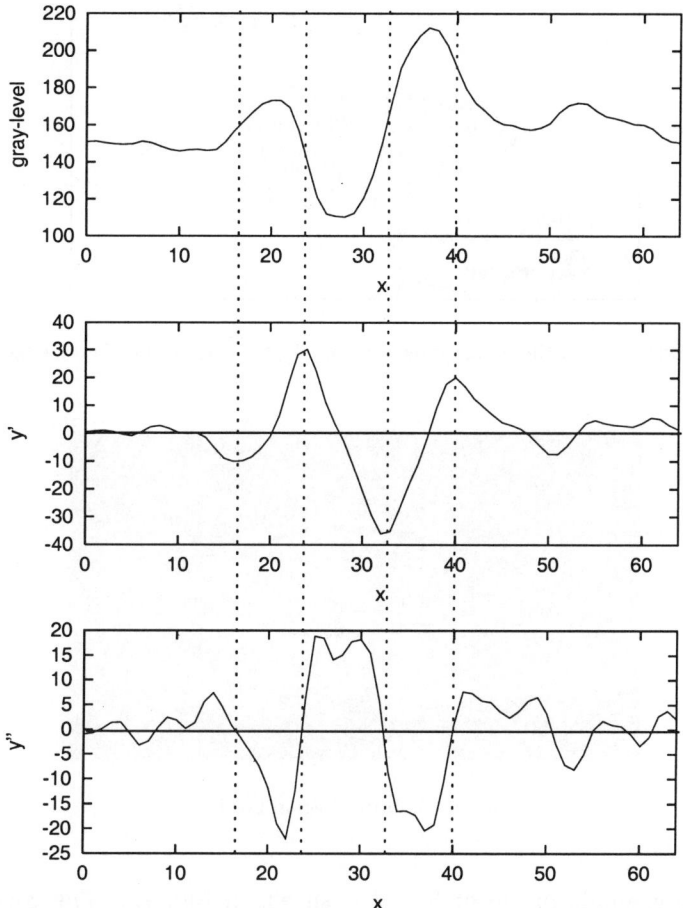

Figure 10. Top: one-dimensional gray value profile of the left edge of the diatom shown in Fig. 7; middle: first derivative; bottom: second derivative. The vertical dashed lines indicate the most important zero crossings of the second derivative.

and top-to-bottom scan, until a pixel of a new region is found. This pixel is taken as a starting point of the new region's border. The pixels of the 3 × 3 neighborhood are searched anti-clockwise to find a pixel with the same value as the current pixel. This search continues until the starting point of the contour is reached again. All pixels of the contour are then labelled as "visited" and the search for a new region is resumed. This procedure continues until the lower right corner of the image has been reached.

Labelled contours are post-processed by evaluating their straightness. In the case of significant curvatures, which imply contour deformations by noise or debris, it is checked whether there is a better contour candidate with a smaller curvature. Contours counting less pixels than the minimum diatom size are rejected. The final

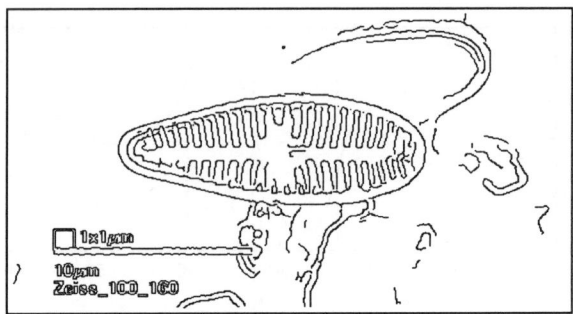

Figure 11. Edge map of the image shown in Fig. 7 obtained by the Canny edge detector.

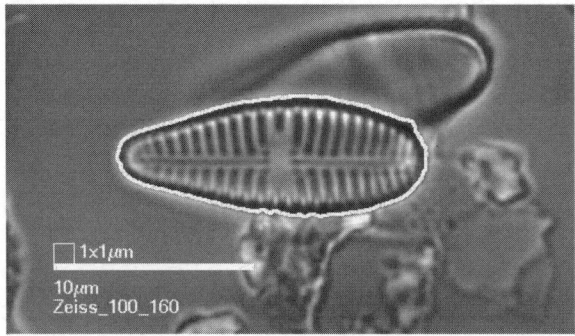

Figure 12. Initial region border.

result for the example image of Fig. 7 is shown in Fig. 12. The contour of the central diatom has been separated from the other, unsharp diatom.

4 Results

The contour extraction procedure proposed in this chapter has been evaluated on a subset of the mixed genera data set (see Chapter 4). Because this evaluation was done before the end of the project, this subset consisted of 808 different diatom images. For most diatoms there is only a single image, but for a subset multiple images have been taken at different focal planes. In the case of multiple images, the contour extraction was applied to the image in which the outline is in sharp focus. In most cases, the contour could be detected completely automatically by at least one of the described methods. Only for 2% of the images the contour had to be extracted semi-automatically. In these cases parts of the contour with a low contrast were manually marked in the image. In another 2% of the cases, it was necessary to change the automatically selected threshold in order to obtain good quality contours. Our results showed that only for diatoms with low contrast, weak poles, or a large diffraction halo, interactive optimization of the selected threshold

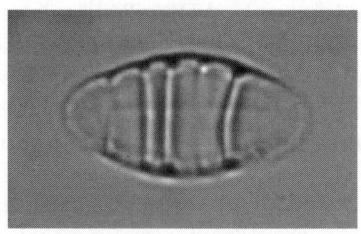 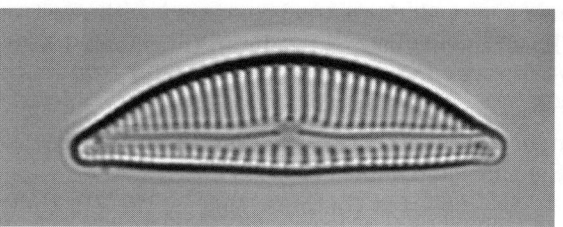

Figure 13. Problematic diatom images. Left: low contrast and weak outline; right: very dark outline with light pole areas.

may be required.

The results of the different contour detection methods are summarized in Tab. 1 according to the quality of the detected contours. As can be seen in the second column the threshold-based contour extraction method yielded 652 good contours without any user intervention. In 73 cases the contour was not smooth enough because threshold values were slightly too high or too low. In 46 cases, characterized by a very dark outline and light pole areas, the contour was broken into two parts. In another 28 cases, the contour was surrounded by some debris very close to the diatom, which caused deformations. An appropriate contour could not be extracted in only in 9 cases.

Using the edge detection-based contour extraction method (column 3), 660 good contours were obtained. In 129 cases the contours were only partly detected by the edge following. In 4 cases the contour was not smooth, and in 5 cases the contours were deformed by debris. Only in 10 cases no appropriate contour could be extracted.

When the results of both methods were combined (column 4), good contours could be extracted from 776 diatoms. Still 8 contours were deformed by debris, 12 were not smooth, and 12 were only partly detected because of low contrast or weak poles. Nevertheless, in 16 of these cases it was possible to extract a good contour by manually adjusting the gray level threshold. In the remaining 16 cases the contour had to be corrected by hand. Two examples of problematic images are shown in Fig. 13. The diatom on the left has very low contrast at the poles and the prominent ornamentation leads to a wrong contour. The diatom on the right

Table 1. Results of the contour detection.

quality of the detected contour	threshold based	edge detection based	both methods
good	652	660	776
not smooth	73	4	12
partly detected	46	129	12
deformed by debris	28	5	8
no contour	9	10	0

has a very dark contour compared with most other images in ADIAC's database. This type of contour results in thresholds which are too low, i.e. they do not allow detection of the contour at the poles. Such images are also difficult for the edge-based contour detection. The contour of the diatom on the left needs to be extracted manually and that of the diatom on the right requires interactive optimization of the threshold value.

All extracted contours, that is, the 776 that were extracted automatically plus the 16 that required threshold optimization as well as the 16 extracted manually, are part of the mixed genera data set used for feature extraction and identification tests as described in other chapters. But *automatic* contour extraction is only possible when the images captured on the microscope have good contrast across the entire diatom outline. As shown here, the vast majority of all contours needs to be inspected visually only once, and only very few require more user intervention.

5 Conclusion

In this chapter a robust scheme for the detection of diatom contours in microscopic brightfield images has been presented. This includes a pre-segmentation step for images containing multiple diatoms, and two different methods for finding the exact location of a contour. The first of these is based on a global gray level thresholding, whereas the second employs an edge detector. The method to be used depends on the quality of the input image. In very few cases user intervention is still necessary to optimize a threshold value, such that a closed contour of acceptable quality is obtained. Smooth contours could be detected completely automatically from 776 diatom images out of a total of 808 images, that is, in 96% of all cases.

References

Anoraganingrum, D., Kröner, S. and Gottfried, B. (1999) Cell segmentation with adaptive region growing. Proc. Int. Conf. on Image Analysis and Processing, Venice (Italy), IEEE Computer Society, Los Alamitos, CA, pp. 1043-1046.

Canny, J. (1986) A computational approach to edge detection. IEEE Trans. Pattern Analysis and Machine Intelligence, Vol. 8, pp. 679-698.

Fu, K.S. and Mui, J.K. (1981) A survey on image segmentation. Pattern Recognition, Vol. 13, pp. 3-16.

Haralik, R.M. and Shapiro, L.G. (1992) Computer and robot vision. Addison-Wesley, Reading, MA.

Marr, D. and Hildreth, E. (1980) Theory of edge detection. Proc. Royal Society of London, Vol. 207, pp. 187-217.

Otsu, N. (1979) A threshold selection method from gray-level histograms. IEEE Trans. Systems, Man, and Cybernetics, Vol. 9, pp. 62-66.

Pal, N.R. and Pal, S.K. (1993) A review on image segmentation techniques. Pattern Recognition, Vol. 26, pp. 1277-1294.

Papamarkos, N. and Atsalakis, A. (2000) Gray-level reduction using local spatial features. Computer Vision and Image Understanding, Vol. 78, pp. 336-350.

Reed, T.R. and du Buf, J.M.H. (1993) A survey of recent texture feature extraction and segmentation techniques. Computer Vision, Graphics, and Image Processing: Image Understanding, Vol. 57, pp. 359-372.

Sahoo, P.K., Soltani, S. and Wong, K.C. (1988) A survey of thresholding techniques. Computer Vision, Graphics, and Image Processing, Vol. 41, pp. 279-295.

Sonka, M., Hlavac, V. and Boyle, R. (1999) Image processing, analysis, and machine vision. Brooks/Cole, Pacific Grove, CA, (2nd edition).

CHAPTER 7

IDENTIFICATION
USING CLASSICAL AND NEW FEATURES
IN COMBINATION WITH DECISION TREE ENSEMBLES

STEFAN FISCHER AND HORST BUNKE

This is the first chapter in this book to deal with automatic diatom identification. We present many different feature sets that explore the shape of valve contours and the ornamentation patterns. Many features are classical, e.g. Fourier descriptors and moment invariants, but we also developed new features specific to diatoms. We present an introduction to decision tree classifiers, including ensembles, which are used here and in other chapters for identification tests. Our features, in combination with decision trees, bagging and cross-validations with the leave-one-out strategy, yielded excellent identification results: 92.5% correct identification in the case of the *Sellaphora pupula* data set, and 95% in the case of the mixed genera data set. These numbers increase to 99% when the matches ranked one to three are considered. When only 27 interpretable features of all 171 features are used, the results are worse, but only a few percent.

1 Introduction

Many diatomists start an identification by using the type of symmetry of the diatom's contour (Barber and Haworth, 1981). Subsequently other characters, such as shape and ornamentation properties, are used. Apart from the difficulty of detecting sometimes minute shape differences, this is relatively easy because the human visual system is so efficient. Unfortunately, often it is very difficult, if not impossible, to devise computer algorithms that extract the same features that are used by experts. For example, the spatulate shape of *Gomphonema augur* is difficult to model mathematically. However, fortunately there are many morphological diatom characters that are related to features used in pattern recognition and image processing. One example is the density of a valve's striation, i.e. the number of stripes per 10 microns, which is easier to measure by computer than by eye.

Because there is no single character or feature that allows the identification of all taxa, many different features or even many different feature sets have to be used. Each of the features or feature sets can describe different characteristics. Some features may be directly related to characters used by experts, whereas others have a purely mathematical background. In addition to the symmetry type mentioned above, many direct properties of a valve's shape, such as length and width, will be used in this chapter for automatic identification. Mathematically inspired features, like Fourier descriptors and moment invariants, will also be investigated. The ornamentation of a valve will be exploited by using texture descriptors based on gray level co-occurrence matrices and Gabor wavelets.

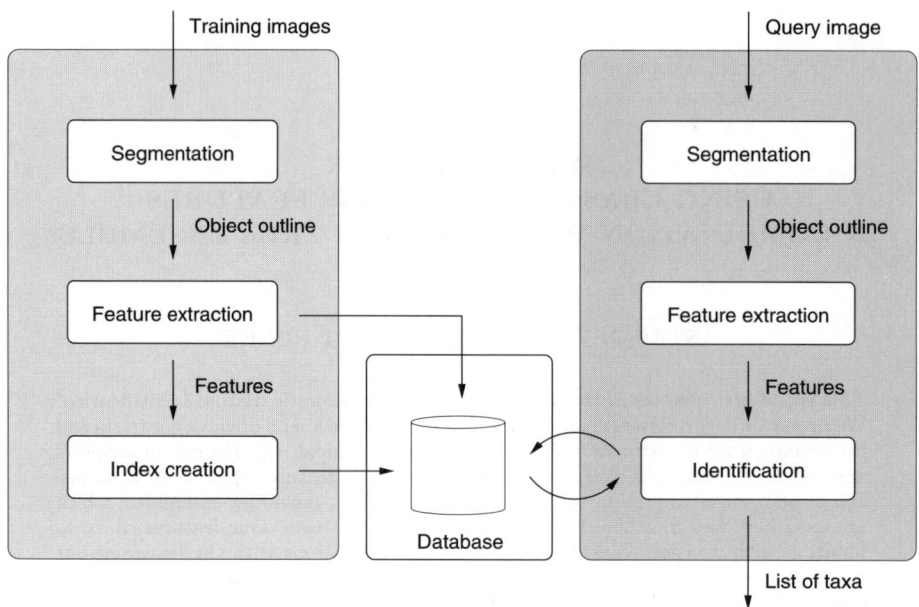

Figure 1. Overview of automatic identification system.

Features are extracted from a set of training images and stored in a database. A classifier is trained using the features in conjunction with taxon names determined by expert taxonomists. After training, the classifier can be used to identify new specimens: features of the new specimen are extracted and compared with the features in the database—indirectly, that is, using the rules established by the classifier during the training—and the classifier computes the best matching taxon or taxa. This procedure is visualized in Fig. 1. The training phase of the identification system is shown on the left. Segmentation refers to contour extraction as described in Chapter 6. The identification or query phase, shown on the right, employs the same procedures, except that the features are not stored in the database; they are only used for identification.

In the next section we describe the features to be used. These address the shape of a contour (Section 2.1), general texture properties of the ornamentation (Section 2.2) and specific morphological properties like the striation density (Section 2.3). The identification approach outlined above is described in detail in Section 3. In Section 4 the diatom test sets, used to evaluate the identification system, and the test protocol are described. In Section 5 experimental results are presented. Results obtained with an identification system that consists of a single decision tree are compared to those obtained with a decision forest. Final conclusions are drawn in Section 6.

Chapter 7: Identification using classical and new features 111

2 Feature extraction

2.1 Contour features

Many but not all diatoms can be distinguished by evaluating properties of the valve outline. In this section various methods for extracting such properties will be presented. These include measures for symmetry, global and local shape characteristics, as well as geometric properties, such as length and width, which are all easy to interpret by expert taxonomists. Also, some heuristic shape descriptors, like rectangularity, compactness and circularity, are introduced. Finally, widely used, but mathematically inspired features, such as Fourier descriptors and moment invariants, are described.

At first glance, the extraction of a property like a shape's symmetry seems to be simple. However, the problem is that natural objects are almost never really symmetric in the strict mathematical sense. They often satisfy some criterion to a certain degree, i.e. there are always deviations. This implies that thresholds need to be used in order to decide whether a specific criterion is met or not. Another problem is caused by the image capturing process. Most diatom valves are not flat but have a rim so that they settle more or less level on a cover slip during the preparation of slides. In these cases their images correspond to an optimum orthogonal projection due to the horizontal focusing plane. But this happens not always. The viewing axis may not be *exactly* orthogonal to the valve face, in some cases it may be slightly oblique, such that geometrical distortions may occur. This also complicates symmetry detection. Moreover, diatoms with curved valve faces cannot be photographed with the entire face in focus, especially when using high-magnification lenses. In such cases, parts of the ornamentation appear sharp, whereas other parts are blurred; this will affect the extraction of striation features, but it can also hamper the extraction of a good quality contour.

Below, a method for determining the symmetry class of a contour will be introduced. Then, the computation of geometric characteristics, such as length and width, will be described, as will be properties of the global shape and the valve poles. This is followed by heuristic descriptors and mathematically motivated features, i.e. Fourier descriptors and moment invariants.

2.1.1 Symmetry analysis

As already mentioned above, the type of symmetry is one of the main characters used by experts. Figure 2 shows some examples. The vast majority of diatoms is bilaterally symmetrical about two orthogonal axes (*symmetrical about both axes*). Other diatoms are symmetrical about only one axis. This can either be the long axis (*heteropolar*) or the short axis (*dorsiventral*), although some of these can be *almost* symmetrical about both axes. Other diatoms have no axes of symmetry, but are rotationally symmetric (*sigmoid*), and a large group of genera are more or less *circular* (Barber and Haworth, 1981; see also Bayer, 1999).

Before we describe the method for analyzing the symmetry, we will introduce some basic terminology. A straight line through the centroid of a two-dimensional shape is called a *symmetry axis* if the figure remains identical after a reflection

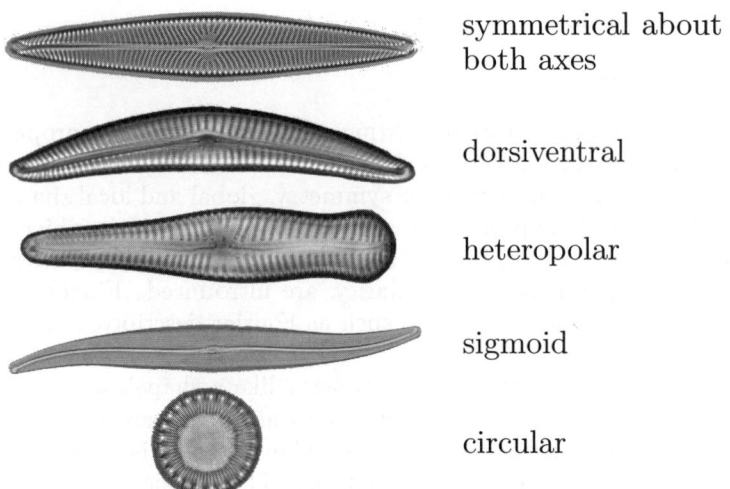

Figure 2. Examples of diatoms with different types of symmetry.

about this straight line. A shape S is called *reflectionally symmetrical* with degree m if it has m symmetry axes. S is called *rotationally symmetrical* with degree $m > 1$ if there are m different angles, such that S remains identical after rotations with $\alpha = k \cdot (360°/m), k = 1, 2, \ldots$ If S preserves its shape after a rotation of $180°$ it is called *point symmetric*.

Fischer et al. (2000a) describe different methods for symmetry detection for the purpose of indexing a diatom database. Here, a similar technique based on the Hausdorff distance will be used. The Hausdorff distance (Huttenlocher et al., 1993) between two sets A and B is defined as

$$H(A, B) = \max\left(h(A, B), h(B, A)\right), \tag{1}$$

where the directed Hausdorff distance $h(A, B)$ is defined as

$$h(A, B) = \max_{a \in A} \min_{b \in B} ||a - b||. \tag{2}$$

Basically, this distance can be used to measure the degree of overlap between two areas in an image. For our purpose, the set A consists of all pixels inside a diatom contour, and set B consists of the same pixels after the diatom has been rotated by some angle. This procedure is visualized in Fig. 3. The contour has been rotated $180°$ around its centroid (i.e. the center of gravity marked with a cross) and the Hausdorff distance is measured between the rotated object (light gray) and the original one (dark gray, most of it occluded). Rotational symmetry will be detected when the Hausdorff distance is below a certain threshold value. For further details see Fischer (2002).

A similar approach is used to detect point symmetry. In this case the object is split into two parts, along an axis through its centroid. One of the parts is rotated by $180°$ degrees and the Hausdorff distance is computed between both parts. Finally, the distance value is compared with a threshold value.

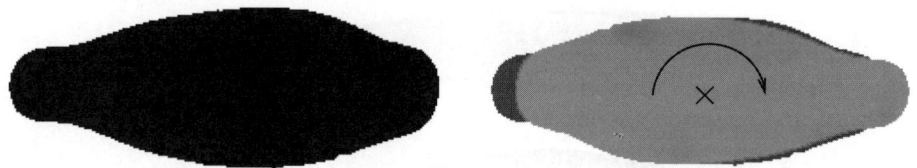

Figure 3. Measuring rotational symmetries using the Hausdorff distance. Left: original object; right: object rotated by 180° and superimposed on the original. The cross indicates the position of the centroid.

Using the number of detected reflectional symmetry axes, and knowing whether the object is point symmetric or not, objects can be classified into distinct symmetry classes. This is done using a small set of rules. If no symmetry axis is detected the object is classified as asymmetrical. If one symmetry axis is detected the object is classified as either dorsiventral or heteropolar. The direction of the symmetry axis is used in order to distinguish these. If the direction of the symmetry axis is parallel to the direction of the major axis, the object is classified as heteropolar, but if these axes are perpendicular it is dorsiventral. In the case of two symmetry axes which are nearly perpendicular to each other, the object is classified as symmetrical about both axes, and if there are more than two axes the object is circular. Only point symmetry is required for sigmoid. Table 1 summarizes the rules.

For circular objects a further post-processing is applied. An object is only classified as being circular if it has more than 2 reflectional symmetry axes *and* the circularity—defined below—is greater than 99%. In all other cases no further post-processing procedure is necessary. If an object does not satisfy any of the rules, it is classified as having an unknown symmetry.

The assignment of a certain diatom taxon to a single symmetry class is often not possible, because some diatoms show different symmetry characteristics during their life cycle. Hence, the symmetry feature has to be applied carefully in the diatom identification process, although it is often used by experts as the first character.

Table 1. Classification of shapes into different symmetry classes. The direction of the symmetry axis is relative to the major axis of pennate diatoms.

symmetry class	number of reflectional symmetry axes	point symmetric	direction symmetry axis
asymmetrical	0	no	
dorsiventral	1	no	$\approx 90°$
heteropolar	1	no	$\approx 0°$
both axes	2	not used	
circular	> 2	not used	
sigmoid	0	yes	

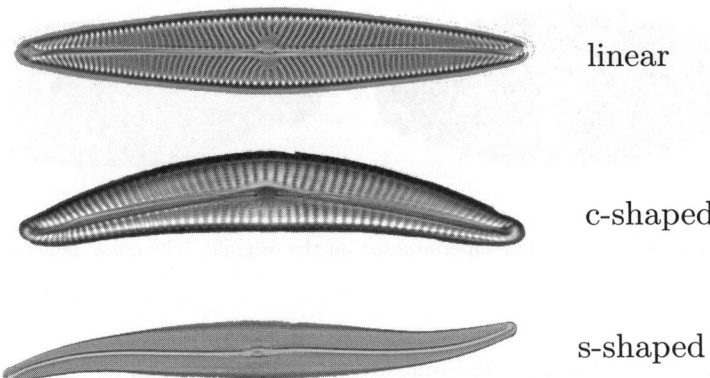

Figure 4. Examples of different types of shape.

In order to reduce the influence of the changes during the life cycle, the strength of the symmetry axis in apical and transapical direction is used together with the symmetry class. The strength of the symmetry axis is measured as the degree of overlap between the original contour and the rotated copy.

2.1.2 Geometric properties

Features that can be easily computed from a contour are the length, width, size and the length-width ratio. The length and width are calculated using a contour's principal axes. The direction of the minor axis is selected perpendicular to the major axis. The length L that we use is the maximum distance between the intersections of the contour and the major axis. The width W is calculated in the same way, but using the minor axis. The length-width ratio is then $R = L/W$. The size S of a contour corresponds to the number of pixels it encloses. Finally, the length and width are converted to μm and the size to μm^2, using the pixel size of the image that is available in the database.

2.1.3 Global shape properties

Most diatoms can be grouped into different types of shape. Figure 4 shows examples. While most diatoms are linear, there are also curved ones. Resembling the shape of the letter *c*, curved diatoms are named *c*-shaped. In this context, sigmoid diatoms are named *s*-shaped.

Mathematically, the global shape of an object can be described by using the form of the object's skeleton (Shih and Pu, 1995). All short branches are removed from the skeleton, such that only the center line remains. To obtain the characteristics of the center line, it is approximated by a third-order polynomial

$$y = ax^3 + bx^2 + cx + d. \qquad (3)$$

Given the coefficients of this polynomial, the shape can be classified. Third order polynomials have only one inflection point and either no or two extrema. Additional

Table 2. Decision rules for different global shapes.

Rule 1:   if $|a| \leq 0.001084$
            and $\Delta \leq 0.000034$
            then   linear

Rule 2:   if $\Delta > 0.000034$
            then c-shaped

Rule 3:   if $|a| > 0.001084$
            and $\Delta > -0.000677$
            then s-shape

information can be derived from the discriminant

$$\Delta = 3ac - b^2 \qquad (4)$$

of the derivative. For example, if $\Delta \leq 0$ there is one inflection point, if $\Delta = 0$ this inflection point is a saddle point, and if $\Delta > 0$ there is no extremum (Bronstein and Semendjajew, 1987).

Because of noise and distortions, the given relations cannot be directly applied to digital images. Suitable threshold values for the polynomial coefficients and the discriminant Δ are found automatically by means of the C4.5 decision tree classifier (see Section 3.1). We tested this by applying C4.5 to a training set of 808 images from the mixed genera data set (see Chapter 4). All 808 images were assigned manually to one of the three shape types. The resulting decision rules are shown in Table 2.

2.1.4 Shape of the poles

Another important characteristic of many diatoms is the shape of the valve endings, called poles. For example, the left and the right poles of the diatom shown in Fig. 5 (left) are different. Not only the shapes of the two poles can be different, also the direction in which they point can be. The diatom in Fig. 5 (right) has poles that point down- and upward.

In order to characterize the shape of a pole, the relevant part of the contour is approximated by a second-order polynomial

$$y = ax^2 + bx + c. \qquad (5)$$

Similar to the global shape analysis described above, the value of the first coefficient (a) is used to describe the sharpness (pointedness) of the pole. In addition, the sign of this coefficient is used to distinguish between convex and concave poles. The direction of the poles is evaluated relative to the long and short axes of the diatom.

A similar approach is applied to characterize the upper and lower contour parts that are separated by the poles. For these parts, only the first polynomial coefficient is used, i.e. the absolute value and the sign. The 10 features listed in Table 3 are used to characterize the poles and contour parts.

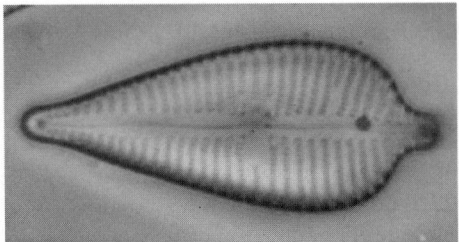 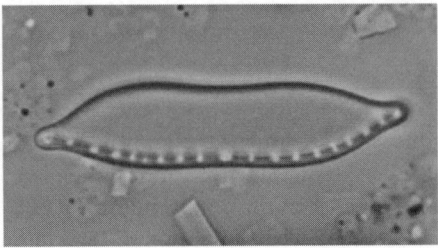

Figure 5. Diatoms with different types of pole. Left: *Gomphonema augur*; right: *Nitzschia brevissima*.

Table 3. Features used for poles and contour halves.

	abs(a)	sign(a)	direction
left pole	x	x	x
right pole	x	x	x
upper contour	x	x	
lower contour	x	x	

2.1.5 Heuristic shape descriptors

Many shape descriptions rely on simple heuristics. They yield acceptable results in the case of simple shapes. Heuristic descriptors are, for example, rectangularity, compactness, ellipticity, etc. (Sonka et al., 1999). These descriptors cannot be used for a reconstruction, and they do not work well for complex shapes, but experts can easily interpret the values. Below we list those that we use in identification experiments.

Rectangularity: The standard approach to measure the rectangularity of an object is to use the ratio of the area of the object to the area of its minimum bounding rectangle. This ratio R_k is given by

$$R_k = \frac{\text{area}_{\text{object}}}{\text{area}_{\text{bounding box}_k}}, \quad (6)$$

where k is the orientation of the bounding box. The bounding box is rotated in discrete steps and for each angle the ratio R_k is computed. The rectangularity is then defined as the maximum of the R_k values, i.e.

$$\text{rectangularity} = \max_k(R_k). \quad (7)$$

Compactness: The compactness of an object is given by

$$\text{compactness} = \frac{(\text{contour length})^2}{\text{area}}. \quad (8)$$

The most compact shape in the Euclidean plane is the circle. All other shapes have a larger compactness, depending on the complexity of the shape. In general, compactness ranges from 1 for a single pixel or a circle to arbitrarily large values. In the case of the diatoms in ADIAC's database, the range starts with values around 10 and goes up to about 80.

Ellipticity: Any ellipse can be obtained by applying an affine transform to a circle (Rosin, 2000). Therefore, it can be characterized by an affine moment invariant of the circle (Flusser and Suk, 1993):

$$I'_1 = \frac{\mu_{20}\mu_{02} - \mu_{11}^2}{\mu_{00}^4}. \tag{9}$$

The moments for the unit circle are

$$m_{pq} = \int_{-1}^{1} \int_{-\sqrt{r^2-x^2}}^{\sqrt{r^2-x^2}} x^p y^q dy dx, \tag{10}$$

which allows us to calculate the value of the invariant of a perfect circle as $I'_1 = 1/(16\pi^2)$. Based on this value, the measure of ellipticity is defined as

$$E = \begin{cases} 16\pi^2 I'_1 & \text{if } I'_1 < 1/(16\pi^2) \\ I'_1/16\pi^2 & \text{otherwise.} \end{cases} \tag{11}$$

This measure ranges from 0 to 1, assuming its maximum value in the case of a perfect ellipse.

Triangularity: The same approach can be used to characterize triangles. Any triangle can be considered to be a simple right-angled triangle aligned with the x- and y-axes after an affine transform has been applied. The moments are

$$m_{pq} = \int_0^x \int_0^1 x^p y^q dy dx, \tag{12}$$

which result in $I'_1 = 1/108$. The triangularity measure is defined as

$$T = \begin{cases} 108 I'_1 & \text{if } I'_1 < 1/108 \\ I'_1/108 & \text{otherwise.} \end{cases} \tag{13}$$

Circularity: An accurate method to compute the circularity of an object is to compute the best-fitting circle and to measure the distance between the contour and the circle. The computation of the best-fitting circle can be done by a standard procedure using a combination of golden section search and successive parabolic interpolation (Forsythe et al., 1977). From the detected circle the distance to the object's contour is computed as

$$d = \frac{1}{N} \sum_i \left(r - \sqrt{(x_i - x_c)^2 + (y_i - y_c)^2} \right)^2, \tag{14}$$

where N is the number of contour points (x_i, y_i) and r is the radius of the circle with center (x_c, y_c). The measure for circularity is

$$\text{circularity} = 1 - d. \tag{15}$$

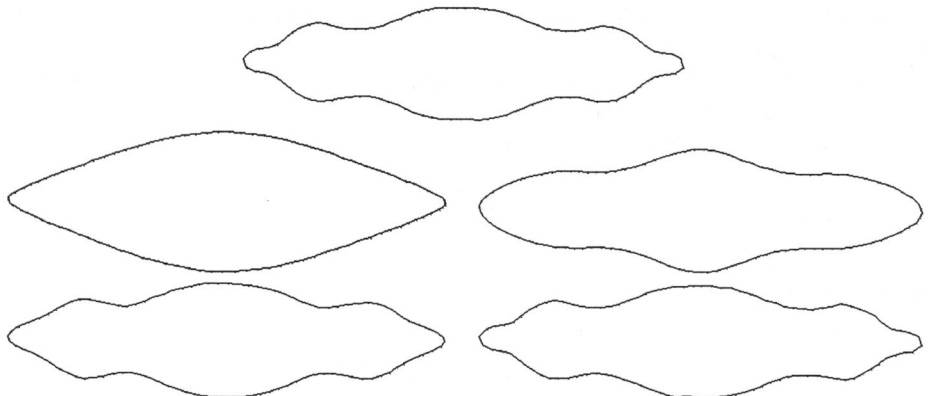

Figure 6. Original contour (top) and reconstructions using (left-to-right and top-to-bottom) 6, 14, 30 and 100 Fourier descriptors from a total of 128.

2.1.6 Fourier descriptors

Two mathematically motivated methods for the description of two-dimensional objects are described below. These are Fourier descriptors and moment invariants. The basic idea behind Fourier descriptors is to see a closed curve—a diatom contour—as a periodic function and to represent this by a set of Fourier coefficients (Arbter et al., 1990). Following a contour clockwise, N coordinate pairs $(x_0, y_0), (x_1, y_1), \ldots (x_{N-1}, y_{N-1})$ are recorded. These coordinate pairs can be seen as a sequence of complex numbers

$$z_k = x_k + \mathrm{i} \cdot y_k, \quad k = 0, 1, \ldots N - 1. \tag{16}$$

Applying the discrete Fourier transform to the sequence z_k leads to complex coefficients

$$F_n = \frac{1}{N} \sum_{m=0}^{N-1} z_m \exp(-\mathrm{i} 2\pi n m / N), \; n = 0, \ldots N - 1. \tag{17}$$

In general the original coefficients

$$a_n = \frac{1}{N} \sum_{m=0}^{N-1} x_m \exp(-\mathrm{i} 2\pi n m / N), \; n = 0, \ldots N - 1 \tag{18}$$

$$b_n = \frac{1}{N} \sum_{m=0}^{N-1} y_m \exp(-\mathrm{i} 2\pi n m / N), \; n = 0, \ldots N - 1 \tag{19}$$

are not invariant to translation, rotation and scaling (Sonka et al., 1999). Translation and rotation invariance are obtained by the transformation

$$r_n = \sqrt{|a_n|^2 + |b_n|^2}. \tag{20}$$

In order to obtain also scale invariance, the descriptors r_n must be normalized with respect to the first coefficient r_1, i.e. $w_n = r_n / r_1$.

Chapter 7: Identification using classical and new features

Fourier descriptors allow for a perfect reconstruction of the original curve. If, however, only low-frequency components are used in the reconstruction, sharp details such as corners are lost, but the global shape of the object is preserved. This is demonstrated in Fig. 6, which shows a contour (top) and reconstructions with different numbers of descriptors. The quality improves when more decriptors are used. One can also see that not all, in this case 128, Fourier descriptors are really required to approximate the contour. During evaluations we found that the first 30 descriptors are sufficient to distinguish between most shapes.

2.1.7 Moment invariants

Another approach to describe contours is based on moments. An image object can be seen as a two-dimensional density function $f(x,y)$. This allows us to compute geometric moments given by

$$M_{pq} = \int_{-\infty}^{\infty} \int_{-\infty}^{\infty} x^p y^q f(x,y) dx dy, \quad p,q = 0,1,2,\ldots, \quad (21)$$

where M_{pq} is the $(p+q)$th-order moment of the continuous image function $f(x,y)$. In the case of digital images, the integrals are replaced by sums and m_{pq} becomes

$$m_{pq} = \sum_{x=0}^{N-1} \sum_{y=0}^{M-1} x^p y^q f_{xy}, \quad (22)$$

where $N \times M$ is the size of the image and f_{xy} are the gray levels. A simplified version is often applied in pattern recognition. Since we are only interested in the shape of an object—a contour—only a binary function f_{xy} is considered, which is 1 if pixel (x,y) is inside the contour and 0 if it is outside.

General moments are not invariant to translation, rotation and scaling. Translation invariance is obtained by using central moments, defined as

$$\mu_{pq} = \sum_{x=0}^{N-1} \sum_{y=0}^{M-1} (x - x_c)^p (y - y_c)^q f_{xy}, \quad (23)$$

where x_c and y_c are the coordinates of the centroid given by

$$x_c = \frac{m_{10}}{m_{00}} \quad \text{and} \quad y_c = \frac{m_{01}}{m_{00}}, \quad (24)$$

in which m_{00} is the number of pixels inside the contour. Using central moments, the orientation of a contour can also be obtained. The angle of the major axis is given by

$$\phi = \frac{1}{2} \tan^{-1} \left(\frac{2\mu_{11}}{\mu_{20} - \mu_{02}} \right). \quad (25)$$

Moments are made scale invariant, as proposed by Hu (1962), by normalizing them:

$$\nu_{pq} = \mu_{pq} / \mu_{00}^{(p+q+2)/2}, \quad p+q = 2,3,\ldots \quad (26)$$

Rotation invariance is achieved by combining the moments using the theory of algebraic invariance. For example, Hu (1962) proposed the following seven moment invariants:

$$I_1 = \nu_{20} + \nu_{02}, \tag{27}$$
$$I_2 = (\nu_{20} - \nu_{02})^2 + 4\nu_{11}^2,$$
$$I_3 = (\nu_{30} - 3\nu_{12})^2 + (3\nu_{21} - \nu_{03})^2,$$
$$I_4 = (\nu_{30} + \nu_{12})^2 + (\nu_{21} + \nu_{03})^2,$$
$$I_5 = (\nu_{30} - 3\nu_{12})(\nu_{30} + \nu_{12})$$
$$\cdot [(\nu_{30} + \nu_{12})^2 - 3(\nu_{21} + \nu_{03})^2]$$
$$+ (3\nu_{21} - \nu_{03})(\nu_{21} + \nu_{03})$$
$$\cdot [3(\nu_{30} + \nu_{12})^2 - (\nu_{21} + \nu_{03})^2],$$
$$I_6 = (\nu_{20} - \nu_{02})[(\nu_{30} + \nu_{12})^2 - (\nu_{21} + \nu_{03})^2]$$
$$+ 4\nu_{11}(\nu_{30} + \nu_{12})(\nu_{21} + \nu_{03}),$$
$$I_7 = (3\nu_{21} - \nu_{03})(\nu_{30} + \nu_{12})$$
$$\cdot [(\nu_{30} + \nu_{12})^2 - 3(\nu_{21} + \nu_{03})^2]$$
$$- (\nu_{30} - 3\nu_{12})(\nu_{21} + \nu_{03})$$
$$\cdot [3(\nu_{30} + \nu_{12})^2 - (\nu_{21} + \nu_{03})^2].$$

These functions are invariant under translation, rotation and scaling of the object, and have been used in various pattern recognition tasks (Dudani et al., 1977; Belkasim et al., 1991; Bigun and du Buf, 1993, 1994).

2.2 Ornamentation features

An important character of diatoms is the ornamentation of the valve face, which is a specific type of texture. Texture (and texture segmentation) is one of the main topics in pattern recognition. There are statistical methods that measure variance, entropy or energy. Moreover, perceptual techniques are able to identify the orientation and regularity of textures (Pitas, 1993). Here we use two well-known feature sets: features derived from gray level co-occurrence matrices and features based on Gabor wavelets. The application of these methods to diatom identification is especially useful for the description of morphological properties. Such properties are often more or less constant over a big part of a valve. A striation pattern can be present within different areas, within which texture measures can be applied. Because a fixed number of features is required for the identification approach presented below, a fixed number of areas will be selected. The morphology is then described by averaging feature values within these areas.

Figure 7 shows an example in which the five areas to be used are marked by white squares. Inside these five areas local texture properties will be computed. One of the areas is located in the center of the diatom. The four other areas are located to the left, right, above, and below the center of the diatom. The position of the areas is computed by using the shape of the contour. Hence, even for nonlinear diatoms all areas will always be inside the contour. The size of the areas is adapted

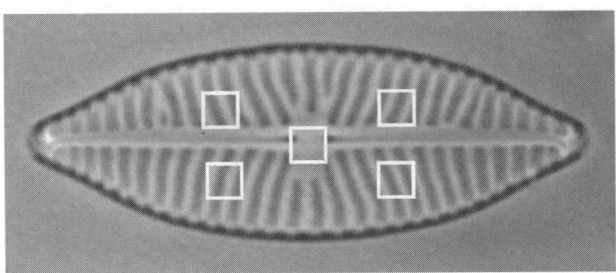

Figure 7. Examples of textured areas to describe the striation.

to the resolution of the image, and is therefore constant for each object. For each of the five areas, we compute 13 features of the gray level co-occurrence matrix proposed by Haralick et al. (1973), as well as the mean and standard deviation of the responses of 4 Gabor filters with different orientations.

2.2.1 Gray level co-occurrence

One statistical method for describing texture is based on the gray level co-occurrence matrix (GLCM). This method, first proposed by Haralick et al., (1973), is based on statistically sampling the way that certain gray levels occur in relation to other gray levels. This means that spatial relations of texture properties are estimated by using second-order statistics.

By using a displacement vector $d = (\delta_x, \delta_y)$, a matrix or histogram $P_d(i,j)$ is filled that counts the number of times that a pixel with gray level i occurs at distance d from a pixel with gray level j. Formally, this can be expressed by

$$P_d(i,j) = \{((r,s),(t,v)) : f(r,s) = i, f(t,v) = j\}, \qquad (28)$$

where (r,s) is a coordinate in the image and $(t,v) = (r + \delta_x, s + \delta_y)$ is the displaced coordinate. The gray-level co-occurrence matrix C_d is obtained by normalizing P_d with respect to the total number of pixels. Subsequently, different features, such as the maximum value, entropy and uniformity, are computed from the GLCM. The description of all 13 features that we use can be found in Haralick et al. (1973).

2.2.2 Gabor wavelets

Apart from statistical methods like the GLCM, there are methods which are motivated by the human visual system. Gabor filters are based on neurophysiological findings indicating that the primate brain applies a frequency and orientation analysis in the visual cortex, i.e. there are neurons that are tuned to specific frequencies and orientations. This approach is followed in the wavelet model which will be described below. For a detailed introduction to Gabor wavelets we refer to Tuceryan and Jain (1998).

The Fourier transform analyzes the global frequency spectrum of a signal. A localization in the spatial domain is obtained by introducing spatial dependencies into the Fourier analysis. The classical way of doing this is by windowing. The

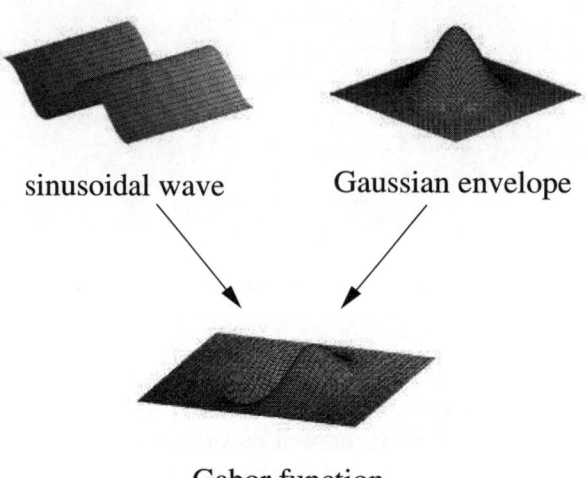

Figure 8. Combination of a sinusoidal wave and a Gaussian envelope to create a Gabor function.

windowed Fourier transform of a one-dimensional signal $f(x)$ is defined by (see Tuceryan and Jain, 1998):

$$F_w(u) = \int_{-\infty}^{\infty} f(x)\omega(x - \psi)e^{-i2\pi u x} dx. \tag{29}$$

If the window function $\omega(x)$ is Gaussian, then the transform becomes a Gabor transform.

A two-dimensional Gabor function in the spatial domain consists of a sinusoidal wave of a certain frequency and orientation that is modulated by a Gaussian envelope. The impulse response of the Gabor filter is given by (see Jain and Farrokhnia, 1991)

$$h(x, y) = \exp\left\{-\frac{x^2 + y^2}{2\sigma^2}\right\} \cdot \cos(2\pi\lambda x), \tag{30}$$

where λ is the frequency of the sinusoidal wave and σ is the size of the Gaussian envelope; see also Fig. 8. Filters with arbitrary orientation can be obtained by a simple rotation of the coordinate system, whereas a scaling allows to make filters with arbitrary frequency.

Motivated by the approach of Jain and Farrokhnia (1991), a bank of Gabor filters with multiple scales and orientations is used to compute image features. These are the means and standard deviations of the filter responses within windows. Although the window size can be determined automatically, using the central frequency, we use the same size as used for the GLCM approach. Gilomen (2001) studied different values for the orientation, the width of the Gaussian envelope, and the frequency. Best results were obtained by using four filter orientations (0, 45, 90 and 135°), a standard deviation $\sigma = 2.6$, and a frequency $\lambda = 5$.

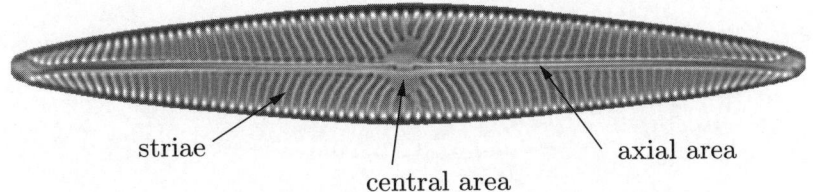

Figure 9. Example showing striae, axial area and central area.

2.3 Other morphological features

After having discussed various features of the valve shape and general texture features, morphological features of the ornamentation will be introduced in this section. While the features mentioned so far can be applied to any kind of object, the features introduced below are specific for diatoms. Figure 9 shows some of the ornamentation characters (see Chapter 2 for more detail). The first group of features describes the striation, i.e. the density of the *striae* and their mean orientation. Closely related are properties of the axial area with a *raphe*. The next feature, not depicted in Fig. 9, is specific for *costae*, which are ribs running from one side of a valve to the other. Finally, a density measure will be introduced for characterizing diatoms that have neither striae nor costae, but large individual pores.

2.3.1 Features of the striae

In order to be able to measure properties such as the density or mean orientation, the striated regions must be detected first. To this end we apply an edge detector to obtain the vertical edges in the image. Edge points are connected to lines, and the lines are approximated by curves. These curves are used to compute the following three properties:

Density: A stable value for the striation density is obtained by measuring the distance between adjacent vertical edges in different areas around the centroid of the contour. Outliers are removed and the average of the remaining values is taken as the final result.

Orientation: The orientation of the detected edges is measured separately in four quadrants around the centroid. An example is shown in Fig. 10 (top) where different orientations are indicated by white lines.

Changeover point: this feature indicates where the striae switch from left-leaning to right-leaning. Figure 10 (top) shows that the orientation of the striae does not need to be constant inside the different quadrants. Most striae run towards the center, but before the poles their orientation can change. This is indicated in Fig. 10 (top) by the detected changeover points, marked by the vertical black lines.

2.3.2 Width of the axial area

After masking out the regions where vertical edges have been detected, only the axial area around the raphe, the central area and some areas along the contour

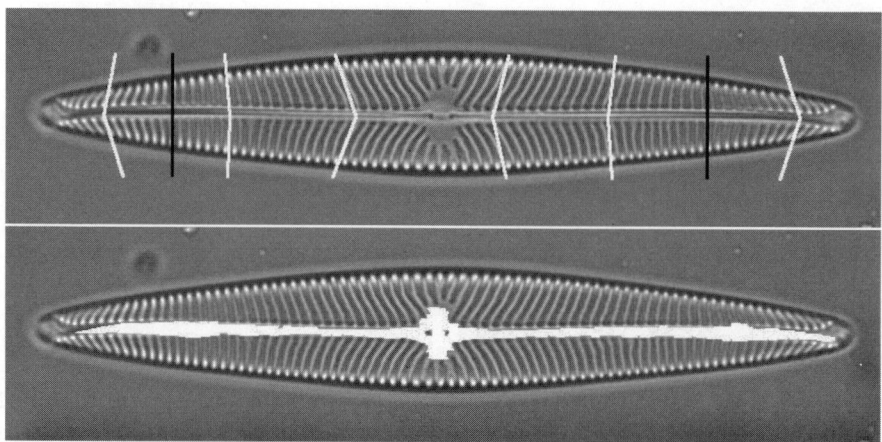

Figure 10. Examples of ornamentation features of *Navicula radiosa*. Top: white markers for the direction of the striae, black markers for the changeover points. Bottom: extracted axial and central areas.

remain. If, furthermore, all pixels at a certain distance from the contour are neglected, only the central and axial areas can be obtained; see Fig. 10 (bottom). From this area the mean vertical width is measured to characterize the width of the axial area.

2.3.3 Features of the costae

Some diatoms do not have a fine striation pattern, but instead have thick ribs, called costae, that run from one side of the valve to the other. The features introduced for the striation density and orientation cannot be used for costae. Hence, a different procedure has been developed. First, a gray value profile is computed along the central line of the diatom. This profile is slightly smoothed in order to suppress noise, after which its second derivative is computed. Similar to an edge detector, zero crossings of the second derivative are located, and, using a threshold on the amplitude of the first derivative, the presence of a rib is detected at the position of the zero crossing. Because of possible blurring at poles, ribs detected at the poles are rejected, and only the ribs in the central part are used to compute the relative density. Figure 11 illustrates this procedure. The central, horizontal line is marked in white in the image. The gray level profile and its derivative are shown to the right. The detected ribs are marked by vertical lines in the image, after which their distances can be averaged.

2.3.4 Horizontal frequency

The frequency of the ornamentation in horizontal direction is mainly related to the stria density, but there are also diatoms that do not have striae that are visible as continuous lines; these diatoms have big individual pores. Such pores appear as aligned points rather than solid lines, and it is difficult to detect the individual

Chapter 7: Identification using classical and new features

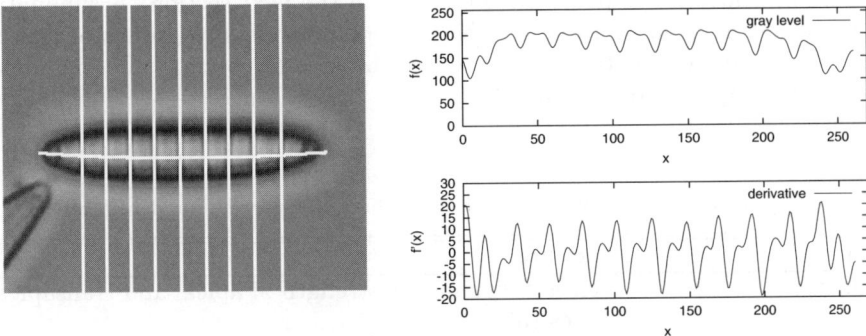

Figure 11. Feature extraction of costae (*Denticula subtilis*). Left: original image with positions of ribs marked. Right: gray level profile along the center line (top) and derivative of the profile (bottom).

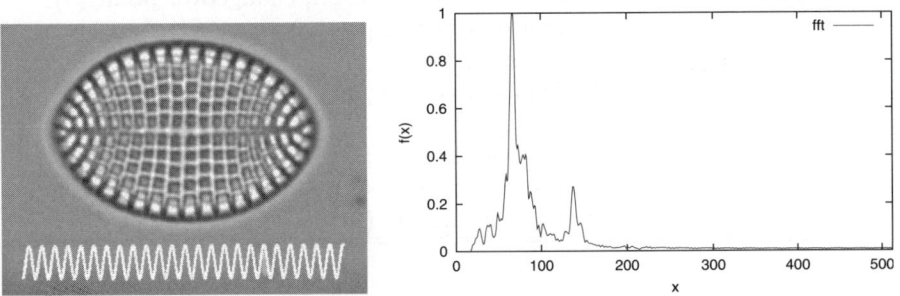

Figure 12. Example of the detection of the horizontal frequency in the case of a diatom with big pores (*Cocconeis stauroneiformis*). Left: original image with sinewave showing the computed main frequency. Right: frequency spectrum of a selected region across the center line.

lines. In order to compute the frequency, we use a method which is similar to the one used for costae.

Instead of only considering the gray values along a center line, a broader rectangular area around the center line is used to build a gray level profile. The main frequency is computed by applying the Fourier transform to the new profile, and then extracting the Fourier coefficient with the highest amplitude. Figure 12 (right) shows the frequency spectrum of the gray level profile from the diatom shown at the left. The corresponding sinewave is shown in the lower part of Fig. 12 (left). The frequency of the sinewave corresponds to the main pore frequency near the center line.

2.4 Summary

In the previous sections various methods for the extraction of features from the contour and the ornamentation have been presented. These features are used for automatic identification experiments described in the next section. A summary of all 171 features is given in Table 4. The first column refers to groups of features, whereas the individual features are listed in the second column. The number of features per group is given in brackets.

Table 4. Complete list of features and their numbers used for diatom identification.

group	feature
symmetry	class of symmetry, strength of apical and transapical axes (3)
shape descriptors	rectangularity, circularity, ellipticity, triangularity, compactness (5)
shape properties	global shape properties (1)
	shape of the poles (10)
geometric properties	length, width, length-width ratio, size (4)
diatom-specific features	stria density, orientation, changeover point (3)
	axial area width (1)
	costa density (1)
	horizontal frequency (1)
texture properties	GLCM (65)
	Gabor wavelets (40)
other features	Fourier descriptors (30)
	moment invariants (7)

3 Classification by decision trees

The core of the automatic identification system as described in this chapter, but also used in some other chapters, is the classification module. In order to implement this module, a scheme is necessary that maps images of unidentified specimens to labels of trained taxa, using the extracted features. From the various classification methods that are available in machine learning and pattern recognition, we selected decision trees. In this approach, a decision tree is induced given a set of training samples. Similar to other learning systems, the accuracy of the system depends on the quality and amount of available training data. Different methods have been proposed to predict the accuracy of an identification system. Results have shown that decision trees have a low error rate, especially in cases where the training data is sparse. But this low error rate can still be improved by building ensembles of decision trees, rather than using a single decision tree (Fischer and Bunke, 2001, 2002).

The theory of decision trees will be reviewed in Section 3.1. Then, in Section 3.2, the validation of the accuracy of decision trees is described. Finally, in Section 3.3, strategies for ensemble learning will be introduced.

Chapter 7: Identification using classical and new features

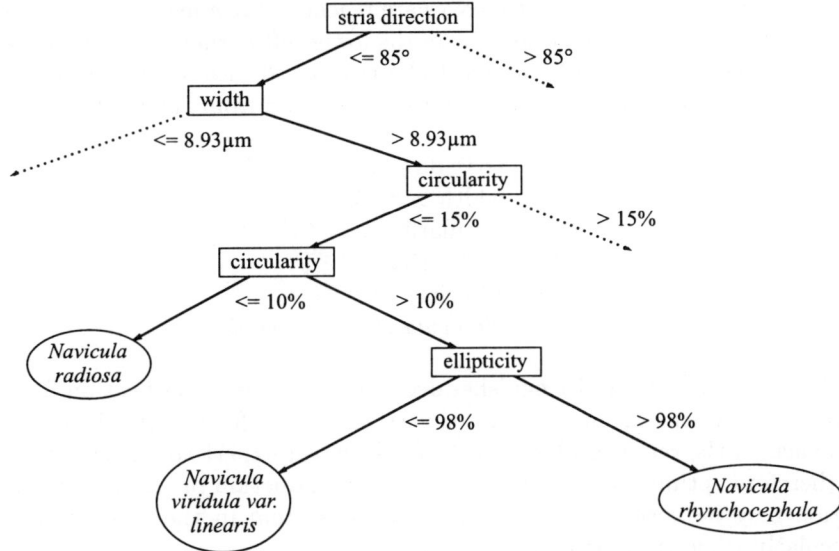

Figure 13. Example of a decision tree for diatom identification.

3.1 Decision trees

There are many different classification methods in neural-network and statistical-decision theory (Michie et al., 1994). Within learning and reasoning approaches, decision trees are among the most popular ones (Janikow, 1998). For the problem considered here, diatom identification using many different features, a decision tree-based approach is very suitable, because it resembles the way that experts identify diatoms. Moreover, in view of the large number of classes[a] (taxa) involved, a single-level decision procedure, as employed by a neural network or a statistical classifier, seems less appropriate. An additional problem with applying a statistical classifier or a neural network is the lack of sufficient training data.

Popular decision tree algorithms like C4.5 (Quinlan, 1993) or CART (Breiman et al., 1984) originate from Hunt's concept for learning systems (Hunt et al., 1966). The basic idea is to split a set of training samples into subsets, so that the data in each of the descending subsets are "purer" than the data in the parent subset. With "purer" we mean that, in the ideal case, the final subsets will contain only samples that belong to a single class.

A method based on a decision tree is a supervised learning strategy that builds a decision tree from a set of training samples. The result of the learning procedure is a tree in which each leaf carries a class name and each interior node specifies a test on a particular feature, with one branch corresponding to each possible feature value or range of values. Figure 13 shows an example of a part of a decision tree. In the root node the feature "stria direction" is chosen for the first split. The different

---

[a]The term "class" in this context is used in the general sense, without any relation to the taxonomic category "class."

subsets are recursively split, until leaves are reached. Each leaf carries a class label, for example *Navicula viridula linearis* in the lower left corner. A decision tree, such as the one shown in Fig. 13, can easily be translated into a set of rules which we can interpret. For example, one of the rules obtained from the above tree is:

        if   stria direction $<= 85°$
        and   width $> 8.93 \mu m$
        and   circularity $<= 15\%$
        and   circularity $> 10\%$
        and   ellipticity $> 98\%$
        then *Navicula rhynchocephala*

This property, which distinguishes decision tree-based classification from neural networks and statistical approaches, is very important for taxonomic identification tasks such as the one considered here. It allows an expert to change or to extend and therefore optimize a tree obtained from the induction procedure. Such a modification may be necessary to obtain a robust system in the case of many taxa, particularly when training data is sparse.

3.1.1 Decision tree induction

A decision tree is constructed during the induction process. Using a data set—that is, samples represented by feature sets with class labels—a decision tree is grown top-down in the following way. Starting with the entire data set, represented by the root node, a feature search is done in order to construct recursively the subsequent layers of the tree. The basic idea is to split the data set into subsets in such a way that each subset holds only samples belonging to one class. Here we will use the well-known C4.5 algorithm (Quinlan, 1993). This algorithm builds subsets by using a criterion based on information theory. At each node in the decision tree, the data set is split into subsets by choosing the feature that maximizes the information gain.

The probability that a sample of the data set T belongs to class c_i is

$$P(c_i) = \frac{\text{number of samples in } T \text{ belonging to } c_i}{\text{number of all samples in } T}. \tag{31}$$

Using the probabilities of all classes $c_1, \ldots, c_k$, the information of the data set is measured by the entropy

$$I(T) = -\sum_{i=1}^{k} P(c_i) \log_2 P(c_i). \tag{32}$$

If a feature a is selected and the data set is split into subsets according to all possible values of that feature, the entropy for the split can be expressed as the weighted sum over all subsets $T_1, \ldots, T_n$:

$$B(T, a) = \sum_{i=1}^{n_a} \frac{|T_i|}{|T|} I(T_i), \tag{33}$$

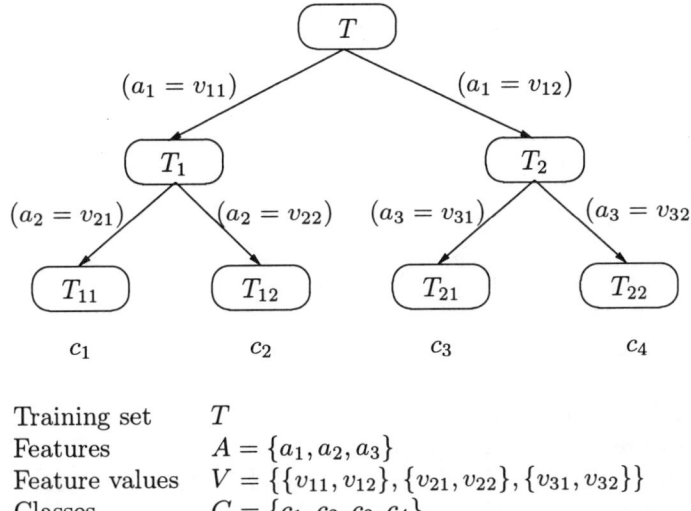

Training set T
Features $A = \{a_1, a_2, a_3\}$
Feature values $V = \{\{v_{11}, v_{12}\}, \{v_{21}, v_{22}\}, \{v_{31}, v_{32}\}\}$
Classes $C = \{c_1, c_2, c_3, c_4\}$

Figure 14. Schematic decision tree.

where a indicates the selected feature and n_a the number of possible values of that feature. The entropy of a split is small if and only if the subsets T_i hold samples from different classes. Based on this criterion, for each feature a the information gained by splitting T according to the possible outcomes is measured by

$$G(T, a) = I(T) - B(T, a). \tag{34}$$

Consequently, when splitting the data set at each node of the tree, the feature that maximizes the information gain according to Eqn. (34) will be selected. The set of training samples that is associated with the node is divided into subsets according to the different values of the feature a, and each subset represents a child node. This procedure is applied recursively for all child nodes until the associated subsets hold only samples belonging to a single class.

This procedure is visualized in Fig. 14. During initialization, the complete set of samples T from the data set is assigned to the root node. For each class $c_1, \ldots, c_4$ in the dataset, the probability $P(c_i)$ is computed using Eqn. (31). For all features $a_i, i = 1, \ldots, 3$, the gain criterion $G(T, a)$ is evaluated. Let us assume that the best split at the root node is done using feature a_1. Then T is split into two subsets T_1 and T_2 according to the two feature values v_{11} and v_{12}. The subset T_1 will contain all samples with $a_1 = v_{11}$, and T_2 will contain all samples with $a_1 = v_{12}$. Now, both subsets are regarded as new data sets T and the procedure is continued until "pure" subsets for classes $c_1, \ldots, c_4$ are obtained.

By applying the gain criterion no constraints are imposed on the size of the individual subsets T_i of T. In some cases it can happen that splits are chosen such that unbalanced subsets are obtained. This can be avoided by integrating the entropy of the size of the subsets into the split criterion. The information of the

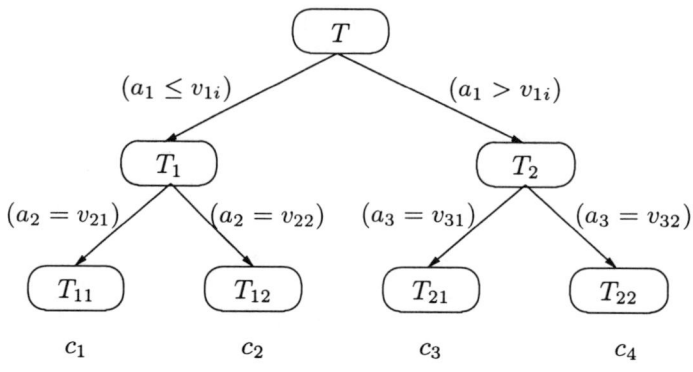

Training set	T
Features	$A = \{a_1, a_2, a_3\}$
Feature values	$V = \{\{v_{11}, \ldots, v_{1m}\}, \{v_{21}, v_{22}\}, \{v_{31}, v_{32}\}\}$
Classes	$C = \{c_1, c_2, c_3, c_4\}$

Figure 15. Schematic decision tree for feature a_1 with continuous values.

split is computed in the same way as the information of the data set (see Eqn. 32):

$$S(T,a) = -\sum_{i=1}^{n} \frac{|T_i|}{|T|} \cdot \log_2\left(\frac{|T_i|}{|T|}\right). \qquad (35)$$

The new split criterion is then taken as the ratio between the split info and the gain criterion:

$$G'(T,a) = \frac{G(T,a)}{S(T,a)}. \qquad (36)$$

Below, the decision tree approach will be extended in order to work with continuous feature values, rather than ordinal numbers as considered so far.

3.1.2 Features with continuous values

Most classification problems involve features with continuous values. In the classical top-down induction process of decision trees, a binary decision is taken at each step. Therefore, features with continuous values have to be discretized before they can be selected. This is normally done by partitioning the range of each feature into at least two subranges, and the main question is then where to cut the range of values. One solution is to introduce a threshold, like τ_a for feature a. Then, for example, the condition $a \leq \tau_a$ is assigned to the left branch of a node and $a > \tau_a$ to the right branch. To determine the value of the threshold, again, the entropy is used to determine τ_a, such that it provides a maximum information gain.

To select a threshold, all samples are sorted using the values $v_1, \ldots, v_m$ of a certain feature. Each value v_i can be a possible threshold; hence, $m-1$ tests are necessary to find the best split. Each threshold v_i then separates the samples into

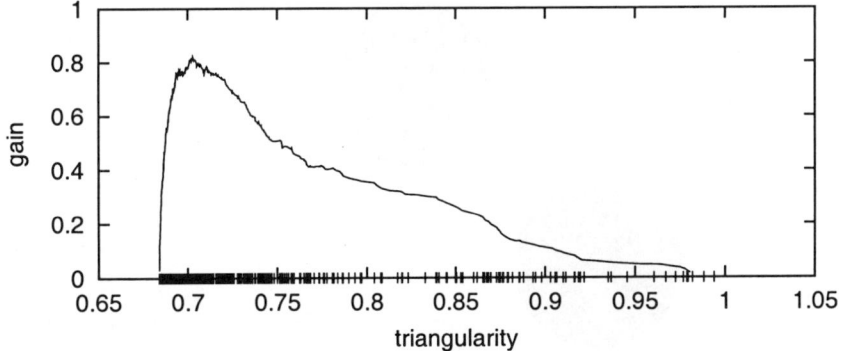

Figure 16. Selection of a threshold for the feature "triangularity." The gain information is plotted as a curve and feature values are marked by vertical ticks. In this case a threshold of 0.703 will be obtained.

two groups, i.e. those with feature values $\{v_1, \ldots, v_i\}$ and those with $\{v_{i+1}, \ldots, v_m\}$. If the feature a_1 of the previous example (see Fig. 14) becomes a feature with continuous values $v_{11}, \ldots, v_{1m}$, the condition for the left branch becomes "less or equal" and that for the right branch becomes "greater than." The new schematic decision tree is shown in Fig. 15.

Below, a real data set with continuous feature values will be considered. One of the features is the triangularity as described in Section 2. Before the gain criterion for that feature can be evaluated, the threshold τ has to be selected. Basically, for each value of the triangularity that occurs in the data set, the information gain is computed and the "best" value is selected as a threshold. Figure 16 illustrates this procedure as applied to a big database. As can be seen, the maximum gain occurs at a triangularity value of 0.703. Hence, a threshold value of 0.703 will be used for triangularity, and the corresponding gain value (0.824) will be used to compare this feature with all other features in the data set.

In this entire process there is an important source of variance. At each node the induction procedure must choose a single value for splitting a set into subsets. This choice can be affected by the addition or removal of a single training sample. Furthermore, all subsequent splits at descendent nodes in a tree are influenced by the split at the root node. Hence, one training sample can produce a cascade of changes in subsequent splits and can alter the entire tree (Dietterich and Kong, 1995). The decision forest approach as outlined in Section 3.3 provides one solution to this problem.

3.2 Validation

In this section it will be shown how the error rate of decision trees can be predicted for new samples that have not been used for training (induction). Traditionally, cross-validation tests, which provide a nearly unbiased estimate of error rates, are used (Efron and Tibshirani, 1995).

In the following it will be assumed that a tree $\mathcal{T}_T$ can be constructed from a

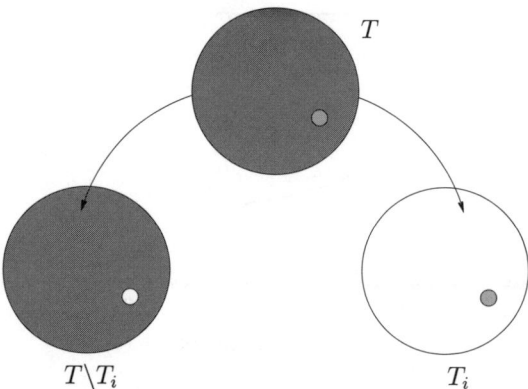

Figure 17. The leave-one-out cross-validation trains on all samples except one; the latter is used for testing. This procedure is applied until all samples have been tested exactly once.

training set $T = \{s_1, s_2, \ldots, s_n\}$. Each sample, or diatom specimen, $s_i = (\mathbf{x_i}, y_i)$ is represented by a feature vector $\mathbf{x_i}$ and a class label y_i. Using this information, the error rate for new samples s can be predicted by cross-validation. Cross-validation methods range from the leave-one-out method to dividing the data set into non-overlapping and equally-sized training and test sets. In the case of leave-one-out there are as many tests as there are samples in the data set. In each test the classifier is trained on all samples except one, and that sample is used for testing. This procedure is visualized in Fig. 17, where the entire data set T is divided into a subset $T \backslash T_i$ and a set T_i which holds only the sample s_i (the symbol \ means excluding). A decision tree is built using $T \backslash T_i$, after which the tree will be used to classify T_i. This procedure is repeated until each sample s_i $(1 \leq i \leq n)$ has been used exactly once for testing. In k-fold cross-validation the dataset T is randomly split into k mutually exclusive subsets $T_1, T_2, \ldots, T_k$ of approximately equal size. The classifier is trained and tested k times; each time i $(1 \leq i \leq k)$ it is trained on $T \backslash T_i$ and tested on T_i. Leave-one-out is a special case of k-fold cross-validation. It is also called n-fold cross-validation, because all possible n subsets are being considered.

The cross-validation accuracy estimate is the number of correct classifications divided by the number of samples in the dataset. Formally, this is (see Kohavi, 1995):

$$A_{\text{c-v}} = \frac{1}{n} \sum_{(\mathbf{x_i}, y_i) \in T} \delta\left(\mathcal{I}\left(T \backslash T_{(i)}, \mathbf{x_i}\right), y_i\right) \qquad (37)$$

where δ is 1 if the decision tree induced on $T \backslash T_i$ assigns $\mathbf{x_i}$ to class y_i and 0 otherwise. The cross-validation estimate is a random number that depends on the divisions into subsets or folds. Repeated cross-validations, using different random folds with the same number of samples, provides a better estimate, but at an added cost.

3.3 Ensemble learning

In a previous study (Fischer et al., 2000b) we found that decision trees trained on a diatom feature database are very specific for the training data. This phenomenon, known as overfitting, is a persistent problem (Ho, 1998). In existing approaches, a fully trained tree is often pruned in order to improve the generalization accuracy, even if the error rate on the training data increases (Quinlan, 1996). During the last decade, different approaches have been studied to prevent overfitting. Promising results were achieved by using ensembles of multiple classifiers. With respect to decision trees they are often called decision forests. Such methods construct a set of classifiers, and new samples are classified by combining the "votes" of all classifiers. This strategy is based on the observation that decision tree classifiers can vary significantly when a small number of training samples is added to or removed from the training set. This instability not only affects the structure of the decision trees, but also the decision rules in the trees. This means that decision trees that have been trained on slightly different data sets will disagree in the classification of some test samples. In the context of automatic diatom identification, a decision forest can be used to identify unknown specimens with a much higher generalization accuracy than a single decision tree could achieve.

3.3.1 Conditions for creating good ensembles

Ensemble learning combines the individual decisions from a set of classifiers, for example by a majority vote, to classify new samples. As already mentioned above, an ensemble is often much more accurate than its individual classifiers. Nevertheless, there are some conditions which need to be met in order to obtain good results. A necessary and sufficient condition is that the classifiers are accurate and diverse (Dietterich, 2000). A classifier is called *accurate* if it has an error rate which is better than randomly guessing the class of new samples. Two classifiers are called *diverse* if they make different errors on new samples. If these conditions are fulfilled, the error rate of the ensemble often decreases.

In general, a learning algorithm searches a space $\mathcal{H}$ of hypotheses in order to identify the best hypothesis in the space. A statistical problem arises when the amount of available training data is too small when compared to the size of the hypothesis space. Without sufficient data, the learning algorithm can find many different hypotheses in $\mathcal{H}$ that all give the same accuracy on the training data. This problem can be overcome by constructing an ensemble of classifiers. In this case the algorithm can average the votes and find a good approximation of the best target function f.

Many learning algorithms, including decision trees, work by performing a kind of local search that may get stuck in a local optimum, rather than the global one. An ensemble constructed from multiple local search processes, with different starting conditions, can, in many cases, provide a better approximation of the target function f than any individual classifier can attain.

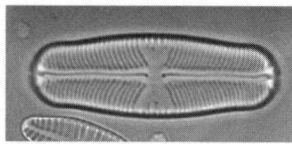 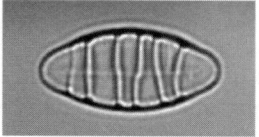 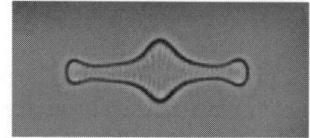

Figure 18. Example images, from left to right: *Sellaphora pupula*, *Diatoma mesodon*, and *Tabellaria flocculosa*

3.3.2 Construction methods

Various methods have been proposed for constructing ensembles. In general, the intention is that each individual classifier should learn a different aspect of the problem. This can be satisfied, for example, by forcing each classifier to train on different subsets of the training set, but there are also other approaches available. Here we will review a few examples of different construction methods.

The most straightforward way to construct an ensemble is *bagging* (Breiman, 1996). In each run, a bootstrap replicate of the original training set is presented to the learning algorithm. Given a training set T of m samples, a replicate T' is constructed by drawing m samples uniformly with replacement from T. A decision tree is constructed for each set T'. Given n decision trees, the final classification is obtained by majority voting. Another popular approach to construct ensembles is *boosting* (Freund and Schapire, 1996). A boosting algorithm maintains a set of weights over the original training set T, and adjusts these weights after each classifier has been trained by the basic learning algorithm. In each iteration i, the learning algorithm is invoked in order to minimize the weighted error on the training set. The weighted error of the hypothesis h_i is computed and applied to update the weights on the training samples. The weights of samples that are misclassified by h_i are increased, and the weights of samples that are correctly classified are decreased. The final classifier is constructed by a weighted vote of the individual classifiers h_i. New training sets T' are constructed by drawing samples, with replacement, from T, with a probability that is proportional to the samples' weights. A disadvantage of boosting is the overweighting of errors, which is sometimes a problem in real-world classification problems that contain noise. Dietterich (1999) has shown that bagging performs better than boosting in such situations. Therefore, only bagging will be used in our identification tests.

4 Data sets used

During the ADIAC project thousands of diatom images were captured and integrated into different databases (see Chapter 4). Two of these databases are used here and in other chapters to test identification algorithms and feature sets. The first data set consists of 120 images of *Sellaphora pupula* diatoms; see Fig. 18 (left) for an example image. The specimens are from 6 different demes, which are morphologically very similar but still with distinct shapes. For example, one deme is almost rectangular, whereas others are more elliptical or blunt. For each deme there

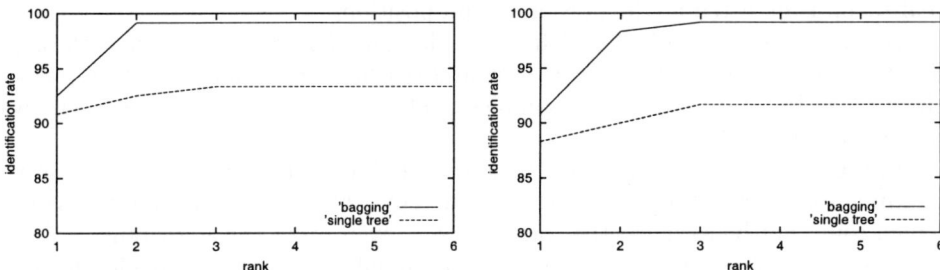

Figure 19. Identification rates for the *Sellaphora pupula* set, using cross-validation with single trees and bagging. Left: entire feature set; right: reduced feature set.

are exactly 20 specimens. This data set was created to analyze the performance of identifiers on samples with fine-grain morphological differences.

The second data set consists of a subset of 808 images from the mixed genera set (see Chapter 4). These images belong to 38 different taxa. For most taxa there are exactly 20 images available, but there are a few taxa represented by more images, up to a maximum of 29. The diatoms in this set vary not only in shape but also in ornamentation; two examples are shown in Fig. 18 (middle and right). This data set is used to analyze the performance of identifiers on samples with coarse-grain morphological differences, but it should be stressed that all specimens from each taxon cover the entire life cycle, that is, the size and shape variations that occur for each taxon.

5 Identification results

The feature sets and the decision-tree classifier can be used to identify diatoms. One goal of the complete system is to assist a diatomist in identifying a wide range of different diatoms. Instead of only presenting a final identification of an unknown diatom, a list of most likely taxon matches will be presented to the user. The user must then decide which of the suggested taxa is the correct one. This is an example of semi-automatic identification; the system suggests but the user decides.

To evaluate the performance of decision tree classifiers on the two data sets introduced in the previous section and in Chapter 4, two different tests were carried out. In the first test only single decision trees were built. In the second test 20 bootstrap replicates of the training set were built, following the bagging approach. In this case the classification result is the majority class of all 20 classifiers. Ties were solved randomly. All results were cross-validated following the leave-one-out approach. This means that each sample in the data set was used once for testing and all other samples were used for training. This procedure was repeated until each sample was used exactly once for testing.

The results obtained in the case of the *Sellaphora pupula* set are visualized in Fig. 19. Each diagram shows the results of both experiments, i.e. using single trees (dotted lines) and bagging (solid lines). The x-axis represents the ranking. For example, rank 2 represents the cumulative identification rate for all samples whose real taxon is detected at the first or second position.

As can be seen in Fig. 19 (left), the identification rate obtained with single decision trees starts slightly above 90% and reaches a maximum of 93.33% at the third rank. In total there were 8 samples of all 120 for which the correct class was not among the first three ranks.

In the case of bagging, the identification rate for the first rank is 92.5% and the maximum is reached on the second rank already (99.17%). There was one sample which could not be assigned the correct taxon, even up to the sixth rank.

In order to evaluate the influence of the number of features on identification rates, we ran a second experiment with a reduced feature set. In the first experiment all 171 features, including Fourier descriptors and moment invariants, were applied. In the second experiment we used only "interpretable" features. This is very important if, for example, a diatomist wants to judge the decisions made by an automatic procedure. Hence, from the complete set of 171 features we selected all features except Fourier descriptors, moment invariants and global texture features, which are all difficult to interpret. Furthermore, we decided *not* to use triangularity and compactness, because these are closely related to other features. Table 5 lists the remaining 27 features of the reduced feature set.

As can be seen in Fig. 19 (right), the identification rates obtained with the reduced feature set are almost the same as the results obtained with all 171 features. The curves are somewhat flatter and the maximum is reached at a higher rank, but in the case of bagging the same maximum ID rate of 99.17% could be obtained—at rank 3 instead of rank 2. Hence, a decision forest (bagging procedure) is capable of yielding excellent results, even when using far fewer features. These results, not forgetting those obtained with single decision trees, also indicate that many of the 171 features are correlated, not providing much additional information.

The same experiments were applied to the mixed genera data set of 808 images, belonging to 38 different taxa, with single trees and bagging, using the entire and reduced feature sets. As can be seen in Fig. 20 (left), the ID rate, when using single decision trees and all 171 features, starts slightly above 86% and reaches a maximum of 88.36% at rank 2. In total there were 94 samples that could not be

Table 5. Reduced feature set for evaluation of interpretable features.

group	feature
symmetry	class of symmetry, strength of apical and transapical axes (3)
shape descriptors	rectangularity, circularity, ellipticity (3)
shape properties	global shape properties (1)
	shape of the valve endings (10)
geometric properties	length, width, length-width ratio, size (4)
diatom-specific features	stria density, orientation, changeover point (3)
	axial area width (1)
	costa density (1)
	horizontal frequency (1)

Chapter 7: Identification using classical and new features

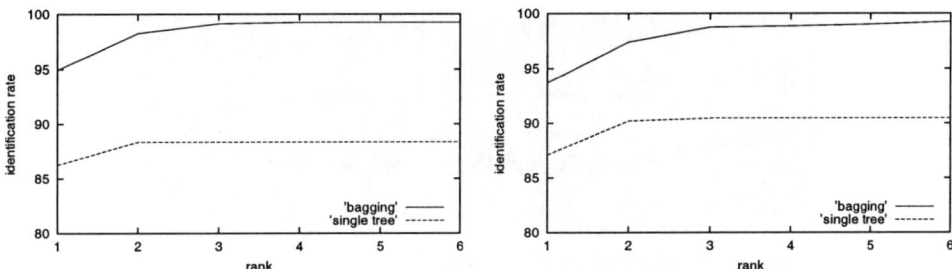

Figure 20. Identification rates obtained with the mixed genera data set, using single trees (dotted curves) and bagging (solid curves). Left: entire feature set; right: reduced feature set.

identified at ranks 1 and 2. In the case of bagging, the ID rate at rank 1 is already 94.9%, and the maximum, slightly more than 99%, is reached at rank 3. There were only four samples that could not be correctly identified.

As can be seen in Fig. 20 (right), the ID rates obtained with the reduced feature set are almost the same. At rank 1 the ID rate is slightly better when using single trees, and slightly worse when using bagging. The improvement for single trees is a typical phenomenon: decision trees tend to make arbitrary decisions if there are too many features; hence using fewer features improves their performance. In the case of bagging the ID curve is slightly flatter than before and the maximum is reached at a higher rank, but still the same maximum ID rate, slightly over 99%, could be obtained.

On average, decision trees belonging to the induced decision forest have a size smaller than 120 nodes. The average depth, or the number of decisions per class, is around 7 (cf. Fig. 13). This means that an expert has to trace about 7 tests, and at most 11 tests, from the root node to a leaf, in order to verify a decision taken by the automatic identification system. From all decision trees that were involved in the identification experiments, the maximum number of 11 tests occurred only for the two taxa shown in Fig. 21; these have a very similar shape and ornamentation.

Our results demonstrate that ensembles of classifiers—in this case decision forests—can achieve much better identification rates than single decision trees. The main reason is that there is no single feature that allows to discriminate all different taxa. This often implies, even in the case of reduced feature sets, that the induction algorithm can create arbitrary decisions when analyzing the available features. This can lead to an unstable decision tree, but it can be corrected by employing an ensemble that "averages" the instabilities of the individual trees in order to obtain a more stable behavior.

6 Conclusions

In this chapter we have presented a large set of features suitable for automatic diatom identification. Many features are "classical" in pattern recognition, but we also developed many new features specific to diatoms. We introduced two different identification approaches that employ decision trees. We have compared single deci-

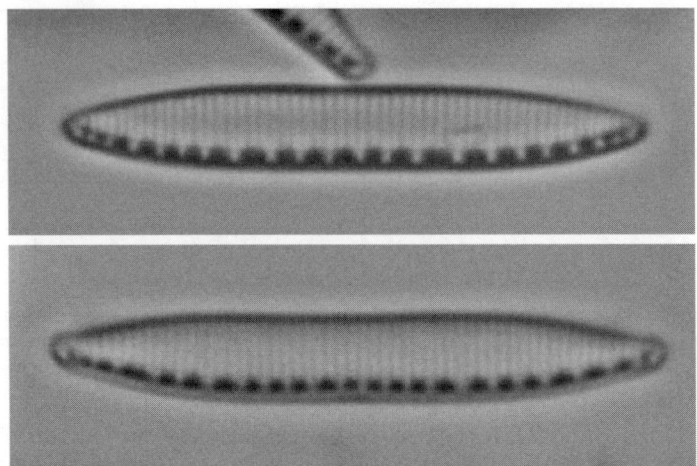

Figure 21. Examples of diatoms belonging to two different species of the same genus: *Nitzschia* sp. 2 (top) and *Nitzschia hantzschiana* (bottom).

sion trees and decision forests, using the leave-one-out cross-validation strategy, by applying them to two different diatom data sets: the *Sellaphora pupula* and mixed genera sets (Chapter 4). The results show that much better identification rates can be obtained by using decision forests rather than single decision trees. Even in the case of the very difficult mixed genera data set, which contains 38 different taxa and specimens that cover the size and shape variations that occur during diatoms' life cycles (Chapters 2 and 4), identification rates above 99% were achieved when taking into account the first three ranks. The misidentification of certain specimens is due to the variability and/or the lack of specific features that can distinguish these taxa. Nevertheless, the identification rates are impressive compared to those achieved by the experts that participated in the "groundtruthing" tests (Chapter 5).

References

Arbter, K., Snyder, W.E., Burkhardt, H. and Hirzinger, G. (1990) Application of affine-invariant Fourier descriptors to the recognition of 3-d objects. IEEE Trans. Pattern Analysis and Machine Intelligence, Vol. 12, pp. 640-647.

Barber, H.G. and Haworth, E.Y. (1981) A guide to the morphology of the diatom frustule. The Freshwater Biological Association, The Ferry House, Ambleside, Cumbria, UK.

Bayer, M.M. (1999) Introduction to diatom identification. Royal Botanic Garden Edinburgh, UK. See http://www.rbge.org.uk/ADIAC/intro/intro.htm

Belkasim, S.O., Shridhar, M. and Ahmadi, M. (1991) Pattern recognition with moment invariants: a comparative study and new results. Pattern Recognition, Vol. 24, pp. 1117-1138.

Bigun, J. and du Buf, J.M.H. (1993) Symmetry interpretation of complex moments and the local power spectrum. J. of Visual Communication and Image Representation, Vol. 6, pp. 154-163.

Bigun, J. and du Buf, J.M.H. (1994) N-folded symmetries by complex moments in Gabor space and their application to unsupervised texture segmentation. IEEE Trans. Pattern Analysis and Machine Intelligence, Vol. 16, pp. 80-87.

Breiman, L., Friedman, J., Olshen, R. and Stone, C. (1984) Classification and regression trees. Wadsworth, Belmont (CA), USA.

Breiman, L. (1996) Bagging predictors. Machine Learning, Vol. 24, pp. 123-140.

Bronstein, I.N. and Semendjajew, K.A. (1987) Taschenbuch der Mathematik. Harri Deutsch, Thun, Germany (23rd Edition).

Dietterich, T.G. and Kong, E.B. (1995) Machine learning bias, statistical bias, and statistical variance of decision tree algorithms. Technical report, Dept. of Computer Science, Oregon State Univ., Corvallis (OR), USA.

Dietterich, T.G. (1999) An experimental comparison of three methods for constructing ensembles of decision trees: bagging, boosting, and randomization. Machine Learning, Vol. 40, pp. 139-158.

Dietterich, T.G. (2000) Ensemble methods in machine learning. In: Lecture notes in computer science, J. Kittler and F. Roli (eds), Springer Verlag, Heidelberg, Germany, Vol. 1857, pp. 1-15.

Dudani, S.A., Breeding, K.J. and McGhee, R.B. (1977) Aircraft identification by moment invariants. IEEE Trans. on Computers, Vol. 26, pp. 39-45.

Efron, B. and Tibshirani, R. (1995) Cross-validation and the bootstrap: estimating the error rate of a prediction rule. Technical report TR-477, Dept. of Statistics, Stanford Univ., Stanford (CA), USA.

Fischer, S., Binkert, M. and Bunke, H. (2000a) Symmetry based indexing of diatoms in an image database. Proc. 15th Int. Conf. on Pattern Recognition (ICPR 2000), Barcelona (Spain), September 2000, Vol. 2, pp. 899-902.

Fischer, S., Binkert, M. and Bunke, H. (2000b) Feature based retrieval of diatoms in an image database using decision trees. Proc. Advanced Concepts for Intelligent Vision Systems (ACIVS 2000), Baden-Baden, Germany, pp. 67-72.

Fischer, S. and Bunke, H. (2001) Automatic identification of diatoms using decision forests. Proc. 2nd Int. Workshop Machine Learning and Data Mining in Pattern Recognition (MLDM 2001), Leipzig (Germany), P. Perner (ed.), Springer Verlag, Heidelberg, Germany , pp. 173-183.

Fischer, S. (2002) Automatic identification of diatoms. PhD thesis, University of Bern, Switzerland (to be submitted).

Fischer, S. and Bunke, H. (2002) Automatic identification of diatoms using visual human-interpretable features. Int. J. of Image and Graphics (in press).

Flusser, J. and Suk, T. (1993) Pattern recognition by affine moment invariants. Pattern Recognition, Vol. 26, pp. 167-174.

Forsythe, G.E., Malcolm, M.A. and Moler, C.B. (1977) Computer methods for mathematical computations. Prentice-Hall, Englewood Cliffs (NJ), USA.

Freund, Y. and Schapire, R.E. (1996) Experiments with a new boosting algorithm. Proc. 13th Int. Conf. on Machine Learning, Morgan Kaufmann, San Mateo (CA), USA, pp. 148-156.

Gilomen, K. (2001) Texture based identification of diatoms. MSc thesis, University of Bern, Switzerland (in German).

Haralick, R.M., Shanmugam, K. and Dinstein, I. (1973) Textural features for image classification. IEEE Trans. Systems, Man, and Cybernetics, Vol. 6, pp. 610-621.

Ho, T.K. (1998) The random subspace method for constructing decision forests. IEEE Trans. Pattern Analysis and Machine Intelligence, Vol. 20, pp. 832-844.

Hu, M.K. (1962) Visual pattern recognition by moment invariants. IEEE Trans. Information Theory, Vol. 8, pp. 179-187.

Hunt, E.B., Marin, J. and Stone, P.J. (1966) Experiments in induction. Academic Press, New York (NY), USA.

Huttenlocher, D.P., Klanderman, G.A. and Rucklidge, W.J. (1993) Comparing images using the Hausdorff distance. IEEE Trans. Pattern Analysis and Machine Intelligence, Vol. 15, pp. 850-863.

Jain, A.K. and Farrokhnia, F. (1991) Unsupervised texture segmentation using Gabor filters. Pattern Recognition, Vol. 24, pp. 1167-1186.

Janikow, C.Z. (1998) Fuzzy decision trees: issues and methods. IEEE Trans. Systems, Man, and Cybernetics, Vol. 28, pp. 1-14.

Kohavi, R. (1995) A study of cross-validation and bootstrap for accuracy estimation and model selection. Proc. 14th Int. Joint Conf. on Artificial Intelligence, Morgan Kaufmann, San Mateo (CA), USA, pp. 1137-1143.

Michie, D., Spiegelhalter, D.J. and Taylor, C.C. (1994) Machine learning, neural and statistical classification. Ellis Horwood, Chichester, UK.

Pitas, I. (1993) Digital image processing algorithms. Prentice Hall, London, UK.

Quinlan, J.R. (1993) C4.5: programs for machine learning. Morgan Kaufmann, San Mateo (CA), USA.

Quinlan, J.R. (1996) Bagging, boosting, and C4.5. Proc. 13th National Conf. on Artificial Intelligence, AAAI Press/MIT Press, Cambridge (MA), pp. 725-730.

Rosin, P.L. (2000) Measuring shape: ellipticity, rectangularity, and triangularity. Proc. 15th Int. Conf. on Pattern Recognition (ICPR 2000), Barcelona (Spain), September 2000, Vol. 2, pp. 952-955.

Shih, F.Y. and Pu, C.C. (1995) A skeletonization algorithm by maxima tracking on Euclidean distance transform. Pattern Recognition, Vol. 28, pp. 331-341.

Sonka, M., Hlavac, V. and Boyle, R. (1999) Image processing, analysis, and machine vision. Brooks/Cole, Pacific Grove (CA), USA (2nd Edition).

Tuceryan, M. and Jain, A.K. (1998) Texture analysis. In: The handbook of pattern recognition and computer vision, C.H. Chen, L.F. Pau and P.S.P. Wang (eds), World Scientific, Singapore (2nd Edition).

CHAPTER 8

IDENTIFICATION BY CURVATURE
OF CONVEX AND CONCAVE SEGMENTS

ROBERT E. LOKE AND HANS DU BUF

In this chapter we describe a new contour feature set. A contour is segmented into convex, concave and straight segments, after which length and curvature features are computed. A symmetry analysis allows the detection of the number of elementary segments. Using only four features, a simple nearest-mean classifier yielded a perfect identification rate of 100% on the *Sellaphora pupula* test set. Using six features, the procedure of training/testing and reversed training/testing still resulted in an average ID rate of 97.5%. The same procedures applied to the mixed genera data set resulted in ID rates of 83.5 and 82.9% when using 10 features.

1 Introduction

There are many different feature sets for describing contours: the "classical" morphometric features and Fourier descriptors (Chapter 7), and Legendre polynomials (Chapter 9). More recent methods are often based on a multiscale approach, such as Gabor event space (Chapter 10). We wanted to test the performance of a relatively simple, fast and single-scale approach that is inspired by the way that diatomists look at valve outlines: the entire shape consists of one or more convex parts, plus concave and even straight parts. Straightforward processing with methods that are common in e.g. computer graphics allows the partitioning of a contour into elementary parts and the computation of curvature features of these.

This idea may seem simple, but its realization is not quite trivial for two reasons: (a) Extracted contours are discrete and include some noise, even small deformations. This can be solved by applying a lowpass filtering. However, this will affect the shape and therefore the convex and concave parts. (B) Computers can process data with tremendous accuracy, but there are limits to this, like for human observers: our visual system is characterized by "just noticeable differences" in all percepts, such as brightness, color and shape. We can only distinguish two shapes if they differ by more than a minimum amount. We may assume that diatomists are well-trained observers, but this implies that computer algorithms must also be "trained" (optimized) to achieve the same performance.

In this chapter we first describe the fully automated (unsupervised) feature-extraction methods. Using simple techniques we extract inflection points, convex and concave segments, and compute local curvature. The latter can be thresholded in order to detect straight segments, after which small and insignificant segments are merged with neighboring significant ones. Similar segments are grouped into *elementary* segments on the basis of symmetry, and length and curvature features are computed. In Sections 3 and 4 the classification methods and implementation

Table 1. Feature types at different levels.

features	level	examples
global	high	mean curvature (θ_μ), number of segments (n_S), number of convex, concave, straight and elementary segments ($n_{cv}, n_{cc}, n_{st}, n_{el}$), mean segment length ($l_\mu$), mean of max. elementary-segment curvature ($\Theta_{\mu,max}$), min. of mean segment curvature ($\theta_{min,\mu}$), min. segment noncircularity ($\theta_{min,q}$)
elementary segment	intermediate	length (L_s), mean curvature ($\Theta_{\mu,s}$), max. curvature ($\Theta_{max,s}$), min. curvature ($\Theta_{min,s}$), noncircularity ($\Theta_{q,s}$)
segment	intermediate	length (l_s), mean curvature ($\theta_{\mu,s}$), max. curvature ($\theta_{max,s}$), min. curvature ($\theta_{min,s}$), noncircularity ($\theta_{q,s}$)
point	low	convexity (C_p), curvature (θ_p), inflection (I_p)

details are presented, followed by experimental results obtained on different contour sets (Section 5). In Section 6 the results are discussed and conclusions are drawn.

2 Feature extraction

Figure 1 illustrates the entire feature extraction process. First, the coordinates of the contour points are filtered to reduce quantization noise and deformations introduced by the contour tracking (see Chapter 6). Thereafter, *point features* are determined, taking into account the neighborhood of each contour point (Dong and Hillman, 2001). These are (a) a point's local convexity and (b) its local curvature. Both contour signals are filtered to further reduce the noise. We then identify different segment types on the contour: (1) by determining the inflection points (zero crossings of the convexity signal) we partition the contour into segments such that all points of each segment are either convex or concave; (2) by thresholding the curvature signal we obtain the straight segments. Then we fuse the straight segments with the convex and concave ones and determine *segment features* for each segment based on its length and curvature. Thereafter, a symmetry analysis is performed by correlating the features of the segments in order to derive more robust *elementary-segment features*. Finally, *global features* are computed. Table 1 lists all contour features and their level of analysis. Below, the computation of the different features is described in detail.

Chapter 8: Curvature of convex and concave segments

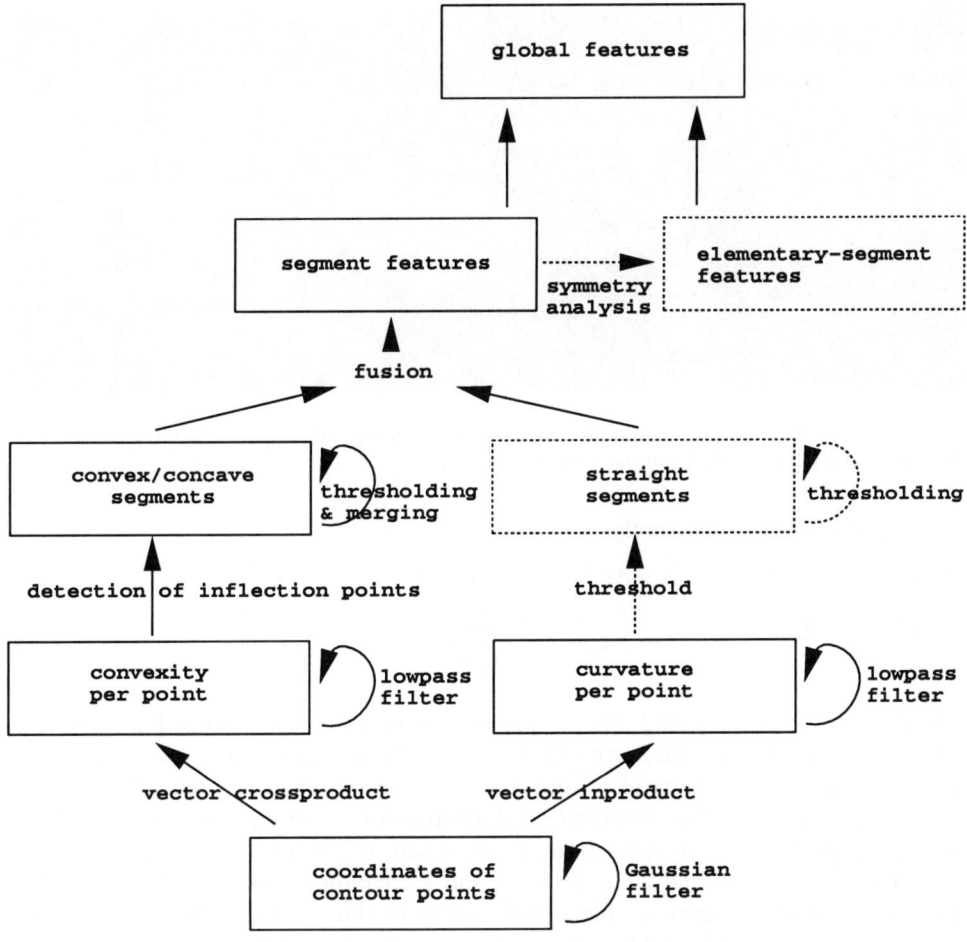

Figure 1. Feature extraction scheme.

2.1 Contour filtering

We assume that a contour is sampled clockwise and consists of n_C points P_i, $i \in [0,..,n_C - 1]$, each with coordinates (x_i, y_i). The x and y coordinates are filtered in floating point using a Gaussian kernel:

$$\bar{x}_i = \sum_{j=-m_0}^{m_0} x_{i+j} \exp(-\frac{j^2}{2\sigma^2}), \qquad \bar{y}_i = \sum_{j=-m_0}^{m_0} y_{i+j} \exp(-\frac{j^2}{2\sigma^2}),$$

with $m_0 := m_0 n_C$ being the kernel size and σ equal to $m_0/3$. We deliberately implement the Gaussian filter size (σ) to be a function of the contour size (n_C) in order to make the feature extraction scale invariant. Figure 2 (top) shows an example. The local shape has been deformed somewhat, but this will apply to all contours in a systematic way.

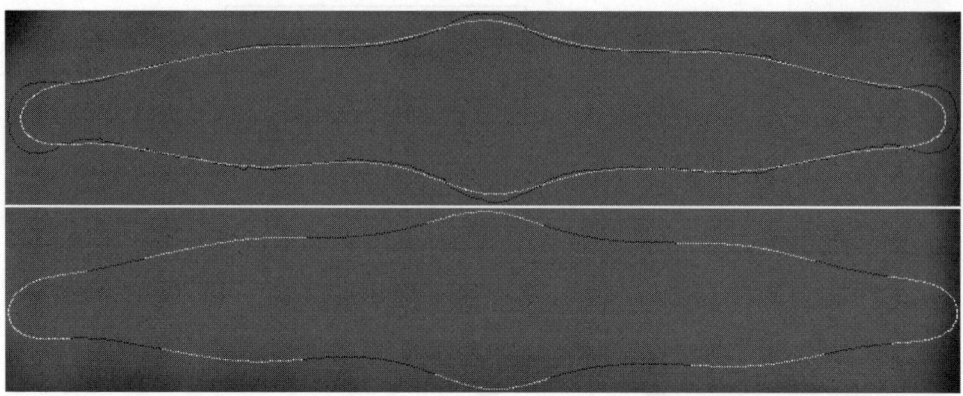

Figure 2. A *Nitzschia sinuata* contour. Top: original (black) and filtered (white) coordinates. Bottom: convex (white) and concave (black) points.

2.2 Local convexity/concavity

For each point P_i the convexity is computed by

$$C_i = \frac{1}{m_1} \sum_{j=1}^{m_1} \vec{a}_j \times \vec{b}_j, \qquad (1)$$

with $\vec{a}_j = P_{i-j}P_i$ and $\vec{b}_j = P_iP_{i+j}$ being vectors defined as shown in Fig. 3, $\vec{a}_j \times \vec{b}_j$ the vector crossproduct computed by $(x_i - x_{i-j})(y_{i+j} - y_i) - (y_i - y_{i-j})(x_{i+j} - x_i)$ and m_1 a mask size. Instead of obtaining a noisy convexity by considering only P_{i-1}, P_i and P_{i+1}, a *robust* convexity is determined by averaging m_1 crossproducts, each calculated symmetrically around P_i for a pair of 2 points. The result of Eq. (1) is a vector with zero x and y components and a z component of varying magnitude and sign. Any reference to C_i below refers only to the z component. A point P_i is considered convex with respect to its neighboring points if C_i has a negative sign and concave if it has a positive sign. Figure 2 (bottom) shows an example. In order to further reduce the noise the resulting signal C is once more lowpass filtered using

$$C_i := \frac{1}{2m_2 + 1} \sum_{j=-m_2}^{m_2} C_{i+j}, \qquad (2)$$

where m_2 is again a mask size. Inflection points are obtained as zero crossings of C:

$$I_i = \begin{cases} \text{TRUE,} & \text{if } (C_{i-j} < 0 \land C_{i+k} > 0) \lor (C_{i-j} > 0 \land C_{i+k} < 0) \\ \text{FALSE,} & \text{elsewhere} \end{cases}$$

where j ($1 \leq j \leq n_c - 1$) is the smallest integer for which $C_{i-j} \neq 0$ and k ($1 \leq k \leq n_c - 1$) the smallest integer for which $C_{i+k} \neq 0$. In the case that multiple inflection points are grouped together only the central inflection point is set to TRUE. All other adjacent inflection points are set to FALSE and the values at the corresponding positions in C are set to 0 in order to stabilize the signal (see below).

The contour has now been divided into segments. In order to remove irrelevant, small (noisy) segments we perform a length-thresholding process which consists of

Chapter 8: Curvature of convex and concave segments

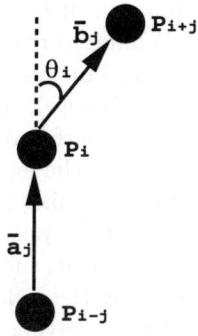

Figure 3. Contour points and vectors in the (x,y)-plane for convexity and curvature computations.

two steps. First, each inflection point I_p located closely between two other inflection points I_i and I_j ($i < p \wedge j > p$), such that the relative distance to each is smaller than t_0 ($|p-i|/n_C < t_0 \wedge |p-j|/n_C < t_0$), is set to FALSE. This eliminates redundant inflection points between the significant ones. We deliberately set the values of C to 0 at the positions between subsequent inflection points in order to stabilize the sign of C, which will facilitate the identification of the different segment types (see Eq. (4)). Second, insignificant pairs of inflection points, where one inflection point is located at a relative distance smaller than t_0 from the other one, are eliminated as follows: a) the two inflection points are set to FALSE; b) the values of C at and between the positions of the two inflection points are set to 0, again for stabilization; c) a new inflection point, located at the center of the two deleted ones, is set to TRUE. This effectively replaces each insignificant pair of inflection points (a small segment) by a single inflection point. Because adjacent segments may now have the same sign of C, we perform a merging in which inflection points between identical segments are set to FALSE.

The inflection points are ordered according to their position on the contour: I_o precedes I_p if $o < p$. In the following n_{ip} denotes the number of inflection points.

2.3 Local curvature and straightness

For each point P_i, $i \in [0, .., n_C - 1]$, the bending angle θ_i, which represents the local curvature (see Fig. 3), is computed by

$$\theta_i = \frac{1}{m_3} \sum_{j=1}^{m_3} \arccos \frac{\vec{a}_j \cdot \vec{b}_j}{|\vec{a}_j||\vec{b}_j|}, \qquad (3)$$

where $\vec{a}_j$ and $\vec{b}_j$ are defined as above, $\vec{a}_j \cdot \vec{b}_j$ is the vector inproduct given by $(x_i - x_{i-j})(x_{i+j} - x_i) + (y_i - y_{i-j})(y_{i+j} - y_i)$, $|\vec{a}_j| = \sqrt{(x_i - x_{i-j})^2 + (y_i - y_{i-j})^2}$, $|\vec{b}_j| = \sqrt{(x_{i+j} - x_i)^2 + (y_{i+j} - y_i)^2}$, and m_3 is a mask size. Like Eq. (1), Eq. (3) yields a *robust* angle by averaging m_3 angles, each calculated for a pair of 2 points arranged symmetrically around P_i. The resulting signal θ is lowpass filtered as described above (Eq. (2)), using a mask of size m_4.

Straight segments are detected by thresholding the local curvature θ_i (in degrees) at each point P_i, using a threshold t_1 which is set as proportional to the value of 360/(number of contour points): if $\theta_i < (360/n_\mathrm{C})t_1$ then P_i is straight. The result of the thresholding is a 1D binary array in which 1's denote points which are straight and 0's points which are not. In this array straight parts which are too small are removed (using the same t_0 as before). Thereafter, each straight segment, indicated by a series of 1's in the array, is fused with the previously determined convex/concave ones by searching, around the endings of the straight segment, for closely located inflection points. For each ending (at position p_e) an inflection point (at position p) is sought (using I), first at the positions inside the straight segment, and then at the positions outside the segment, such that $|p_e - p|/n_c < t_0$. Then the straight segment is fused by: a) unsetting the inflection points within the new segment and, at each ending, the one at p; b) setting, for each ending, a new inflection point at the center between p_e and p to TRUE; and c) setting C, at all positions between the found inflection points and within the detected straight segment, to 0 (for stabilization; see above). If no inflection point is found around an ending, a new inflection point at p_e is set to TRUE. If the inflection point searched for at one ending is identical to the one searched for at the other ending, only the position of the closest ending is moved towards it. After the fusion all segments with relative lengths smaller than the threshold t_0 are again removed and equal neighboring segments are merged. See Fig. 4 for an example of the fusion process.

2.4 Segment features

After the processing described above we know the local curvature as well as the inflection points between significant convex, concave and straight segments of a contour. For all segments we need to compute features using the convexity C and the curvature θ. Let n_S denote the number of segments defined as

$$n_\mathrm{S} = \begin{cases} 1, & \text{if } n_\mathrm{ip} = 0 \\ n_\mathrm{ip}, & \text{otherwise.} \end{cases}$$

The segments are numbered in clockwise order, the first segment starting at the first inflection point and the last one starting at the last inflection point. Each pair of subsequent inflection points (I_i, I_j) $(0 \leq i < j < n_\mathrm{C}; j \neq i+1)$ defines a segment S_k $(0 \leq k < n_\mathrm{S})$ whose contour points range between position i and j. The value of k is equal to the rank order of I_i on the contour. For each segment S_k the following features are determined:

$$l_k = (j - i)/n_\mathrm{C},$$

which is the relative length,

$$\theta_{\mu,k} = \frac{1}{j-i-1} \sum_{l \in (i,j)} \theta_l, \quad \theta_{\max,k} = \max_{l \in (i,j)} \theta_l, \quad \theta_{\min,k} = \min_{l \in (i,j)} \theta_l, \quad \theta_{q,k} = \frac{\theta_{\max,k}}{\theta_{\mu,k}},$$

being the mean, maximum, minimum and "noncircular" curvature, as well as

$$type_k = \begin{cases} \text{convex,} & \text{if } n > 0 \wedge p = 0 \\ \text{concave,} & \text{if } n = 0 \wedge p > 0 \\ \text{straight,} & \text{if } n = 0 \wedge p = 0, \end{cases} \tag{4}$$

Chapter 8: Curvature of convex and concave segments

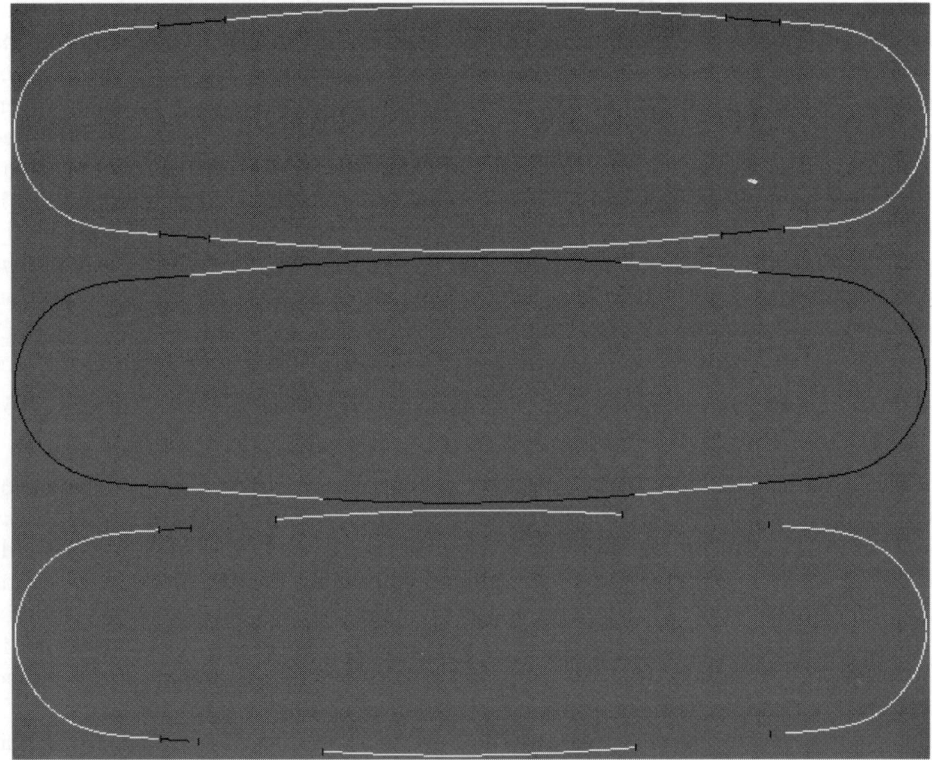

Figure 4. A *Sellaphora pupula* deme "large" contour. Top: detected convex (white) and concave (black) segments. Middle: straight segments (white). Bottom: the fusion of all segments. The small bars orthogonal to the contour indicate inflection points. Gaps surrounded by bars denote straight segments and gaps without bars thresholded segments. On the right, the fused concave segments were removed by the length-thresholding process.

in which n and p are the number of contour points with negative or positive convexity C in S_k. Feature $\theta_{q,k}$ represents the *noncircularity* because only for circular segments its value will be 1, in all other cases $\theta_{q,k} > 1$ (apart from straight segments).

2.5 Elementary-segment features

Having the segmented contour and all segment features, we perform a segment-based symmetry analysis in order to find similar segments. Thereafter, the feature space can be reduced by deriving robust elementary-segment features for each set of similar segments. This symmetry analysis does not depend on the diatom center or centroid, as opposed to other approaches (see e.g. Chapter 7).

The analysis is performed three times, yielding a similarity matrix for each of three different features (l, θ_μ and $\theta_{\max}$). First, given the ordered segments S_i, $i \in [0, n_S - 1]$, a two-dimensional correlation matrix M of size $n_S \times n_S$ is determined by correlating the features in the neighborhood of each segment with the features in the neighborhood of all other segments. Because correlations can only be computed for a set of numbers, "in the neighborhood" above and below refers to a segment with its 2,4,... neighboring segments. In the matrix M the lower triangle positions ($i > j$) will represent cross-correlations between the features in the neighborhoods of segments S_i and S_j. The upper triangle ($i < j$) contains cross-correlations between the features in the neighborhood of segment S_i and the features in the *mirrored* neighborhood of segment S_j. The diagonal elements ($i = j$) are autocorrelations which are by definition equal to 1. Two neighborhoods will be considered similar if their corresponding element M_{ij} and/or their mirrored element M_{ji} is sufficiently high; the diagonal positions of M are simply ignored. Only pairs of equal *type* are considered; if types differ the matrix element is set to -1. Each correlation coefficient is determined by (see James, 1999):

$$M_{ij} = \frac{1}{m_5 \sigma_i \sigma_j} \sum_{k=-(m_5-1)/2}^{(m_5-1)/2} (S_{i+k,x} - \mu_i)(S_{j+k,x} - \mu_j), \qquad (5)$$

where $S_{i,x}$ is feature $x \in \{l_i, \theta_{\mu,i}, \theta_{\max,i}\}$ of segment S_i, μ_i is the mean value of segment feature x in the m_5-neighborhood of S_i, σ_i the standard deviation of x about μ_i in m_5, and m_5 an odd mask size (set to 3 in our experiments). Second, the values in M are thresholded: all values higher than t_2 are set to 1; all equal and smaller ones to 0. Finally, the upper triangle of the resulting binary matrix is mapped (mirrored) onto the lower triangle using the logical OR operator: if $M_{ij} = 1 \vee M_{ji} = 1$ then $M_{ij} = 1$; else $M_{ij} = 0$. Now all relevant data are in the lower matrix triangle, with ones indicating similarity and zeros dissimilarity.

The three similarity matrices obtained are merged into one matrix using the logical AND operator, i.e. only if all three features l, θ_μ and $\theta_{\max}$ are 1, the result will be 1. This matrix is mapped onto a one-dimensional array A with values in the interval $[0, n_S - 1]$ in which elements having same integer values indicate similar segments. The number of elementary segments (n_{el}) is equal to the number of different values in this array. Let $n_{\text{el},k}$ be the number of elementary segments with value k. For each elementary segment ES_k, $k \in [0, n_{\text{el}} - 1]$, the following average features are determined: length L_k, mean curvature $\Theta_{\mu,k}$, maximum curvature $\Theta_{\max,k}$, minimum curvature $\Theta_{\min,k}$ and "noncircular" curvature $\Theta_{q,k}$, using

$$L_k = \frac{1}{n_{\text{el},k}} \sum_{i=0}^{n_S-1} l_i, \quad A(i) = k,$$

etc. Hence, these are the same features as used for all segments but averaged over all similar and (anti-)symmetrical segments. See Fig. 5 for two examples.

2.6 Global features

In the previous two subsections we have introduced two feature sets: one set for all segments and an averaged one for all elementary segments. In both cases there are five features plus the segment type, but also the numbers of all segments and the different elementary segments. These features follow directly from the idea to split a contour into different segments and to reduce these into elementary ones. However, it is also possible to compute additional features for the entire contour

Chapter 8: Curvature of convex and concave segments

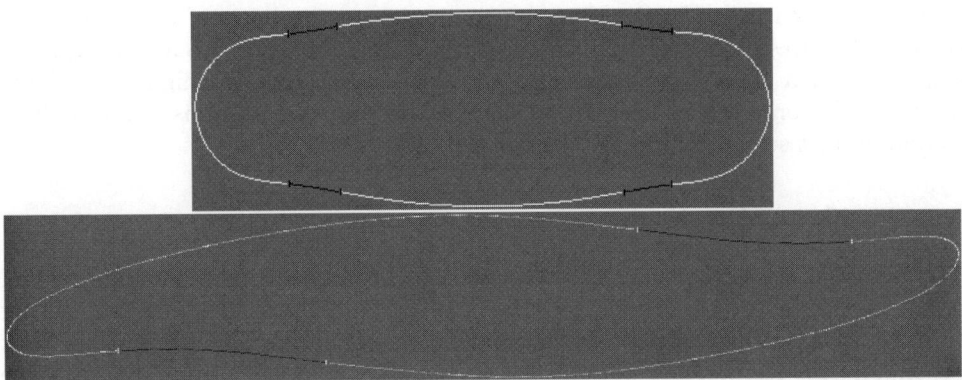

Figure 5. Two diatoms with similar contour segments. Top: a *Sellaphora pupula* deme "pseudocapitate" with 3 elementary segments (2 convex, 1 concave) from a total of 8 (4 convex, 4 concave). Bottom: a *Gyrosigma acuminatum* with 2 elementary segments (1 convex, 1 concave) from a total of 4 (2 convex, 2 concave).

taking into account (a) all contour points, (b) all segments and (c) all elementary segments. We call these *global* because they characterize the entire shape. First, we can compute curvature features on the basis of all n_C contour points:

$$\theta_\mu = \frac{1}{n_C} \sum_{i=0}^{n_C-1} \theta_i, \qquad \theta_{\max} = \max_{i \in [0, n_C-1]} \theta_i, \qquad \theta_{\min} = \min_{i \in [0, n_C-1]} \theta_i,$$

i.e. the average, maximum and minimum curvature.

On the basis of all segment features we can compute the number of convex (n_{cv}), concave (n_{cc}) and straight (n_{st}) segments, the average, maximum and minimum segment lengths:

$$l_\mu = \frac{1}{n_S} \sum_{i=0}^{n_S-1} l_i, \qquad l_{\max} = \max_{i \in [0, n_S-1]} l_i, \qquad l_{\min} = \min_{i \in [0, n_S-1]} l_i,$$

as well as 10 curvature and "noncircularity" features:

$$\theta_{\mu,\mu} = \frac{1}{n_S} \sum_{i=0}^{n_S-1} \theta_{\mu,i}, \qquad \theta_{\max,\mu} = \max_{i \in [0, n_S-1]} \theta_{\mu,i}, \qquad \theta_{\min,\mu} = \min_{i \in [0, n_S-1]} \theta_{\mu,i},$$

$$\theta_{\mu,\max} = \frac{1}{n_S} \sum_{i=0}^{n_S-1} \theta_{\max,i}, \qquad \theta_{\min,\max} = \min_{i \in [0, n_S-1]} \theta_{\max,i},$$

$$\theta_{\mu,\min} = \frac{1}{n_S} \sum_{i=0}^{n_S-1} \theta_{\min,i}, \qquad \theta_{\max,\min} = \max_{i \in [0, n_S-1]} \theta_{\min,i},$$

$$\theta_{\mu,q} = \frac{1}{n_S} \sum_{i=0}^{n_S-1} \theta_{q,i}, \qquad \theta_{\max,q} = \max_{i \in [0, n_S-1]} \theta_{q,i}, \qquad \theta_{\min,q} = \min_{i \in [0, n_S-1]} \theta_{q,i}.$$

Likewise, 13 features $(L_i, \Theta_{i,j})$ can be computed on the basis of all elementary-segment features. We note that on the basis of these features we also compute a $\Theta_{\min,\min}$ and a $\Theta_{\max,\max}$. This very large feature set contains highly-correlated features. The only way to find out which ones are the most useful is to apply them to diatom test sets and to study the results.

3 Classifier and test sets

The principal classifier we use here is based on the traditional offline training and online testing approach (Duda and Hart, 1973; Fukunaga, 1990). First, during training, the classifier is built by calculating cluster centers for all diatom taxa. Second, during testing, new diatoms are classified by applying a minimum-distance criterion between each cluster center and the extracted diatom features to determine the closest class (taxon). The Mahalanobis metric can be used for poorly scaled and/or highly correlated features (Duda and Hart, 1973), but here we use the simpler Euclidean distance, together with a scaling of the features by the standard deviation.

If $\vec{X}_c^i$ is the feature vector of diatom i and k_c is the number of diatoms in class c, cluster centers are calculated by averaging all feature vectors of diatoms in the training set which belong to the same class c:

$$\vec{A}_c = \frac{1}{k_c} \sum_{i=0}^{k_c-1} \vec{X}_c^i.$$

If $\vec{X}_d$ is the feature vector of a diatom d in the test set and $\vec{A}_i$ is the cluster vector of class i, $i \in [0, n_{\rm C} - 1]$, with $n_{\rm C}$ being the number of classes, then d is attributed class j ($j \in [0, n_{\rm C} - 1]$) for which

$$|\vec{A}_j - \vec{X}_d|$$

is smallest. If multiple classes yield the same minimum distance, the class is randomly selected from these. This procedure is also referred to as nearest-mean or Bayes minimum-distance classifier. In additional experiments we will apply decision trees (C4.5/C5.0) which are explained in Chapter 7.

The feature sets were examined using two diatom sets: (a) the set of 120 *Sellaphora pupula* specimens with 6 demes, and (b) the mixed genera set, with 781 diatoms belonging to 37 different taxa. These data sets have been described in Chapter 4. In both cases the sets have been split randomly into separate training and test sets with equal numbers of images. Partner RBGE created the training and test sets of the *Sellaphora pupula* data set.

4 Implementation, CPU times and parameter selection

The feature extraction and identification software was developed in ANSI C on an SGI Origin 200QC server using *cc* compiler options *ansi* and *fullwarn*. In order to achieve a good performance, we used also the options *Ofast=ip27* and *n32*. A UNIX (IRIS 6.3) interface was developed in order to perform the parameter selection tests.

Computation times, using only 1 of the 4 available processors, were 10.8 seconds for the feature extraction for all 120 *Sellaphora pupula* diatoms, 0.1 second for the classifier construction using 60 training contours, and 74 minutes for the identification of 60 test contours using *all* single features and *all* feature pairs, triplets, quadruplets, quin- and sextuples. For the mixed genera these times were 44.8 seconds (for 740 diatoms), 0.7 second (for 370 training contours) and 659 minutes (up to quintuples), respectively. We note that the Origin has MIPS R10000 processors at 180 MHz, and that a normal Pentium III computer at 733 MHz is faster by a factor of 2.2.

There are 8 parameters (not including m_5, which was always set to 3): five for the coordinate, convexity and curvature filtering (m_0; m_1 and m_2; m_3 and m_4), one for the length thresholding (t_0), one for the straight-line detection (t_1) and one for the symmetry analysis (t_2). The mask sizes m_i range between 0 or 1 (no filtering) and $(n_c - 1)/2$ (the filter mask spans the entire contour). The threshold t_0 ranges between 0 (no segment removal) and 1 (all segments are removed), t_1 between 0 (no straight points) and $\theta_{\max}$ (a point P_i is straight if $\theta_i < \theta_{\max}$), and t_2 between -1 (all segments are correlated, $n_{el} = 1$) and 1 (all segments are uncorrelated, $n_{el} = n_s$). Because a parameter tuning experiment, in which all parameter combinations are considered, is computationally very expensive, we divided these parameters into 5 groups of importance: a) m_0, m_1, m_2, b) m_3, m_4, c) t_1, d) t_0 and e) t_2.

For *Sellaphora pupula*, different parameter combinations were applied to the 60 training contours by varying the parameters in the following intervals: $m_0 \in [0, 0.01, 0.02, \ldots, 0.07, 0.08]$, $m_1 \in [1, 10, 20, \ldots, 70, 80]$, $m_2 \in [0, 10, 20, \ldots, 70, 80]$, $m_3 \in [1, 5, 10, \ldots, 65, 70]$, $m_4 \in [0, 5, 10, \ldots, 65, 70]$, $t_0 \in [0, 0.005, 0.010, \ldots, 0.035, 0.040]$, $t_1 \in [0, 0.002, \ldots, 0.078, 0.080]$ and $t_2 \in [-0.9, -0.8, \ldots, 0.8, 0.9]$. In order to quickly determine values for m_0, m_1 and m_2 all combinations of the values in their intervals were used in a first pass, with the other parameters set to their initial values. In a second pass, the parameters were finetuned by selecting the best values and varying the values again, but only around the selected ones in intervals that were a factor of two smaller. In order to determine values for the other parameters we varied 1) m_3 and m_4, 2) t_1, 3) t_0 and 4) t_2, each time using all combinations of the parameter values, and with the other parameters set to their pre-determined or initial values ($m_0 = 0, m_1 = 1, m_2 = 0, m_3 = 1, m_4 = 0, t_0 = 0, t_1 = 0$ and $t_2 = 1$). For each parameter set, a mean score was computed by averaging the training recognition rates of all one-dimensional nearest-mean classifiers, each one built using another single feature. We selected the parameters with the highest overall score. The resulting parameters were $m_0 = 0.04 n_c, m_1 = 30, m_2 = 10, m_3 = 40, m_4 = 35, t_0 = 0.015 n_c, t_1 = 0.01$ and $t_2 = 0.2$.

The same procedure was applied to the mixed genera set. However, we had to implement two modifications: (1) Since no training and test sets were available, we built these ourselves by randomly selecting first 20 samples per class (taxon), thus using each time 740 of the 781 diatoms, and then 10 per set. This way we generated a large number (750) of different pairs of training and test sets. From these we selected the pair that gave the highest score on the training set, with all parameters set to their initial values. (2) We selected those parameters which

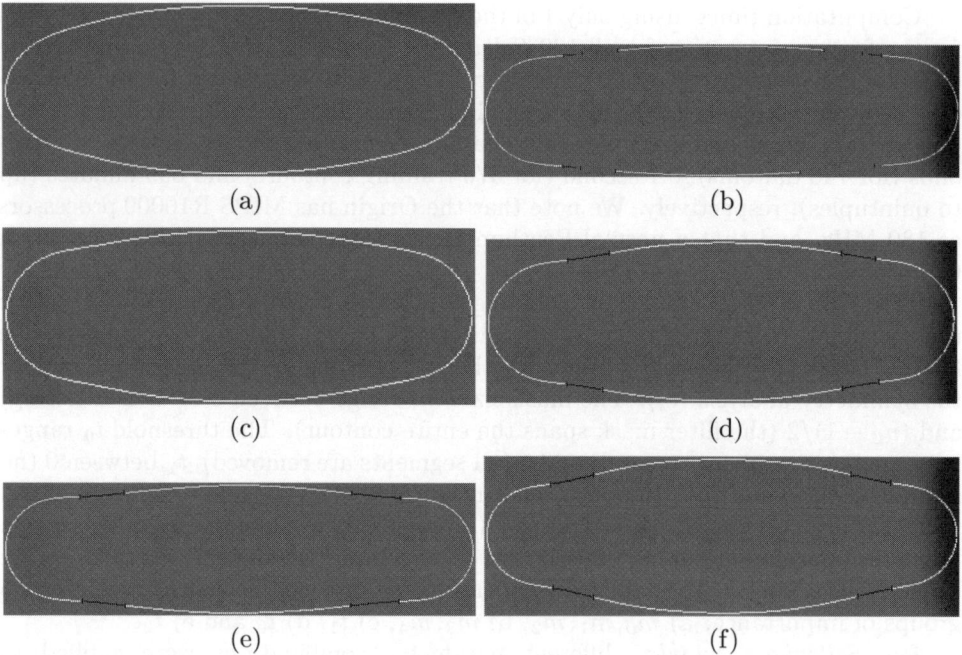

Figure 6. Typical processing results for each *Sellaphora pupula* deme: (a) "elliptical," (b) "large," (c) "pseudoblunt," (d) "pseudocapitate," (e) "cf. rectangular" and (f) "tidy." Note that images are not to the same scale.

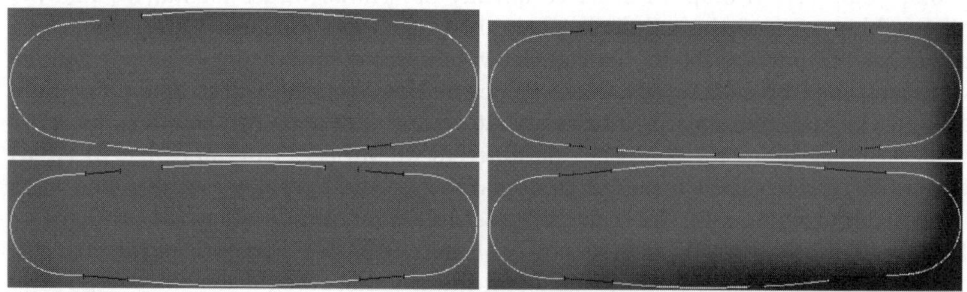

Figure 7. Irregular results: demes "large" (top) and "cf. rectangular" (bottom).

scored the highest mean of the recognition rates of the classifiers built for all feature pairs taken from $n_s, n_{cv}, n_{cc}, n_{st}, l_\mu, l_{max}, l_{min}, l_{cv}, l_{cc}$ and l_{st} (see Section 5.2 for the definition of the latter three features), and not, as above, those which scored the highest mean score of all 1D classifiers, built for all individual features. This increased the variation in the scores obtained for the different parameter sets. The resulting parameters were $m_0 = 0.05 n_c, m_1 = 10, m_2 = 0, m_3 = 45, m_4 = 30, t_0 = 0.015 n_c, t_1 = 0.016$ and $t_2 = 0.9$. Note that $m_2 = 0$ indicates that no lowpass filtering of the convexity is required.

Chapter 8: Curvature of convex and concave segments

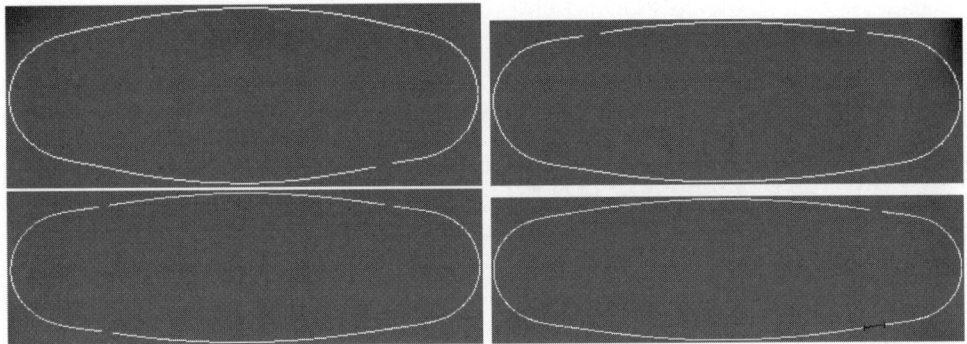

Figure 8. Results of contour processing for 4 "pseudoblunt" diatoms.

5 Identification results

In the following we first present results obtained with the 120 *Sellaphora pupula* contours, and then those for the large set of 781 contours from the mixed genera set. For the *Sellaphora pupula* set we used complete original contours as input to our method and not the subsampled ones, as used in some other chapters. For the mixed genera data we used the subsampled contours.

5.1 Sellaphora pupula *data set*

Figures 6, 7 and 8 show processing results from the test set. The small lines orthogonal to the contour indicate inflection points. The black contour parts denote concave segments, the white ones convex segments. Gaps surrounded by bars denote straight segments and gaps without bars thresholded segments. The number of extracted segments is *normally* equal to 8, except for the "large" deme, where it ranges from 4 to 12, and "elliptical" and "pseudoblunt," where it equals 1. The system always finds "elliptical" contours to be completely convex, but in the case of "pseudoblunt" the segment extraction and analysis is exactly at the limit of the thresholding process. Note that a visual inspection of the "pseudoblunt" contours is not a trivial task, even for trained diatomists, and segment separation is quite difficult (see Figs 6(c) and 8).

Figure 9 illustrates the identification (classification) using one pair of features: it shows feature vectors calculated for all 60 test diatoms of the 6 demes in feature space $(\theta_{max}, \theta_{max,min})$. These two features (being the maximum curvature and the maximum of the minimum segment curvature) adequately distinguish most demes, including "large" and "tidy," but also "elliptical" and "pseudoblunt." We note that, despite the fact that the latter two are completely convex, they have been almost correctly separated. We also note that: a) the 2 features are largely uncorrelated—otherwise all points would form a line—and b) simple linear decision boundaries seem to be sufficient to separate the demes and there are no significant subclasses, except maybe for the "cf. rectangular" deme. Below we will demonstrate that, by using this feature pair, 90% of the test diatoms can be correctly identified, and that by extending the dimension of the feature space from 2 to 4 a 100% identification

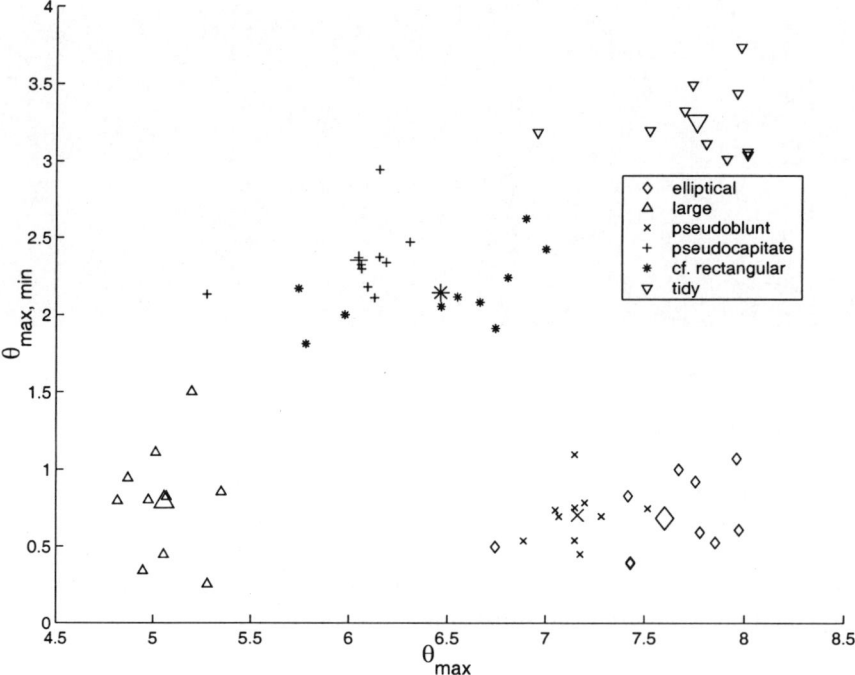

Figure 9. Example of 2D feature vectors of 60 contours in the *Sellaphora pupula* test set (small symbols) and the class centers of the 60 contours in the training set (big symbols).

rate can be achieved.

Tables 2 and 3 show sorted identification results using 1, 2, 3 or 4 features. The single-feature results (Table 2) show that the curvature features rank highest, before the length features and the various segment numbers. Note for example the big difference between $\theta_{\mu,\mu}$ (the average of the segment curvature means) and θ_μ (the contour curvature mean), which indicates the importance of the contour division into segments. Note also the relatively high recognition rate of the "large" deme when feature n_{st} is used, which indicates that "large" contours can be separated from other contours on the basis of their straight segments. For the "elliptical," "pseudoblunt" and "cf. rectangular" demes none of the single features can attain a perfect recognition rate of 100%.

The results with multi-dimensional feature spaces (Table 3) were obtained by: a) normalizing the features; b) building a classifier for each combination of feature pairs, triplets or quadruplets; and c) calculating and sorting the classifiers' identification rates. In the normalization each feature value was divided by the standard deviation along the feature dimension. Apart from the features described above, four additional, independent features have been tested as well: diatom width (d_W); diatom length (d_l); stria density (s_d); and stria orientation (s_O). This gives a total of 40 features. The reasons for including these features are (according to diatomists): (1) "large" and "cf. rectangular" demes are also separated by s_d, (2) "cf. rectangular" and "pseudocapitate" as well as "large" and "pseudoblunt" demes

Table 2. Sorted identification rates (%) of nearest-mean classifiers using individual features for the *Sellaphora pupula* set.

feature	ellipt	large	pseudo blunt	pseudo capitate	cf. rect-angular	tidy	total
$\Theta_{\mu,max}$	90	100	80	70	50	90	80.0
$\theta_{\mu,max}$	90	90	80	90	30	90	78.3
$\theta_{max,\mu}$	70	90	70	80	40	100	75.0
θ_{max}	40	100	90	80	50	80	73.3
$\Theta_{max,min}$	40	90	50	40	70	100	65.0
$\Theta_{max,max}$	30	70	90	90	40	70	65.0
$\theta_{\mu,\mu}$	20	100	70	70	30	90	63.3
d_W	30	90	30	100	50	80	63.3
$\theta_{min,\mu}$	70	100	70	40	0	90	61.7
$\Theta_{min,\mu}$	70	100	70	40	0	90	61.7
$\theta_{\mu,q}$	70	10	90	30	70	80	58.3
$\Theta_{\mu,\mu}$	30	100	70	50	50	30	55.0
$\Theta_{\mu,min}$	20	100	50	30	30	100	55.0
$\theta_{max,min}$	0	40	80	80	20	100	53.3
$\theta_{min,max}$	90	0	80	70	20	60	53.3
$\Theta_{min,max}$	90	0	80	70	20	60	53.3
$\Theta_{\mu,q}$	70	20	90	40	10	90	53.3
$\Theta_{max,\mu}$	10	50	70	50	30	100	51.7
$\theta_{\mu,min}$	10	100	50	30	20	100	51.7
d_l	20	80	70	60	60	10	50.0
$\theta_{max,q}$	70	10	70	20	40	90	50.0
θ_{min}	40	100	30	50	60	20	50.0
$\Theta_{min,min}$	40	100	50	50	50	0	48.3
n_S	50	60	70	100	0	0	46.7
n_{cv}	50	60	70	100	0	0	46.7
$\Theta_{max,q}$	70	0	80	50	0	80	46.7
n_{el}	50	60	60	100	0	0	45.0
$\Theta_{min,q}$	70	10	80	20	10	70	43.3
$\theta_{min,q}$	70	10	80	0	10	80	41.7
θ_{μ}	10	90	70	40	10	20	40.0
s_o	50	20	60	0	40	60	38.3
s_d	30	0	0	70	70	60	38.3
l_{max}	50	10	70	100	0	0	38.3
n_{cc}	50	0	70	100	0	0	36.7
L_μ	50	10	60	100	0	0	36.7
l_μ	50	10	60	100	0	0	36.7
l_{min}	50	0	60	60	10	40	36.7
L_{min}	50	0	60	50	10	50	36.7
L_{max}	50	0	70	80	0	0	33.3
n_{st}	20	90	30	20	0	20	30.0

Table 3. Sorted identification rates (%) of nearest-mean classifiers using different feature pairs, triplets and quadruplets for the *Sellaphora pupula* set. E = elliptical, L = large, PSB = pseudoblunt, PSC = pseudocapitate, CFR = cf. rectangular, T = tidy.

features	E	L	PSB	PSC	CFR	T	total
$\Theta_{\mu,\max}, d_W$	80	100	90	100	100	90	93.3
$\Theta_{\min,\mu}, d_W$	80	90	90	100	100	90	91.7
$L_{\max}, d_W$	70	90	100	100	100	90	91.7
$\theta_{\mu,\max}, d_W$	80	90	90	100	100	90	91.7
$\theta_{\min,\mu}, d_W$	80	90	90	100	100	90	91.7
$\theta_{\max,\min}, d_W$	60	100	100	100	100	90	91.7
$\Theta_{\max,\mu}, d_W$	80	100	100	100	80	90	91.7
$\theta_{\mu,q}, d_W$	80	90	100	100	90	90	91.7
$\theta_{\max,\mu}, d_W$	80	90	90	100	100	90	91.7
$\Theta_{\max,\min}, d_W$	60	100	100	100	90	90	90.0
$\Theta_{\mu,q}, d_W$	80	90	100	100	80	90	90.0
$\theta_{\max}, \theta_{\max,\min}$	90	100	90	90	70	100	90.0
n_{el}, d_W	70	100	100	100	80	90	90.0
$\theta_{\max}, \theta_{\max,\mu}, d_W$	100	100	100	100	100	90	98.3
$n_{CV}, \theta_{\max}, d_W$	90	100	100	100	100	90	96.7
$n_{el}, \theta_{\max}, d_W$	90	100	100	100	100	90	96.7
$\theta_{\max}, L_{\max}, d_W$	90	100	100	100	100	90	96.7
$n_{CC}, \theta_{\max}, d_W$	90	100	100	100	100	90	96.7
$\Theta_{\max,\max}, \Theta_{\mu,\min}, d_W$	100	100	90	100	100	90	96.7
$\theta_{\max}, \Theta_{\max,\mu}, d_W$	100	100	100	100	90	90	96.7
$\theta_{\max}, l_{\max}, d_W$	90	100	100	100	100	90	96.7
$\theta_{\max}, \theta_{\max,\min}, d_W$	90	100	100	100	100	90	96.7
$\theta_{\max}, \theta_{\mu,\min}, \theta_{\mu,q}$	100	100	100	90	70	100	93.3
$\theta_{\max,\min}, \Theta_{\max,\max}, \Theta_{\mu,\min}, d_W$	100	100	100	100	100	100	100.0
$\theta_{\max}, l_{\min}, \theta_{\max,\mu}, d_W$	100	100	100	100	100	90	98.3
$\theta_{\mu,\min}, \theta_{\mu,q}, d_W, s_O$	90	100	100	100	100	100	98.3
$n_{el}, \theta_{\max}, \theta_{\mu,q}, \Theta_{\mu,\min}$	100	100	100	100	70	100	95.0

are also separated by d_W, and (3) "pseudoblunt" and "pseudocapitate" demes are well separated by d_l and d_W. These features are described in Chapters 4, 7 and 10. Table 2 shows that d_W ranks relatively high but not the highest, and Table 3 shows that d_W is quite important for obtaining good results using feature pairs, triplets and quadruplets. The importance of the other three features was not directly confirmed, but we will see later that s_O is important for achieving a 100% identification rate on the reversed test set. We note that Table 3 shows only the best results (pairs, triplets and quadruplets) as well as the best result without feature d_W (triplets and quadruplets) from, respectively, 780, 9880 and 91,390 built classifiers, but in all cases using the unseen *test* set of 60 contours. An error analysis showed that one diatom was often misclassified. The contour of this diatom is shown in Fig. 10; as can be seen the misidentification is not caused by an over-filtering of the contour coordinates. We also note that the contour partition was normal, i.e. comparable to the one shown in Fig. 6(f). This indicates that the identification of this diatom should be considered with extra attention.

Chapter 8: Curvature of convex and concave segments

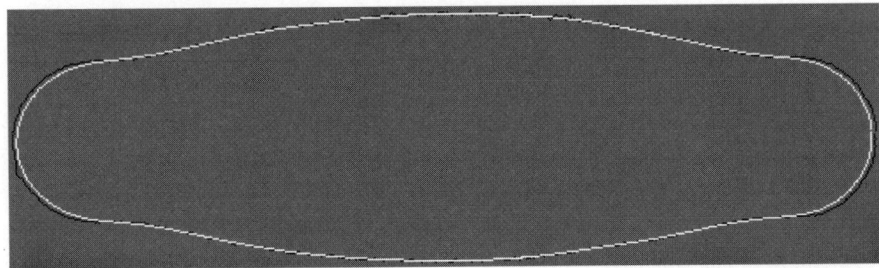

Figure 10. This "tidy" contour was often misclassified as "pseudocapitate." Black denotes the original and white the filtered coordinates.

Table 4. Identification rates (%) using a 4D and 6D feature space on the training, test, reversed training and reversed test sets, and the average training and test rates, for the *Sellaphora pupula* set.

features	train	test	rev. train	rev. test	aver. train	aver. test
$\theta_{\max,\min}$, $\Theta_{\max,\max}$, $\Theta_{\mu,\min}$, d_W	91.7	100.0	96.7	88.3	94.2	94.2
$\theta_{\max}$, $l_{\min}$, $\theta_{\min,\mathrm{q}}$, $\Theta_{\min,\min}$, d_W, s_O	98.3	95.0	96.7	100.0	97.5	97.5

We have seen that by using only four features and a normal nearest-mean classifier with pre-defined training and test sets as created at RBGE, a perfect identification rate of 100% can be obtained. Because in other Chapters (9, 10) results based on swapping the training and test sets, i.e. the reverse procedure, are reported, we will do the same here. Table 4 shows two results in which four and six features were used and the maximum score on the test and the reversed test sets were obtained. Surprisingly, in both cases the test results were better than the training results, and in both cases the result for the test set, other than the one on which 100% was obtained, dropped significantly (88.3 and 95%). The best result with this "set swapping" is obtained using six features: 97.5%. Figure 11 shows that the performance on either set, as a function of the number of features, increases asymptotically to 100%. Hence, we expect that more features, up to a total of 8 or 10, say, may yield perfect results for both training and testing on the two subsets.

Decision trees have been used in the other identification chapters in this book, hence we would like to compare their performance with the Bayes classifier used in this chapter. Table 5 shows identification rates of C5.0 classifiers which were built on the training set using default settings (boosting performed slightly worse). If we compare these results with those of the nearest-mean classifiers (see Tables 2, 3 and 4), we can note that a) a 100% rate is never achieved, b) C5.0 can achieve 98.3% when using only four features, c) the performance for the sextuple, which performed very well on both the test and reversed test set, is much worse, d) the rate obtained when *all* features are used is much lower than when only 2, 3 or 4 features are used, and e) the contribution of the features with a total identification rate smaller than 50% (Table 2) is only very small (the difference between the recognition rates listed in lines 5 and 6 is only 1.7%). Hence, although nearest-mean classifiers are much

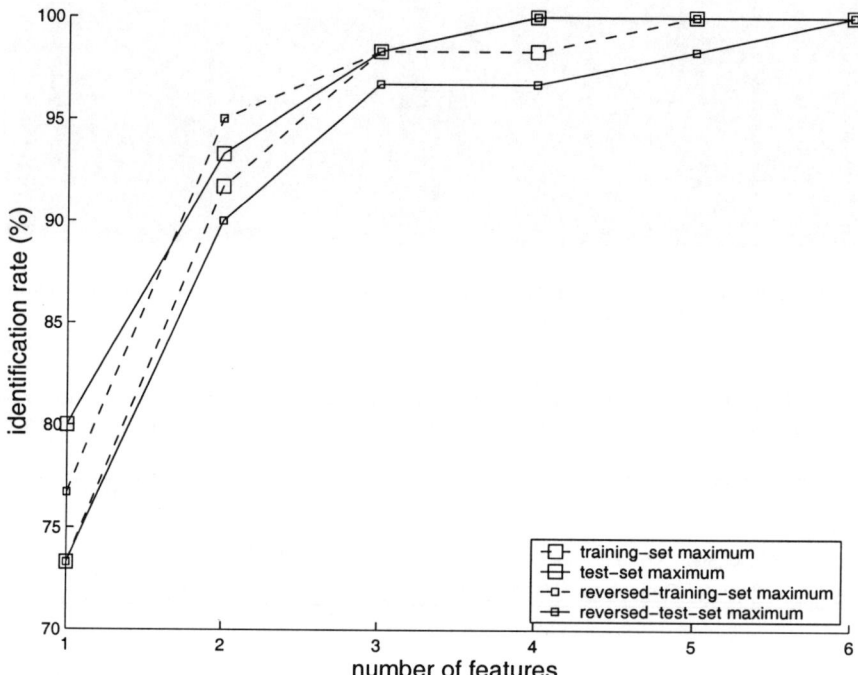

Figure 11. Identification rates for different dimensions of the feature space (*Sellaphora pupula* data set).

Table 5. Identification rates of C5.0 classifiers using selected feature combinations for the *Sellaphora pupula* data set.

features	%
$\Theta_{\mu,\max}, d_W$	96.7
$\theta_{\max}, \theta_{\max,\mu}, d_W$	95.0
$\theta_{\max,\min}, \Theta_{\max,\max}, \Theta_{\mu,\min}, d_W$	98.3
$\theta_{\max}, l_{\min}, \theta_{\min,q}, \Theta_{\min,\min}, d_W, s_O$	83.3
all with a total recognition rate $\geq 50\%$ in Table 2	90.0
all	88.3

simpler, they seem to outperform decision trees on this feature set and these data.

5.2 Mixed genera data set

Figure 12 shows some of the contour processing results obtained with this data set. Obviously, in this set the contour shapes are much more diverse than in the *Sellaphora pupula* set. Initial results were poor due to the diversity of shapes, and we were forced to define three new (global) features for this set, i.e. the sums of the relative lengths of all convex, concave and straight segments (l_{cv}, l_{cc} and l_{st}).

Chapter 8: Curvature of convex and concave segments

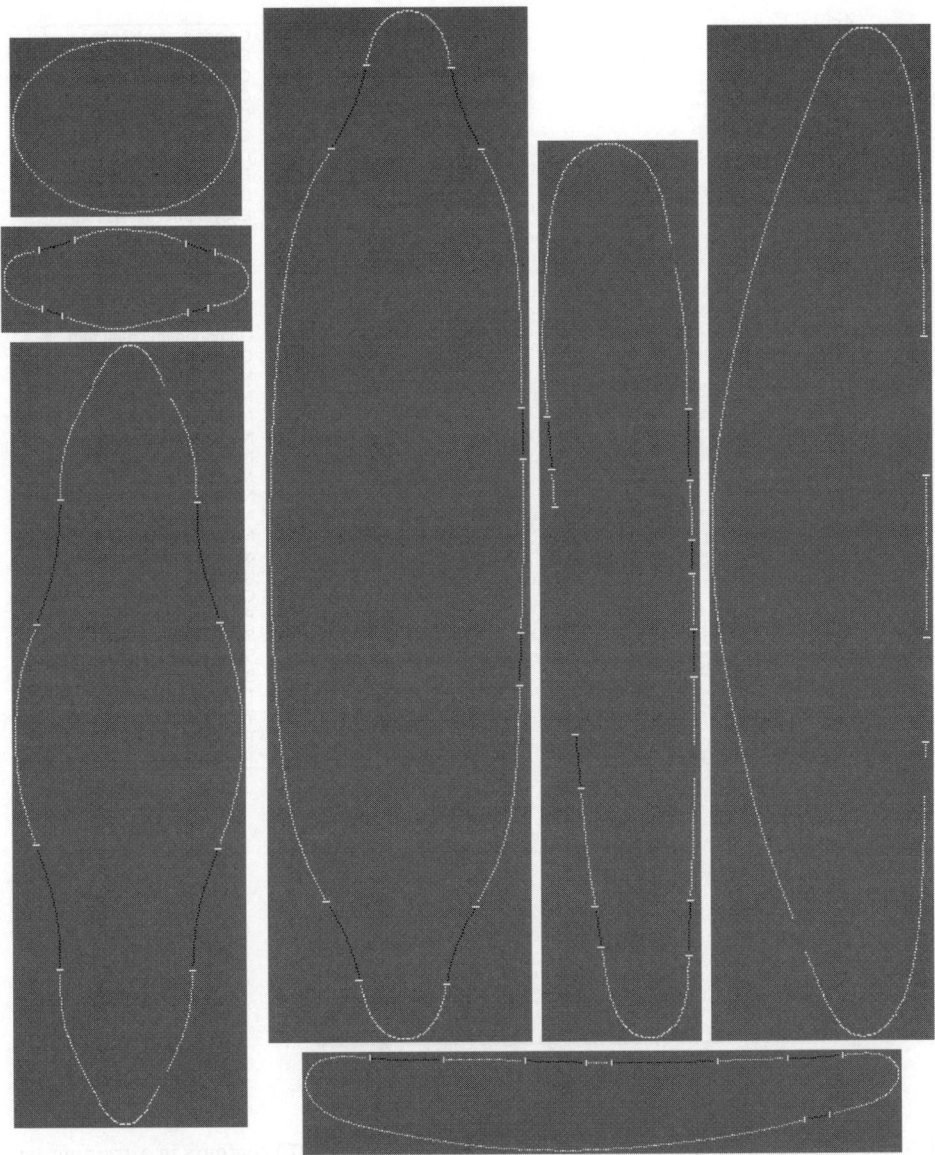

Figure 12. Processing results for some taxa in the mixed genera set, from top to bottom and left to right: *Cocconeis neodiminuta, Staurosirella pinnata, Stauroneis smithii, Cymbella hybrida, Meridion circulare, Cymbella helvetica* and *Eunotia incisa*. Images are not to the same scale.

The basic idea is that these features allow, together with $n_s, n_{cv}, n_{cc}, n_{st}, l_\mu, l_{max}$ and l_{min}, a first rough identification by which, e.g., partly convex/concave contours can be separated from completely convex ones. Then the identification can be improved by using curvature features to separate, e.g., the different completely convex contours.

Figure 13 exemplifies this by showing the computed class centers in feature

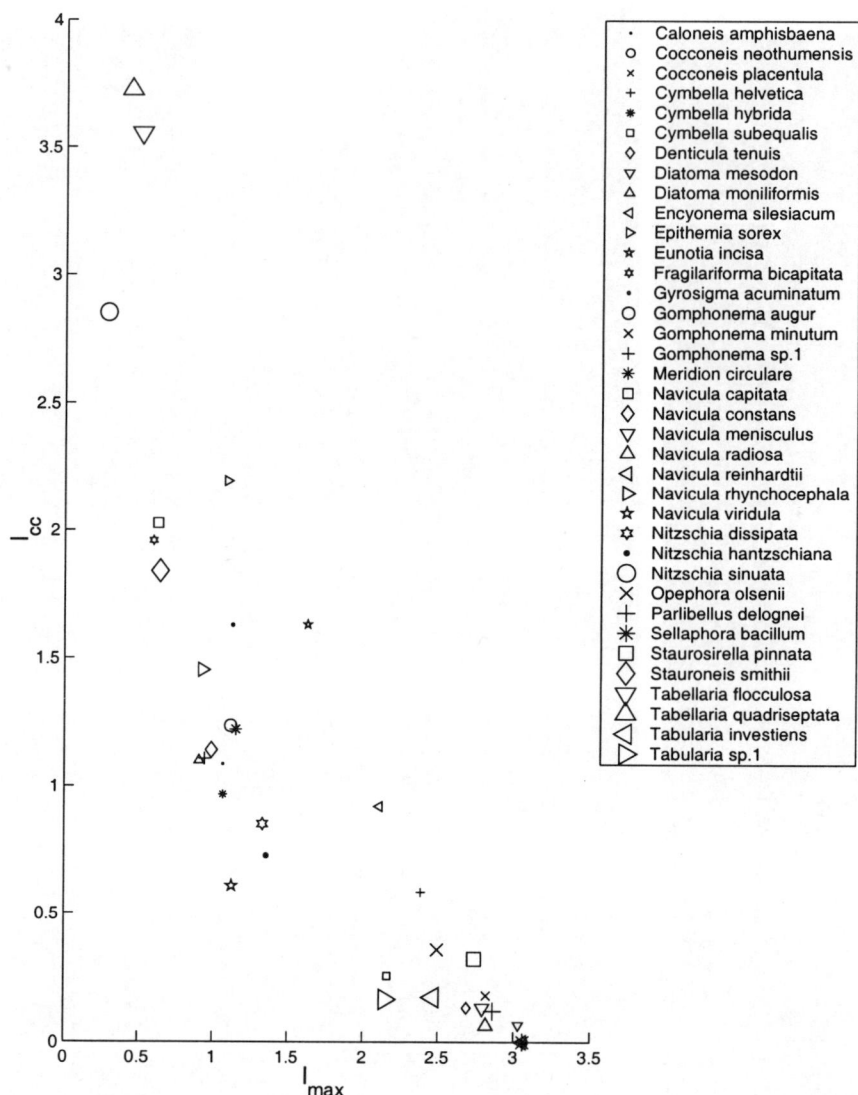

Figure 13. Class centers of features $l_{\max}$ and l_{cc} calculated from 370 contours in a *training* set of the mixed genera data set.

space $(l_{\max}, l_{\mathrm{cc}})$. We note that both features range between 0 and 1, but the figure shows the features after normalization. For clarity we show only the class centers of a training set, and not, as in Fig. 9, the feature vectors of the test samples. Not only is the number of samples and taxa much larger than in the *Sellaphora pupula* set, but as we will see later, the identification rate with feature pairs is also much lower. As can be seen in Fig. 13, there are many taxa for which l_{cc} is (almost) 0; these have contours which are (almost) completely convex. There are also many taxa in the middle of the plot; these have contours with significant concave parts.

Table 6. Sorted identification rates of nearest-mean classifiers, using individual features (mixed genera data set).

feature	%	feature	%
l_{max}	25.7	$\theta_{max,q}$	16.8
θ_{max}	24.9	$\theta_{\mu,q}$	16.8
L_μ	23.8	$\Theta_{\mu,\mu}$	16.5
L_{max}	22.7	$\Theta_{\mu,q}$	16.2
$\Theta_{max,max}$	22.2	$\Theta_{max,\mu}$	16.2
$\Theta_{min,\mu}$	21.4	$\Theta_{max,q}$	15.9
$\theta_{\mu,min}$	20.5	$\Theta_{\mu,max}$	15.7
$\theta_{max,min}$	20.5	s_μ	14.3
l_{cc}	20.5	$\Theta_{min,q}$	13.8
$\theta_{min,\mu}$	20.3	s_{q3}	13.8
$\Theta_{max,min}$	20.0	n_S	13.8
l_{min}	19.7	$\theta_{min,q}$	13.2
l_μ	19.2	$\theta_{\mu,max}$	13.2
$\Theta_{min,max}$	19.2	n_{cc}	13.0
$\Theta_{\mu,min}$	18.4	s_{q1}	12.2
L_{min}	18.4	n_{cv}	11.4
$\theta_{min,max}$	18.4	s_{q4}	10.8
l_{cv}	18.4	s_{q2}	10.5
$\Theta_{min,min}$	18.1	n_{el}	10.5
θ_{min}	18.1	l_{st}	5.9
s_σ	18.1	n_{st}	5.4
$\theta_{\mu,\mu}$	17.8	s_{odd}	4.1
$\theta_{max,\mu}$	17.6	s_{even}	2.4
θ_μ	17.0		

Note that there are only few taxa in the upper part of the plot (*Nitzschia sinuata*, *Tabellaria flocculosa* and *Tabellaria quadriseptata*). Below it can be seen that l_{cc} is indeed important to get good identification results.

Tables 6 and 7 show sorted identification results using up to 10 features. Apart from the described features, eight additional, independent features were used (see Chapter 10): two for the mean and variance of the stria density (s_μ and s_σ), four quadrant means of the stria orientation (s_{q1}, s_{q2}, s_{q3} and s_{q4}) and two for the even and odd striation symmetry (s_{even} and s_{odd}). This makes a total number of 47 features. The single-feature results (Table 6) show that the length features rank highest, before the curvature features and the different segment numbers. Note, however, that l_{cv}, l_{cc} and s_σ rank also relatively high. Note also that the identification rates are much lower than those in Table 2. The features l_{st} and n_{st} rank very low, but their recognition rates are still higher than the statistical guess rate (2.7%). The highest recognition rate of l_{st} was obtained for the *Meridion circulare* (70%).

Table 7. Sorted identification rates of nearest-mean classifiers, using different feature combinations (mixed genera data set).

features	%
θ_{max}, l_{CC}	52.4
θ_{max}, $\theta_{\mu,\mu}$, l_{CC}	64.6
θ_{max}, $\Theta_{\mu,\mu}$, l_{CC}, s_μ	76.2
θ_{max}, $\theta_{\mu,\mu}$, l_{CC}, s_μ	76.2
θ_μ, θ_{max}, $\Theta_{\mu,min}$, l_{CC}, s_μ	78.1
θ_μ, θ_{max}, $\theta_{\mu,\mu}$, l_{CC}, s_μ	78.1
θ_{max}, l_{CV}, l_{CC}, s_μ, s_σ, $\theta_{\mu,min}$	78.4
θ_{max}, l_{CV}, l_{CC}, s_μ, s_σ, θ_{max}, $\Theta_{\mu,min}$	80.8
θ_{max}, l_{CV}, l_{CC}, s_μ, s_σ, θ_{max}, $\theta_{\mu,min}$, $\Theta_{\mu,q}$	82.2
θ_{max}, l_{CV}, l_{CC}, s_μ, s_σ, θ_{max}, $\theta_{\mu,min}$, $\Theta_{\mu,q}$, s_μ	82.7
θ_{max}, l_{CV}, l_{CC}, s_μ, s_σ, θ_{max}, $\theta_{\mu,q}$, $\theta_{max,q}$, $\Theta_{\mu,min}$, s_μ	83.5
θ_{max}, l_{CV}, l_{CC}, s_μ, s_σ, θ_{max}, $\theta_{\mu,min}$, $\theta_{\mu,q}$, $\theta_{max,q}$, s_μ	83.5

One problem when experimenting with the mixed genera set is that much more features are needed than for the *Sellaphora pupula* set and that trying out bigger feature sets is computationally very expensive: when the number of classifier dimensions (d) increases, the number of possible feature combinations grows with a factor of $\binom{47}{d}$. If we want to use eight instead of seven features, for example, the number of combinations grows with a factor of more than 3×10^8! One solution for testing big feature sets is to vary only some features in the classifier input and to fix the rest of the input features. The first part of the multi-dimensional results (the upper part of Table 7) was obtained by trying out all feature pairs, triplets, quadruplets and quintuples. Only the best results are shown. Note that θ_{max} and l_{CC} form part of all combinations and that s_μ forms part of all quadruplets and quintuples. The lower part of Table 7 was obtained by fixing five features and again trying out all combinations of single features, pairs, triplets, quadruplets and quintuples, in order to form 6-, 7-, ... 10-tuples. Again, only the best results are shown. Note that by extending the feature quintuple to a 10-tuple the identification rate increases by only 5.4%. We fixed θ_{max}, l_{CV}, l_{CC}, s_μ and s_σ in the classifier input (l_{CV} and s_σ were important in a similar test in which only three features were fixed in the input). Surprisingly, we found that if θ_{max} or s_μ were used *twice* in the classifier input, the results improved. This may seem strange, but it changes the weighting of the different features. However, it also indicates that the Mahalanobis metric might lead to better results, since it provides for an improved feature scaling and it takes the correlation of the features into account by determining covariance matrices.

Table 8 shows ID rates obtained by applying the reversed training/testing procedure. It shows two feature sets, the one which performed best on the test set and another which performed best on the reversed test set. Surprisingly, the identification rate obtained on the reversed test set (86.2%) was higher than the one obtained on the original test set (83.5%). In contrast to the *Sellaphora pupula* set,

Chapter 8: Curvature of convex and concave segments

Table 8. Identification rates (%) using 10-D feature spaces on the training, test, reversed training and reversed test sets, and the average test rate, for the mixed genera set.

features	train	test	rev. train	rev. test	aver. test
θ_{max}, l_{CV}, l_{CC}, s_μ, s_σ, θ_{max}, $\theta_{\mu,min}$, $\theta_{\mu,q}$, $\theta_{max,q}$, s_μ	87.6	83.5	86.2	80.5	82.0
θ_{max}, l_{CV}, l_{CC}, s_μ, s_σ, θ_{max}, $\theta_{\mu,\mu}$, $\Theta_{min,min}$, s_μ, s_{q4}	86.2	79.5	86.5	86.2	82.9

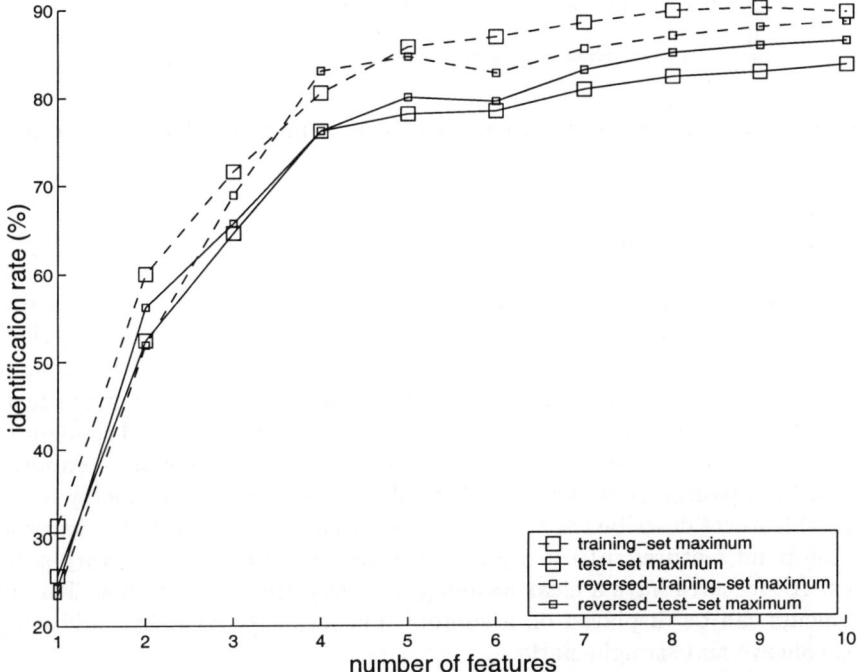

Figure 14. Identification rates for different dimensions of the feature space (mixed genera set).

here the training results were better than the test results. Figure 14 shows that the performance on the training sets, as a function of the number of features, increases asymptotically to 90%. The performance on the test sets is slightly worse. But the slope for 8- to 10-tuples is still positive, which means that better results should be possible when considering more features.

Again, we applied decision trees to the generated feature sets. Table 9 shows recognition rates of C4.5 classifiers, which were applied using windowing. These results are much worse than the results of the nearest-mean classifiers (see Tables 7 and 8). Surprisingly, the performance of the classifier which uses all features is better than those obtained with feature subsets. The contribution of the features listed at the bottom of Table 6 is, again, only very minor (1.9%). As we have already seen in the case of the *Sellaphora pupula* set, nearest-mean classifiers can

Table 9. Identification rates of C4.5 classifiers using selected feature combinations for the mixed genera data set.

features	%
θ_{max}, l_{CC}	51.1
θ_{max}, $\theta_{\mu,\mu}$, l_{CC}	64.1
θ_{max}, $\Theta_{\mu,\mu}$, l_{CC}, s_μ	73.0
θ_μ, θ_{max}, $\Theta_{\mu,min}$, l_{CC}, s_μ	73.0
θ_{max}, l_{CV}, l_{CC}, s_μ, s_σ, θ_{max}, $\Theta_{\mu,min}$	74.9
θ_{max}, l_{CV}, l_{CC}, s_μ, s_σ, $\theta_{\mu,q}$, $\theta_{max,q}$, $\Theta_{\mu,min}$	75.9
all with a total recognition rate > 14% in Table 6	75.7
all	77.6

outperform decision trees, even when identifying a mixture of very different taxa like here.

6 Discussion and conclusions

In this chapter we have presented new features which can be used to describe diatom contours. Shape descriptions are generated at three levels of analysis: at the lowest level, by computing local point features; at intermediate levels, by (elementary) segment features; and at the highest level, by global features. The segment features are computed by processing the point features of convex, concave and straight parts after partitioning the contour. The elementary-segment features are computed by averaging the features of similar segments after performing a symmetry analysis. The global features describe the entire contour shape and are computed by processing a) the point features, b) the segment features and c) the elementary-segment features. All features have a clear meaning, i.e. they are interpretable. The different segments can be displayed on a contour and inspected visually, indicating its convex, concave and straight parts.

The feature extraction method can be applied to noisy contours with convex, concave and straight parts, but also to almost completely convex ones, for shape analysis or identification tasks. We used the method to compute features for two different sets of diatoms, with the aim of building an automated identification system. Therefore, we used a very simple nearest-mean classifier based on the standard Euclidean metric. We explicitly specified which feature combinations should be selected from our feature sets in order to obtain the best identification results (in other chapters mainly results for large feature sets are presented). Results obtained with C4.5/C5.0 decision trees can be comparable, but when an entire feature set is used instead of selected feature combinations, the results are worse.

First, we applied the nearest-mean classifier to different feature combinations in order to identify six demes of *Sellaphora pupula*. This data set contains diatoms with minute differences in shape, which are difficult to identify even for human experts (see Chapter 5). Nevertheless, excellent results were obtained: (1) Using only two features, an ID rate of 93.3% could be obtained. (2) Using only four features, an ID rate of 100% was achieved on the test set. (3) When the training and test sets are swapped, an average of 97.5% was achieved. Despite these excellent results we found that: (a) A recognition rate of 100% on the test set does not guarantee a 100% rate on the training set. (b) The features for which the best test and reversed test results are obtained are not necessarily the same, i.e. it is not a trivial task to find the best classifier input.

We also used our features for the identification of diatoms in the mixed genera set, which, in contrast to *Sellaphora pupula*, contains diatoms with very diverse shapes. Therefore, we defined three additional, global features. Using ten features as input to the nearest-mean classifier, a recognition rate of 83.5% was achieved on the test set. When swapping the training and test sets, an average rate of 82.9% was achieved. We expect that nearest-mean classifiers based on the Mahalanobis metric might lead to better results for this data set.

We note that the numbers presented above have been obtained by the classical 50/50 training/testing method, i.e. only half of all specimens per taxon were used for training. In other chapters decision trees with for example the leave-one-out strategy have been applied. The latter procedure serves to estimate the ID rate for *new* specimens, but uses almost *all* available specimens for training. Since more training data generally yield better identification rates, we expect that the leave-one-out procedure, in combination with the nearest-mean classifier, will improve the results, especially in the case of the mixed genera data set.

All results shown have been obtained by using two sets of parameters, which were selected specifically for the two data sets. These parameter sets were optimized in such a way that the highest recognition averages were obtained by using either single features or feature pairs. Other parameter sets might yield even better results, provided that they are optimized for higher-dimensional feature spaces, with for example four or five features, in order to improve the inter-class separation. Depending on the application at hand, *both* the parameter selection *and* the feature selection can be finetuned.

References

Dong, Y. and Hillman, G.R. (2001) Three-dimensional reconstruction of irregular shapes based on a fitted mesh of contours. Image and Vision Computing, Vol. 19, pp. 165-176.

Duda, R.O. and Hart, P.E. (1973) Pattern classification and scene analysis. Wiley-Interscience, New York.

Fukunaga, K. (1990) Introduction to statistical pattern recognition. Academic Press, New York (2nd Edition).

James, G. (1999) Advanced modern engineering mathematics. Addison-Wesley, Harlow, England (2nd Edition).

CHAPTER 9

IDENTIFICATION BY CONTOUR PROFILING AND LEGENDRE POLYNOMIALS

ADRIAN CIOBANU AND HANS DU BUF

In this chapter we present an improved method for contour profiling which is based on dynamic ellipse fitting. We also apply Legendre polynomials for characterizing upper and lower contour parts, assuming that a contour with a pennate form is in a horizontal position. Three classification methods are used: decision trees, backpropagation neural networks, and a hand-optimized syntactical classifier. Using only contour features of the *Sellaphora pupula* data set with distinct training and test sets, the characteristic profile features outperformed Legendre polynomials and correct identification rates of 91.7 resp. 93.3% were obtained by a neural network and a decision tree. The hand-optimized syntactical classifier could reach 75% on the test set, but a training on all 120 available samples resulted in 99.2% when also width and striation features were used. In the case of the mixed genera data set with 48 taxa and a total of 1009 images, neural networks in combination with Legendre polynomials performed best, yielding 82%. After excluding taxa with either similar shapes, incorrect contours or an insufficient number of images, the remaining 17 taxa with a total of 359 images could be identified with an ID rate of 91.1%.

1 Introduction

There are numerous shape features described in the literature and some chapters of this book introduce even more new ones. In this chapter we will introduce a new contour profiling method based on dynamic ellipse fitting. We will also introduce Legendre polynomials, which have been used to analyze diatom shapes (Stoermer and Ladewski, 1982), but not for diatom identification in the strict sense.

One of the simplest methods for contour matching is based on the centroidal profile (Davies, 1997). First, the centroid of the contour is computed from the coordinates of the contour points. The centroid becomes the origin of a polar coordinate system and the contour is plotted as a one-dimensional angle-radius graph (see Fig. 1). To compare two shapes it is necessary to cross-correlate them, i.e. to shift one graph along the other until the best match is obtained, since there is no fixed starting angle and the rotation of the contours is arbitrary. This method is able to cope with objects of different size by means of a normalization, dividing all radii by the largest one. However, this method has some difficulties when (1) due to occlusions the centroid is moved away from its normal position, (2) the objects are not star-shaped, resulting in a graph with multiple values per angle, and (3) the objects are very elongated, in which case a special smoothing is necessary for the contour parts near the centroid.

Below we will first describe the new profiling method (Section 2) and then

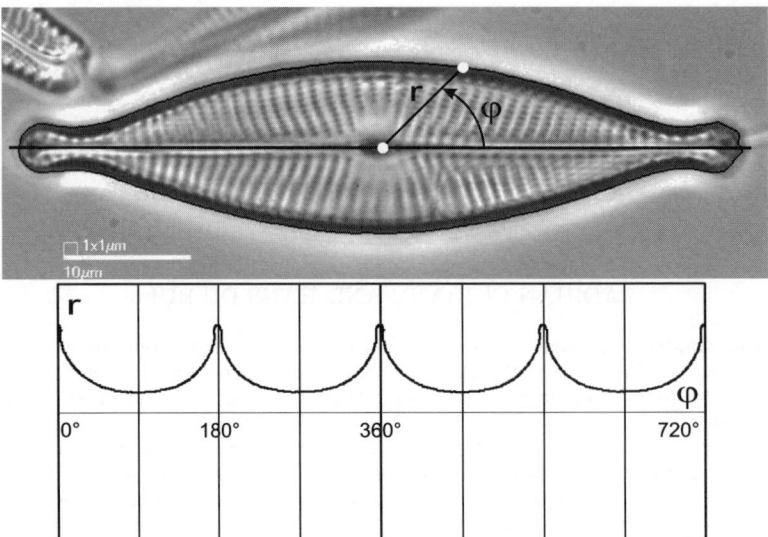

Figure 1. The classical centroidal profile shown for two rotations.

the Legendre polynomial fitting (Section 3). The different feature sets will be tested using neural networks, decision trees as well as a hand-optimized syntactical classifier (Section 4).

2 Dynamic ellipse fitting and contour profiling

Despite the vast amount of different shapes, many diatom contours are star-shaped, i.e., starting from the center or the centroid a ray will intersect the contour only once for any angle. Hence, the centroidal profile can be suitable for the matching task. In order to deal with elliptical (pennate) shapes, we proposed a new method based on dynamically fitting an ellipse to a contour, followed by the extraction of a characteristic profile (Ciobanu et al., 2000). For diatoms with non-pennate shapes, like triangles or squares, it will be necessary to devise other methods. It is also important to note that diatoms of different taxa can have virtually the same shape, being differentiated only by size or valve ornamentation. In addition, diatoms belonging to a same taxon can have different shapes and ornamentations (see Chapter 2). All this complicates automatic identification and suggests that, instead of applying straightforward methods as done here, we need to think about more sophisticated processing in the future.

2.1 Ellipse fitting

The main building block of our method is the fitting of an ellipse to a set of points which represent a contour. The method of direct least-squares fitting (Fitzgibbon et al., 1999) turned out to be appropriate in our case. This method minimizes

Chapter 9: Contour profiling and Legendre polynomials

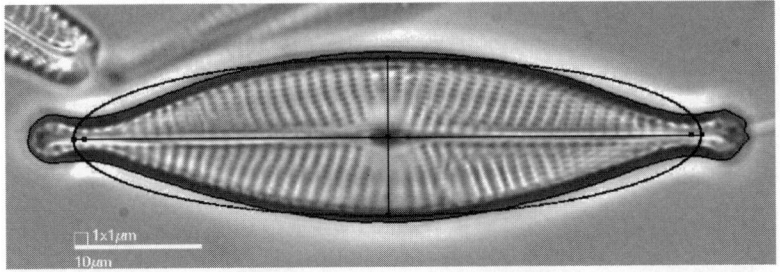

Figure 2. The first-fitted ellipse; the two foci are shown on the major axis.

the sum of squared algebraic distances of a conic to a set of points by using a generalized eigensystem of size 6×6. Starting from the polynomial equation of a conic $ax^2 + bxy + cy^2 + dx + ey + f = 0$, and imposing the condition $4ac - b^2 = 1$ in order to obtain an ellipse, the method finds the coefficients a to f as the normalized components of the eigenvector that corresponds to the smallest eigenvalue.

The method is fast because computations involve matrices of small dimensions. It requires as input at least six points on the contour. In the case of diatoms we may have hundreds to thousands of contour points. By leaving out some of them it is possible to vary the fitting precision according to the needs. With arithmetic formulae the geometrical parameters of the fitted ellipse can be obtained: coordinates of the center and foci, orientation, major and minor axes. Figure 2 shows a typical example when all contour points are used.

In order to obtain good fits the points which represent the contour must be uniformly distributed along it. All the contours extracted by image processing methods have to be preprocessed before applying the ellipse fitting. The contours are thinned, smoothed and subsampled by skipping every second contour pixel. The resulting contour points are used in the first step of the dynamic ellipse fitting presented in the next section. A concise ASCII format for saving and exchanging contours was adopted in the ADIAC project. It consists of a sequence of (x, y) pixel coordinates obtained by following the contour in anti-clockwise order.

2.2 Dynamic ellipse fitting

In general, like in Fig. 2, the first-fitted ellipse does not yield the elliptical part of a diatom, although the human eye can perceive this. The points on the contour belonging to this elliptical part will be closer to the ellipse than some other points, for instance those at the poles of a pennate diatom. An iterative ellipse-fitting procedure has been developed to detect the points on the elliptical parts of contours. The implemented algorithm works as follows: (1) consider all the points of the contour and fit the first ellipse; (2) for all the points on the contour compute the distance to this ellipse; (3) select only those points on the contour for which the distance is less than 60% of the maximum distance (this is the initial selection criterion) and fit a new ellipse to the selected points; (4) repeat step 3 until the new ellipse is stable (10 iterations proved to be sufficient); (5) decrease the selection criterion from its initial value to 1 (the pixel distance), in decrements of 1, while repeating step 3. At the end of this process the last-fitted ellipse overlaps the elliptical part of the contour (see Fig. 3). The two parameters for this fitting, the percentage of the maximum distance (60%), and the number of repetitions (10) for each decreasing step, were selected after extensive trials.

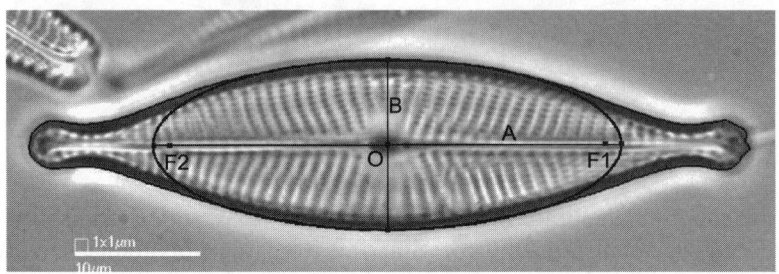

Figure 3. The last-fitted ellipse (O = center, A = major axis, B = minor axis, F1 and F2 = foci).

An essential part of the above algorithm is the computation of the distance from any point on the contour to a fitted ellipse. This distance can be the Euclidean distance expressed in pixels, but for this one has to find the corresponding point on the ellipse. We used an algebraic distance based on the property of an ellipse to be the set of all points in a plane for which the sum of the distances to its foci is a constant and equal to the length of its major axis. Hence, for a point on the contour the two distances to the foci are summed and the length of the major axis is subtracted. This distance is zero for a point *on* the ellipse, positive for a point *outside* and negative for a point *inside* the ellipse. It is this property that makes it suitable for our dynamic ellipse fitting algorithm.

Tests involving a large variety of diatom contours showed that shapes having more than 30% of their contour points on the elliptical part (like pennate diatoms) will have it perfectly detected by the last-fitted ellipse, and the orientation of this ellipse can be used as a reference. If the diatoms are circular or almost circular, their contour is entirely detected as an ellipse with equal or almost equal major and minor axes. In these cases the orientation of the ellipse is arbitrary and cannot be used as a reference. If a contour has elliptic parts that make up less than 30% of the total contour, in some cases the last-fitted ellipse is still usable for further processing, but in most cases the position and orientation make it unusable. For instance, the mixed genera test set used in our identification experiments (see Section 4.1) has 8 problematic taxa out of 48 (almost 17%). Figure 4 shows two examples.

2.3 The characteristic profile

The new characteristic profile is similar to the centroidal one. The polar coordinate system is formed naturally by the center of the last fitted ellipse and its major and minor axes. For any angle the corresponding ray intersects the contour as well as the ellipse. The Euclidian distance d (expressed in pixels) between these two points is taken as the value of the characteristic profile at that angle. If the contour point is outside the fitted ellipse the value is positive, if it is inside the value is negative. Plotting the values for one (or better two) full rotations gives the characteristic profile (see Figs. 5 and 6).

The characteristic profile has the following advantages over the classical centroidal profile:

- Because the major axis of the last-fitted ellipse is used as the 0° reference the characteristic profiles of pennate shapes are already aligned and the matching will be easier. This implies rotation invariance and that no constraints are imposed on the diatom orientation in an input image. This saves time when capturing images with a digital camera on a microscope.

- In the case of occlusions on the contour that leave at least 30% of the elliptical parts intact, the last-fitted ellipse will still have the correct position, assuring robustness of the method.

- If the contours are very elongated, very flat ellipses are obtained and the special smoothing necessary for the parts of the contour near the centroid can be avoided.

The characteristic profile still requires star-shaped contours in order to have only one value for each angle. There is also the possibility of normalizing the characteristic profile, but this is not as straightforward as in the case of the centroidal profile. A normalization should provide some scale invariance. Determining the maximum distance to the fitted ellipse and dividing all the other distances by this will not yield size invariance. The range of distances to the fitted ellipse can be rather large and is not correlated with diatom size. Another option is to divide all

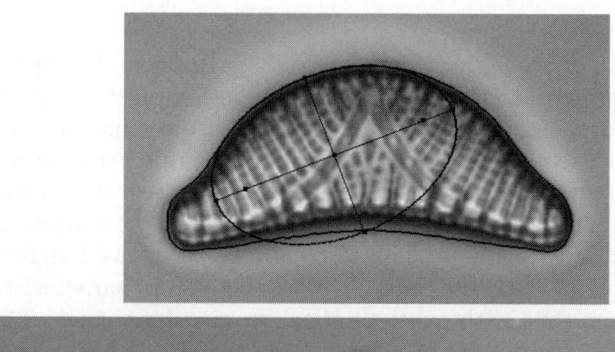
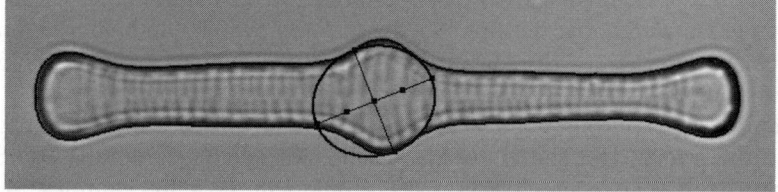

Figure 4. The last-fitted ellipse in the case of non-elliptical contours.

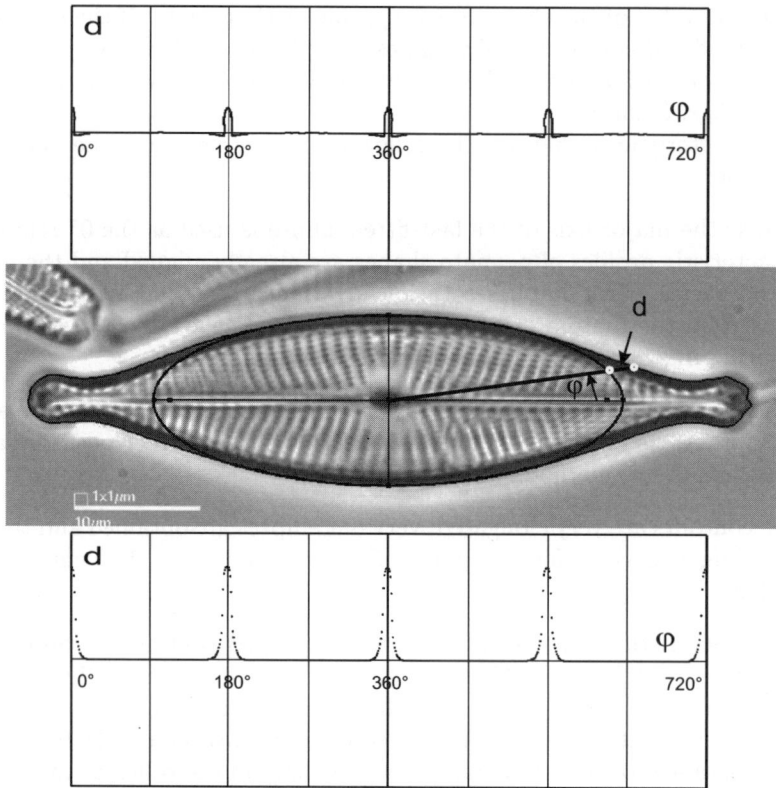

Figure 5. The characteristic profile computed with the first- (top) and last-fitted (bottom) ellipse.

distances by the length of the major axis (A in Fig. 3). Because of the possibility to obtain very flat ellipses with a large major axis, this option must also be discarded. A more convenient normalization factor is the length of the minor axis (B). In the tests done with neural networks and decision trees we verified scale invariance for diatoms that differ moderately in size using this length B (see section 4).

The characteristic profile has one main disadvantage: if the contour is close to a perfect ellipse the characteristic profile will be almost zero for all angles. Ellipses vary from round to elongated, but all will have the same characteristic profile. There are many diatom taxa with quasi-elliptical or -circular shapes. This implies the use of the morphometric feature called *ellipticity*, which can be derived from the geometrical parameters of the last-fitted ellipse, and which can be supplemented by other features like the absolute size.

The number of contour points depends on the size of the images. In order to normalize the size of the characteristic profile, we always sample the contour in steps of one degree, i.e. each characteristic profile will have 360 samples. Although we use the algebraic distance in the ellipse fitting, we use the Euclidean distance between the last-fitted ellipse and the contour for obtaining an accurate profile.

Chapter 9: Contour profiling and Legendre polynomials 173

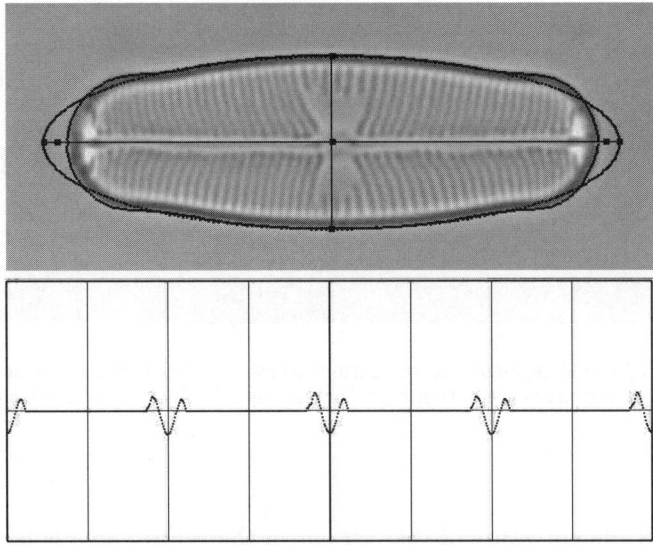

Figure 6. The characteristic profile of a *Sellaphora pupula* diatom.

2.4 Extraction of syntactical features

The 360 values of the normalized characteristic profile can be used directly in a matching task. Since the characteristic profile is translation, rotation and scale invariant, it can be used as the input to any classification method (neural networks, decision trees). Here we extract higher-level contour information in order to build a syntactical classifier.

We use the characteristic profile values without any normalization. First we apply a lowpass (box) filter of size 3 in order to smooth local peaks. Then, in order to eliminate small parts that are close to zero, we force to zero all the values that are smaller than 2 (pixels). This transforms the profile into a sequence of segments that are zero (belonging to the elliptical contour parts), positive (contour parts outside the last-fitted ellipse) and negative (inside this ellipse). The position of each segment is determined by finding all the zero crossings. If, for example, positive values decrease to a zero value followed by at least three consecutive zero values, this is also considered a zero crossing. Then each pair of successive zero crossings is tested. If all the values within a pair are zero then the segment is elliptical. If all the values are positive we have an outside segment and its maxima and minima are determined (always one more maximum). If all the values are negative we have an inside segment and we determine its minima and maxima (always one more minimum).

Concerning the significance of each segment type, let us consider a common pennate diatom. The zero segments correspond to the central elliptical part of the contour that is also part of the last-fitted ellipse. The curvature of the contour in the central area determines the curvature of the fitted ellipse and therefore its horizontal size. The main positive or negative segments correspond to the poles of

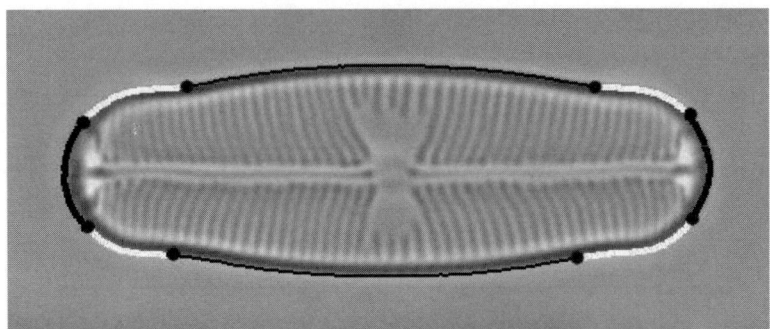

Figure 7. The eight final segments for the diatom whose last-fitted ellipse was already presented in Fig. 6. Elliptical segments are in thin black, outside ones in thick white and inside ones in thick black.

the diatom. If the curvature of the contour in the central area is large, the fitted ellipse will have a small major axis and the poles will be outside it. If the central parts have a low curvature or are close to a straight line, then the fitted ellipse will be elongated and the poles will be found inside it. Other small positive or negative segments can appear along the contour, depending on the shape. Some occur at specific angles and are important for the identification. Others can appear accidentally due to occlusions or debris close to the diatom, i.e. the automatic contour extraction may produce locally false contours.

So far we have cleaned the characteristic profile, and the number of different contour segments can be in the range of 1 to 30 (figures based on tests using the *Sellaphora pupula* set, see Section 4.1). The upper limit is still too high for a syntactical classifier that is simple to implement. In order to reduce the maximum number of segments to a reasonable 12 we assign to each outside or inside segment a relative size (the number of contour points of the segment divided by the total number of contour points). Then we sort these segments in descending size order and keep the biggest six (or seven if the last segments have the same size). Since there cannot exist more than one elliptical segment for each outside or inside segment, the total number or segments will be less than 12 (or 14). In our tests we found that negligible segments were always eliminated, but sometimes small significant positive or negative segments were also eliminated. A segment is eliminated by forcing its profile values to zero, followed by merging it with the neighboring elliptical segments. On the other hand, the most significant segments were always selected for the group of 12.

The next step is to construct the syntactical contour description: this is simply a list of at most 12 (or 14) segments as they appear on the contour. In the list, ordered by ascending angles, each segment is specified by its properties: type (elliptical, outside, inside), starting angle, peak angle, ending angle and the peak amplitude (%) relative to the length of the minor axis B. For outside and inside segments the peak angle is where the maximum or minimum in the profile is found. If a segment has more than one peak, the largest two are selected. If the smallest of the two peaks is at least 90% of the largest one, the peak angle is the midpoint and the peak amplitude equals the average of the two peaks; otherwise the segment is treated as a one-peak segment. Elliptical segments have their peak angle at the middle of the starting and ending angles and they do not have an amplitude since all their values are zero. Table 1 shows an example of a syntactical list (the corresponding diatom image is shown in Figs. 6 and 7).

Table 1. The syntactical description of the *Sellaphora pupula* diatom contour (Figs. 6, 7).

           1 - outside   (10 - 15 - 21), amplitude   13.962%
           2 - elliptical (21 - 88 - 156)
           3 - outside   (156 - 163 - 170), amplitude   16.003%
           4 - inside   (170 - 180 - 190), amplitude -26.234%
           5 - outside   (190 - 195 - 200), amplitude   13.501%
           6 - elliptical (200 - 267 - 334)
           7 - outside   (334 - 344 - 350), amplitude   20.925%
           8 - inside   (351 -   0 - 10), amplitude -24.881%

3 Legendre polynomials

Apart from using the characteristic profile there are other methods that allow to model a contour directly, such as Chebyshev and Legendre polynomials. As shown by Stoermer and Ladewski (1982) one can model the shape of diatom valves by fitting Legendre polynomials. We have adapted this method so that it works for more shape types, i.e. not only for diatoms with apical symmetry, taking advantage of the fact that most pennate diatoms in our database are already in a horizontal position.

Legendre polynomials form an orthogonal set of functions and are defined on the interval $[-1, 1]$. The set can be easily obtained by the following recurrence relation (Abramowitz and Stegun, 1970):

$$(n+1)P_{n+1}(x) = (2n+1)xP_n(x) - nP_{n-1}(x)$$

with $n = 1, 2, \ldots$ and $P_0(x) = 1$ and $P_1(x) = x$. This yields the functions $P_2(x) = (3x^2 - 1)/2$, $P_3(x) = (5x^3 - 3x)/2$, etc. Then, a partial Legendre polynomial approximation like

$$W(x) = \sum_{n=0}^{k} a_n P_n(x)$$

can be used to model one half of a valve contour with sufficient precision. The coefficients $a_0, a_1, \ldots, a_k$ qualify as shape descriptors, like the Fourier coefficients (Chapter 7).

The implementation starts by normalizing the contour. Because the contours are already horizontally oriented, the extreme left and right contour points are

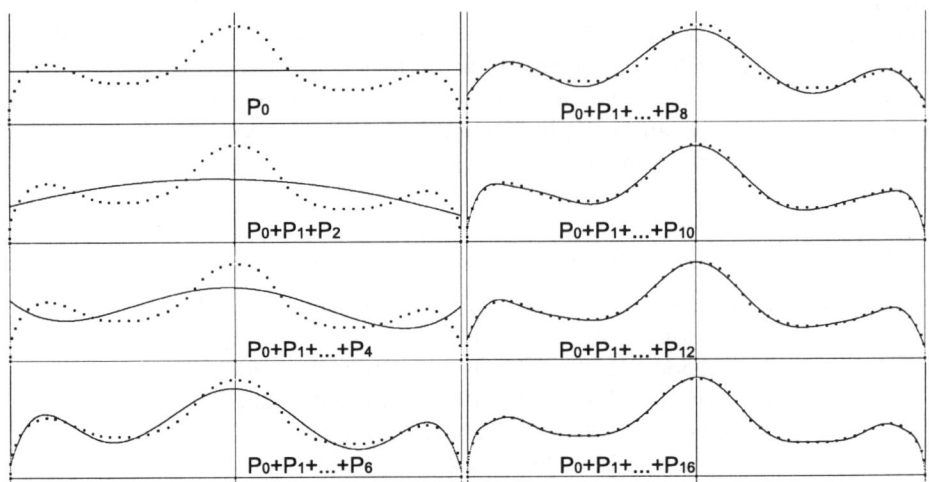

Figure 8. An upper contour part modeled using 17 Legendre polynomials.

detected and the middle of the line segment through these two points becomes the origin of a new Cartesian system. All the coordinates of the contour points are computed relative to this system. Half the length of the line segment mentioned above is used to normalize the contour points, i.e. they will be inside the unit square $[-1 \leq x, y \leq 1]$ and the two extreme points will assume the coordinates $(-1, 0)$ and $(1, 0)$. These two points are the start and the end of two curves, the upper part and the lower part of the contour. This allows the method to be allied to diatoms without apical symmetry.

Next, 60 equidistant points are selected on each of the two curve parts and these are used for the two Legendre approximations. We use Numerical Recipes' (Press et al., 1999) singular value decomposition. The main routine is svdfit, which in turn calls the svdcmp and svbksb routines, whereas the Legendre polynomial values are generated by the fleg routine. Tests using different shapes yielded very good fits (error $\chi^2 < 0.01$) for each half contour curve and as much as 17 shape descriptors, giving a total of 34 parameters. Figure 8 shows the contribution of successive Legendre polynomials to the fitting. We observed that the odd coefficients have always very small values for all contours in our test sets (see Section 4.1), hence we decided not to use them, which leaves 18 shape descriptors for each entire contour.

4 Experimental results

In this section we describe the two diatom test sets, the classifiers used, as well as the results obtained.

Table 2. Subsets of the mixed genera set.

taxa	total no. images	no. images per taxon		
		average	minimum	maximum
48	1009	21	11	29
46	962	21	11	29
23	469	20	11	25
17	359	21	20	25

4.1 Diatom test sets

The first set consists of 120 diatoms belonging to the *Sellaphora pupula* species complex (see Chapter 4). There are 20 diatoms for each of six different demes: "elliptical," "large," "pseudocapitate," "pseudoblunt," "cf. rectangular" and "tidy". The set was randomly divided into equally-sized training and test sets. Since the shapes are very similar, this set is ideal for testing the ability to deal with very small differences in shape and ornamentation. For this set we have very good contours which were extracted semi-automatically at RBGE. Figure 6 in Chapter 4 shows representative examples of the demes.

The second set is more complex. The latest version[a] has a total of 1009 images belonging to 48 taxa (see Chapter 4). Most taxa are represented by about 20 samples, but some have more than 20 whereas one has only 11. In this mixed genera set there is a large diversity in both shape and ornamentation. With respect to shape, there are different taxa with almost identical shapes and there are diatoms with different shapes belonging to the same taxon (due to processes related to life cycle–see Chapter 2). For this reason we analyzed this set in detail and selected subsets. First we selected 46 taxa by eliminating *Nitzschia* sp. 2, which is extremely similar to *Nitzschia hantzschiana*, and *Gyrosigma acuminatum*, which has a sigmoidal shape for which both the characteristic profile and the Legendre polynomials are not well suited. From the remaining 46 taxa we selected 23 in order to get only one type of shape per taxon. These form the third subset. Each taxon not included in this subset has more than one type of shape and should be split into subcategories containing only one shape. This will require extra samples (up to 20 per subcategory) in order to get reliable classification results. We also selected a fourth subset of only 17 taxa. In this subset of the 23-taxa subset, there is only one taxon with an almost elliptical shape (i.e. *Cocconeis placentula* and *Cocconeis stauroneiformis* were eliminated). *Fallacia* sp. 5 was eliminated because it is similar in shape to *Fallacia forcipata*, *Navicula gregaria* because it is represented by only 11 images and *Parlibellus delognei* as well as *Petroneis humerosa* because their contours were not precisely extracted and their shape accidentally resembles shapes found in other taxa. Table 2 presents a summary of the characteristics of the four subsets, whereas Table3 details the taxon names (see Chapter 4 for full names). For these subsets we have only automatically-extracted contours (see Chapter 6). Equally-sized training and test sets for each subset were created randomly.

---

[a] In other chapters a previous, smaller version with 37 taxa and 781 images is used.

Table 3. Taxa in the subsets of the mixed genera set, separated by horizontal lines, and the number of images per taxon. Chapter 4 lists authorities.

nr.	taxon	im.	nr.	taxon	im.
1	*Caloneis amphisbaena*	20	24	*Diatoma mesodon*	26
2	*Cocconeis neothumensis*	20	25	*Diatoma moniliformis*	20
3	*Cymbella hybrida*	20	26	*Gomphonema augur*	20
4	*Cymbella subequalis*	21	27	*Gomphonema* sp.1	20
5	*Denticula tenuis*	22	28	*Navicula reinhardtii*	29
6	*Fallacia forcipata*	24	29	*Sellaphora bacillum*	20
7	*Gomphonema minutum*	25	30	*Fragilariforma bicapitata*	20
8	*Navicula capitata*	20	31	*Opephora olsenii*	20
9	*Navicula constans*	23	32	*Staurosirella pinnata*	20
10	*Navicula menisculus*	20	33	*Tabularia investiens*	20
11	*Navicula lanceolata*	21	34	*Cymbella helvetica*	26
12	*Navicula radiosa*	21	35	*Encyonema silesiacum*	25
13	*Navicula rhynchocephala*	20	36	*Eunotia tenella*	21
14	*Navicula viridula*	20	37	*Eunotia incisa*	20
15	*Stauroneis smithii*	20	38	*Meridion circulare*	20
16	*Surirella brebissonii*	22	40	*Tabellaria quadriseptata*	23
17	*Tabularia* sp.1	20	39	*Pinnularia kuetzingii*	21
18	*Cocconeis placentula*	20	41	*Epithemia sorex*	20
19	*Cocconeis stauroneiformis*	22	42	*Nitzschia hantzschiana*	20
20	*Fallacia* sp.5	17	43	*Nitzschia dissipata*	20
21	*Navicula gregaria*	11	44	*Nitzschia sinuata*	20
22	*Parlibellus delognei*	20	45	*Eunotia denticulata*	22
23	*Petroneis humerosa*	20	46	*Tabellaria flocculosa*	20
			47	*Nitzschia* sp.2	27
			48	*Gyrosigma acuminatum*	20

4.2 Classifiers

We applied three classification methods:

- *Neural Networks:* We used a classical back-propagation neural network with two hidden layers. We did not use a momentum term to accelerate the training, but we changed the gain factor, when necessary, based on the training progress. We imposed training target levels of 0.91 and 0.09, which are less demanding than the ideal 1 and 0. The training process was stopped when the mean-square error of the outputs became less than 0.1. In some cases the training did not reach this condition because of conflicting input patterns. In the case of the *Sellaphora pupula* set we identified the diatoms which produced the conflicts and retrained the networks without them. In the case of the mixed genera set we let the networks train until the mean-square error stabilized at a small value.

In applying the trained networks to the test sets we used the following conventions: (1) if one output exceeds 0.5 and all others are less than 0.5, we either have

Chapter 9: Contour profiling and Legendre polynomials

Table 4. Neural network configurations for the mixed genera set.

taxa	characteristic profile	Legendre polynomials
48	$40 \times 347 \times 193 \times 48$	$18 \times 347 \times 193 \times 48$
46	$40 \times 347 \times 193 \times 46$	$18 \times 347 \times 193 \times 46$
23	$40 \times 273 \times 127 \times 23$	$18 \times 273 \times 127 \times 23$
17	$40 \times 227 \times 103 \times 17$	$18 \times 227 \times 103 \times 17$

a recognized or a misinterpreted contour; (2) if all outputs are below 0.5 we reject the contour; (3) if more than one output exceeds 0.5 we have a misinterpretation.

Because the characteristic profile provides a large number of values (360) we grouped them into only 40 values by averaging each set of 9 values (corresponding to 9 consecutive degrees). The sets of 9 degrees were chosen to preserve important information at the diatoms' poles. Comparative tests using 360 and 40 values showed that there is no decrease in the recognition rate. Hence, we will present in this chapter only results generated by neural networks and decision trees which use 40 input values.

For the *Sellaphora pupula* set we used two configurations: $40 \times 117 \times 67 \times 6$ (40 input neurons, 117 neurons in the first hidden layer, 67 neurons in the second hidden layer and 6 output neurons) for the characteristic profile and $18 \times 97 \times 47 \times 6$ for the Legendre descriptors. For the mixed genera set the configurations are presented in Table 4. In all cases the number of input and output neurons was imposed by the number of features and the number of classes (taxa). The number of hidden neurons was selected after several training tests. The training and test sets were randomly generated for each of the four mixed genera subsets. They are different for the two types of input features.

The neural networks using Legendre shape descriptors trained much slower than those using the characteristic profile values. Reasons for this are the narrow range of values of the shape descriptors, as well as the fact that many characteristic profile values are zero for elliptical contour parts, even in the compressed version with only 40 values.

• *Decision Trees:* We used Quinlan's (1992) C4.5 software package for the mixed genera set and the more recent C5.0 package for *Sellaphora pupula* (see Chapter 7). The mixed genera set and its subsets were again randomly split into training and test sets. C4.5's xval script allows to do this automatically by x-fold cross-validation. We used a 2-fold cross-validation, which splits an entire set into two equally-sized subsets, does trainings and tests on both possible combinations and gives the average test result as the recognition rate. This is similar to the procedure used for the *Sellaphora pupula* set. We also used a 10-fold cross-validation in which case an entire set is split into 10 equally-sized subsets. For each of these subsets the training is done using the other nine subsets and the testing is done on the subset itself. The recognition rate is the average of all 10 trials.

- *Hand-optimized Syntactical Classifier:* In the syntactical or structural approach a certain pattern (class) is expressed as a composition of subpatterns, also called pattern primitives (Fu, 1983). In our case these primitives form a list of contour segments obtained by processing the characteristic profile, as explained in Section 2.4. The list can be seen as a syntactical description or a sentence. The syntactical approach tries to recognize a sentence (contour) by analysing the structure using a set of syntactical rules. The procedure applied to the *Sellaphora pupula* set is as follows.

First, the 60 images in the training set and their associated lists of contour segments were analyzed: the total number of segments per contour, the number of main (pole) segments and their type (in/outside the last-fitted ellipse), position and amplitude, the number of the other, usually smaller, segments, the possible angles where these small segments can occur, the possible amplitudes of such small segments, the symmetry relations between segments of the same type, etc. We found similar features for all demes; these do not allow to define unique, mutually exclusive patterns for each deme. The diatomists involved acknowledged that shape can be an ambiguous character for some demes, in which case a final decision is taken after width measurements or ornamentation analyses. We implemented a classifier using six sets of if-then rules, each set dedicated to one deme. The rules are tested sequentially, starting with higher level features (like the number of segments) and ending with detail features (like the positions of the small segments). If a condition is not verified by the syntactical contour description, the rule set is abandoned with a message containing the particular value of the parameter that broke the rule. If all the conditions for a deme are passed, a message stating that the contour pertains to that deme is issued.

Second, detailed numerical criteria were established for each if-then rule. Each set of rules was tried sequentially using only the training contours from the corresponding deme until all 10 contours were correctly identified. Rule breaking messages made the task easier, pointing out the criteria that had to be adjusted in order to reach this goal.

Third, each rule set was tried also using the training contours belonging to the other demes. Again, using rule breaking messages, the criteria were fine-tuned in order to minimize the number of training contours that pass more than one rule set, i.e. contours that could pertain to two or more demes.

Once the rule sets were established, there were three posibilities for a test contour: (a) the features of the contour pass only one rule set corresponding to one deme, which can be either correct or a misinterpretation; (b) they only partially pass all the six rule sets, in which case the contour is rejected, possibly because of deformations due to the automatic contour extraction (debris or occlusions), and (c) the features of a contour pass two or more rule sets. This is a misinterpretation that can or cannot be solved using features other than the ones contained in the syntactical description.

To deal with situation (c) we used the feedback from the diatomists and added two rules triggered by contours that could belong to more than one deme after parsing the first six rule sets. One rule tests for the diatoms' width and makes a distinction between "pseudoblunt" and "pseudocapitate" demes. The other rule

Table 5. Recognition rates (%) for the *Sellaphora pupula* set, using only contour features.

	neural network		decision tree (C5.0)	
	characteristic profile	Legendre polynomials	characteristic profile	Legendre polynomials
training	100	100	100	100
reverse training	100	100	100	100
testing	91.7	85	93.3	76.7
reverse testing	85	85	85	85
average testing	88.3	85	89.2	80.8

focuses on stria orientation, a helpful feature for the separation of "elliptical" and "pseudoblunt" demes. The width feature corresponds to the absolute diatom width in μm, whereas the orientation feature is the mean stria orientation measured in the upper right quadrant of each contour (see Chapter 10).

The creation of a hand-optimized classifier is quite difficult and costs a lot of time. However, once such a classifier has been developed on the basis of a sufficient number of training samples, it does not require any re-training when adding more classes. Neural networks and decision tree applications like C4.5 have to be re-trained when new classes are added.

4.3 Results for the Sellaphora pupula set

Table 5 shows the neural network and C5.0 results for the *Sellaphora pupula* set. Reverse training and testing rates were obtained by swapping the training and test sets and an average testing rate is also given. The neural networks and the C5.0 decision trees were both able to train perfectly, resulting in 100% discrimination. The average test results are slightly better for the characteristic profile features. But it can be seen that the reverse testing gave the same 85% recognition rate in all four cases, i.e. the particular choice of the training and test sets can be the cause of the differences.

Table 6 presents the recognition results obtained with the hand-optimized syntactical classifier. Using only the syntactical description based on the characteristic profile, a recognition rate of 88.3% was obtained on the training set (see Section 4.2 for details). There were no rejected diatoms, but there were diatoms which passed both the "elliptical" and "pseudoblunt" rule sets or both the "cf. rectangular" and "pseudocapitate" ones. The supplementary rule concerning the width achieved a complete discrimination between the "cf. rectangular" and "pseudocapitate" demes, and the other rule for the stria orientation resolved the "elliptical"/"pseudoblunt" discrimination problem. Hence a 100% discrimination rate could be obtained, but with the help of features not related to shape.

Then the syntactical classifier optimized on the training set was applied to the test set. 75% of the diatoms were correctly recognized. The others were rejected (no misinterpretations were recorded), which implies that the supplementary rules for width and striation were not triggered (hence the dashes in Table 6). It also means that 15 of the 60 test diatoms have characteristics that are not represented in the training set. The syntactical classifier is not able to generalize like a neural network or a decision tree: the generalization has to be explicitly included in the rule sets.

This was corroborated by further fine-tuning the syntactical classifier, taking into account the characteristics of all diatoms (training on the entire set of 120 diatoms). Using all features, i.e. contour, width and striation, only one diatom was misinterpreted. It passed both the "pseudoblunt" and "cf. rectangular" rule sets, and the diatomists agreed that the discrimination between these two demes can be problematic even for experts. The additional features again resolved the discrimination problems for the "cf. rectangular"/"pseudocapitate" and "elliptical"/"pseudoblunt." After training with the entire set of 120 diatoms, the recognition rate approached 100%, indicating that the rules were close to optimal. This also suggests that a near 100% recognition rate for new samples of the *Sellaphora pupula* demes can be expected.

4.4 Results for the mixed genera set

Tables 7 and 8 show recognition results for the mixed genera set, using the 40 values of the characteristic profile and the 18 Legendre shape descriptors. Since this set has shape ambiguities that pose problems for both the characteristic profile method and Legendre polynomials, different subsets with fewer ambiguities were tested (see Section 4.1).

Table 7 summarizes the results obtained with neural networks, where training and testing were carried out by splitting the (sub)sets randomly into two equal parts. By eliminating more and more shape ambiguities, i.e. using the subsets with fewer taxa, the results always improve when the characteristic profile features are used. This is not the case for the Legendre descriptors, although the smallest set with 17 taxa gave the best result (91.1%, compared to 87.8% obtained with the characteristic profile). The Legendre polynomials performed significantly better on

Table 6. Recognition rates (%) of the syntactical classifier for the *Sellaphora pupula* training and test sets, as well as for training on the entire set. C=characteristic profile only, CW=characteristic profile plus absolute width and CWO=characteristic profile plus absolute width and stria orientation.

	features	correct	wrong	reject
training	C	88.3	11.7	0
(60 diatoms)	CW	95	5	0
	CWO	100	0	0
testing	C	75	0	25
(60 diatoms)	CW	–	–	–
	CWO	–	–	–
training on	C	89.2	10.8	0
entire set	CW	95	5	0
(120 diatoms)	CWO	99.2	0.8	0

Table 7. Neural network recognition rates (%) for the mixed genera set.

	48 taxa	46 taxa	23 taxa	17 taxa
	characteristic profile			
training	95.6	96.0	97.4	100
reverse training	94.9	95.3	97.5	96.7
testing	72.2	74.7	77.4	86.9
reverse testing	71.1	72.6	78.4	88.7
average testing	71.7	73.7	77.9	87.8
	Legendre polynomials			
training	97.2	98.1	96.1	100
reverse training	97.7	96.9	97.5	100
testing	81.4	81.9	81.9	90.7
reverse testing	82.5	78.9	79.2	91.5
average testing	82	80.4	80.6	91.1

Table 8. Decision tree (C4.5) recognition rates (%) for the mixed genera set. The testing procedure using 2-fold cross-validation is comparable with the one used to make Table 7.

		48 taxa	46 taxa	23 taxa	17 taxa
		characteristic profile			
training	2-fold	89.3	88.5	93.2	96.1
training	10-fold	90.4	90.1	93.7	96.4
testing	2-fold	60.2	59.4	70.2	72.8
testing	10-fold	66.2	66.6	77.1	84.1
		Legendre polynomials			
training	2-fold	94.6	94.5	96.8	98.3
training	10-fold	95.6	95.8	97.3	98.6
testing	2-fold	71.5	70.2	78.5	87.7
testing	10-fold	72.6	77.6	81.2	89.7

all subsets, i.e. they are more appropriate for dealing with a large variety of shapes, whereas the characteristic profile proved to be better at discriminating between sometimes minute differences in shape (*Sellaphora pupula*, Table 5).

These conclusions are confirmed by the results obtained with the C4.5 decision tree (Table 8), although that classifier performs worse than the neural networks (Table 7). A final conclusion is that the best results for the mixed genera set were obtained by the neural networks in conjunction with Legendre shape descriptors, from 82% for all 48 taxa to 91.1% for the 17-taxa subset. More tests on both contour and striation features are reported in Chapter 10.

5 Conclusions

In this chapter we presented a new method for shape analysis, the characteristic profile, and a method for fitting Legendre polynomials to diatom contours. The feature sets were classified using three different approaches: neural networks, decision trees and a hand-optimized syntactical classifier. Results obtained with two very different data sets revealed that the chosen features have similar capabilities for characterizing diatom shapes. The characteristic profile has a small advantage when fine details have to be discriminated, whereas the Legendre polynomials are clearly better when a large variety of shapes needs to be processed.

With respect to the classification methods, neural networks proved to perform better than decision trees. A possible explanation could be based on the lack of images, only 10 per taxon for training. A larger number of images per taxon will help neural networks to perform better, but the same applies to all classifiers.

The syntactical classifier, i.e. a hand-optimized, rule-based system, was a special case in this context. Such a system is quite elaborate to develop, and can only be applied to a limited number of classes and features. The *Sellaphora pupula* set with 6 demes and a total of 120 images was an interesting test case. After obtaining a recognition rate of only 75% when training and testing on subsets of 60 images each, a further fine-tuning by using all the 120 images yielded almost 100%.

When using only contour features the neural network and decision tree results were not conclusive. The neural network in combination with the characteristic profile features yielded an 88.3% recognition rate for the very difficult *Sellaphora pupula* set. Using Legendre polynomials for a mixed set of very diverse taxa, it resulted in an 82% rate when considering all the 48 taxa, and in a 91.1% rate when problematic shapes were eliminated. However, it must be remembered that these are not final but rather *first* numbers. Further developments will undoubtedly improve the results.

References

Abramowitz, M. and Stegun, I.A. (1970) Handbook of mathematical functions. Dover Publications, New York (9th Edition).

Ciobanu, A., Shahbazkia, H. and du Buf, J.M.H. (2000) Contour profiling by dynamic ellipse fitting. Proc. 15th Int. Conf. on Pattern Recognition, Barcelona, Vol. 3, pp. 758-761.

Davies, E.R. (1997) Machine vision: theory, algorithms, practicalities. Academic Press, London, UK (2nd Edition).

Fitzgibbon, A., Pilu, M. and Fisher, R.B. (1999) Direct least square fitting of ellipses. IEEE Trans. Pattern Analysis and Machine Intelligence, Vol. 21, pp. 476-480.

Fu, K.S. (1983) Applications of pattern recognition. CRC Press, Boca Raton (FL), USA.

Jain, A.K., Duin, P.W. and Mao, J. (2000) Statistical pattern recognition: a review. IEEE Trans. Pattern Analysis and Machine Intelligence, Vol. 22, pp. 4-37.

Pejnovic, P., Markovic, M. and Stankovic, S. (2001) Shape classification by moment and autoregressive invariants. Int. J. of Pattern Recognition and Artificial Intelligence, Vol. 15, pp. 311-327.

Press, H. et al. (1999) Numerical recipes in C: the art of scientific computing. Cambridge University Press, Cambridge, UK (2nd Edition).

Quinlan, J.R. (1992) C4.5: Programs for machine learning. Morgan Kaufmann Publishers, San Francisco (CA) , USA.

Stoermer, E.F. and Ladewski, T.B. (1982) Quantitative analysis of shape variation in type and modern populations of *Gomphoneis herculeana*. Nova Hedwigia, Beiheft 73, pp. 347-386.

CHAPTER 10

IDENTIFICATION BY GABOR FEATURES

LUÍS M. SANTOS AND HANS DU BUF

This chapter presents a new contour feature set and algorithms for the extraction of ornamentation features. After dynamic ellipse fitting of the contour, the characteristic signal is extracted and analyzed by a multiscale Gabor line/edge detection with almost continuous frequency scaling, followed by a local stability analysis in order to eliminate unstable events. Each stable event is characterized by 11 scale-space features. The valve ornamentation is analyzed by improved models of cortical grating and bar cells in order to detect a raphe and striated regions. Within the striated areas averages of the local density and orientation are computed. The Gabor contour and ornamentation features are used for identification by Bayesian and decision-tree classifiers. Best results were achieved by combining the Gabor features with Legendre polynomials. An identification rate of 85% could be obtained in the case of the data set of 120 *Sellaphora pupula* diatoms of six demes. The mixed genera data set of 781 images of 37 different taxa could be identified with a rate of 88%. In both cases entire feature sets were used, but first results obtained with feature reduction showed that better identification rates can be expected by carefully selecting a small set of the best features.

1 Introduction

As for scale-space curvature (see Chapter 11) our contour approach is multiscale but it is inspired by a model of visual perception, i.e. brightness. Although there are different modeling approaches, for example explaining the existence of Mach bands (Pessoa, 1996), the idea is the following: du Buf and Fischer (1995) have developed and tested a one-dimensional brightness model in which a 1D pattern is analyzed by using complex Gabor filters (cortical simple cells) in order to extract basic features (lines and edges, also called events) at different scales. But the visual system does not *reconstruct* the input image; it constructs a symbolic representation that is suitable for a higher-level representation. The image that we see is a virtual one: nature devised virtual reality a long time before the ascent of modern technology. Hence, if a cell in the neural visual substrate fires, at a higher level it will be known that at the corresponding retinotopic position there is a certain event with an associated scale, orientation and amplitude. In other words, there is an *interpretation* that exists either of a generalized line with a Gaussian profile or a generalized edge with a truncated error-function profile. This model can explain Mach bands and other brightness illusions, but it can also be employed in order to characterize 1D signals derived from, for example, diatom outlines.

In another layer in the visual cortex cells have been found that respond either to periodic gratings or to isolated bars. This processing is highly nonlinear because a grating cell does not respond to an isolated bar and a bar cell does not respond to individual bars in a grating. Computational models of such cells can be applied in pattern recognition, i.e. the diatom ornamentation. Because the striation can vary in both density and orientation (Barber and Haworth, 1981), which implies that a normal homogeneity-based segmentation cannot be applied to extract the striated regions, a set of grating cells tuned to different frequencies and orientations can be used for this purpose. Since 2D Gabor filters are employed, local frequency and orientation information can be computed within the striated regions. The bar cell model can be applied in order to detect a raphe.

Our multiscale contour features are related to others, such as those from the well-known Gaussian scale-space approach based on the solution of the diffusion equation, taking into account entire contours or shapes (Abbasi et al., 2000; Lindeberg, 1994). However, these are general methods that create a scale space in which the position of the extrema over scales is represented. In our case we need to be able to link this information back to the original contours, in order to obtain a syntactical meaning of the features. Our approach introduces the notion of a multiscale syntax, but there are more differences: Firstly, instead of using the contour signal itself we use an improved version of the classical centroidal profile on the basis of the dynamically best-fitted ellipse—see Chapter 9 and Ciobanu et al. (2000). Secondly, instead of using a few Gaussian scale-space kernels, the centroidal profile is filtered with a set of almost continuously scaled complex Gabor filters that allows the extraction of line and edge events at all scales. A grouping process eliminates detected events which are not stable over a contiguous range of scales. From the remaining stable events a number of features are selected that describe them over both positions and scales.

With respect to the ornamentation, some normal texture measures have been used to characterize the striation—see Chapter 7 and Fischer et al. (2000). Here our aim is twofold: first we want to detect the striated areas within the valve outline, and then, within the striated areas, we want to estimate the local frequency and orientation to gather their statistics, as well as determining the symmetry type. As far as we know, apart from the texture measures referred to above, ornamentation analysis has never been applied to diatoms. The same holds for the detection of other structures, like a raphe.

The rest of this chapter is organized as follows: in Section 2 we describe the contour processing and feature extraction, and in Section 3 the grating and bar cell models, followed by frequency and orientation estimation. In Section 4, identification results obtained with different feature sets, classifiers and diatom sets are presented. We conclude with a final discussion in Section 5.

2 Contour processing and feature extraction

This section deals with all preprocessing necessary to prepare characteristic contour signals, the Gabor scale space and line/edge detection, as well as the features to be used in the identification experiments. All consecutive steps are explained in the following subsections.

Chapter 10: Gabor features

2.1 Contours and the characteristic signal

After preprocessing an image to correct the inhomogeneous illumination and to normalize the contrast, the contour is extracted by thresholding and, if necessary, postprocessed to reconstruct continuity (see Chapter 6). Then a first ellipse is fitted to all contour points. Since most pennate diatoms are not purely elliptical, the entire ellipse will cover only a small part of the contour. By making the ellipse fitting dynamic, it is forced to cover exactly the elliptical part of the contour. Subsequently, the center of the ellipse is used to sample the contour at integer angles from 0° to 359° relative to the major ellipse axis (no image rotation is required) and for each contour sample the distance to the two ellipse foci minus the length of the major axis is computed. This distance will be zero where the ellipse covers the contour and unequal zero at the other positions, i.e. the valve endings. The maximum or minimum of the characteristic signal (also called characteristic profile) is normalized to obtain size invariance in the shape description, although the absolute diatom size will also be available for the identification. All this processing is described in more detail in Chapter 9 and Ciobanu et al. (2000). As a result, we obtain the characteristic signal

$$-1 \leq S_c(\theta) \leq 1 \ , \ \theta \in [0 \ldots 359]. \tag{1}$$

Figure 1 (left) shows a diatom image with the contour in black and the dynamically best-fitted ellipse in white, as well as its normalized characteristic signal (right). The latter shows the deviations from a pure ellipse, i.e. negative when a contour point is inside the ellipse and positive when outside, around 0 (360) and 180 degrees. This signal provides the improved information that we need for identification (classification) based primarily on the shape of the valve endings. Figure 2 shows another example.

2.2 Gabor filtering and the concept of 1D lines and edges

In our scale space we use a type of filter that has been found in the primate visual system and that can be easily modeled. Complex Gabor filters are quadrature filters with an even-symmetric real part and an odd-symmetric imaginary part. This structure allows the visual system not only to detect positions of lines and edges but also to make a distinction between positive and negative lines as well as edges. It has been shown that the local phase information provides more stable information for position detection than the local maxima of the modulus (du Buf, 1993 and 1994), and that a multiscale line/edge representation allows the modeling of certain visual effects like Mach bands (du Buf and Fischer, 1995). Since we are working with 1D characteristic signals, we use 1D Gabor filters defined by

$$\begin{aligned} G(x|\sigma,\omega,s) &= \exp\left(-x^2/2s^2\sigma^2 + i\omega x/s\right) \\ &= \exp\left(-x^2/2s^2\sigma^2\right) \cdot \left(\cos(\omega x/s) + i\sin(\omega x/s)\right), \end{aligned} \tag{2}$$

where σ is the size of the Gaussian envelope, ω is the frequency, s is a scaling parameter and $i = \sqrt{-1}$. From Eq. 2 we can see that these are Gaussian-modulated periodic functions. If $R_s(x)$ is the complex cyclic convolution (denoted by $*$) of a

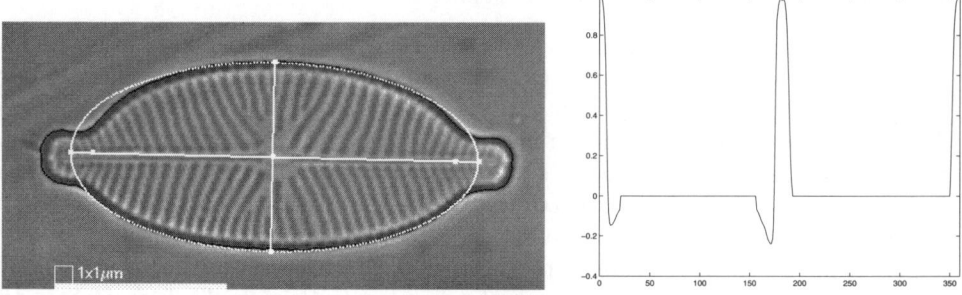

Figure 1. *Navicula constans* with its contour and best-fitted ellipse (left), and its characteristic signal (right).

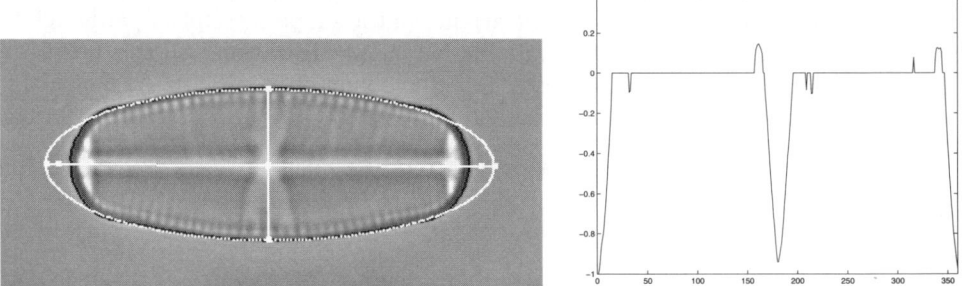

Figure 2. *Sellaphora pupula* deme "pseudoblunt," focused on the outline (left), and its characteristic signal (right).

Gabor filter at scale s with the characteristic signal $S_c(x)$, using x instead of θ, then we can write

$$R_s(x) = S_c(x) * G(x|\sigma, \omega, s). \tag{3}$$

Although presented here in the x domain, the filtering for 256 scales is actually done in the frequency domain because it is much faster: instead of 256 convolutions with increasing kernel sizes it requires one forward FFT (Fast Fourier Transform) and 256 inverse FFTs, plus 256 1D matrix multiplications. Since the size of the characteristic profile is always 360 (not a power of 2), we use a special library (fftw) which implements Rader's algorithm, among others (Frigo, 1999).

The Gabor line/edge analysis will be done in one dimension, which means that we are not really dealing with lines and edges in 2D images but only with 1D

Figure 3. Left to right: ideal line, generalized line, ideal edge, generalized edge (all positive).

Chapter 10: Gabor features

Table 1. Real and imaginary inputs for the grouping operators.

type	a	b	c
positive line (pl)	-Im	Re	Im
negative line (nl)	Im	-Re	-Im
positive edge (pe)	-Re	-Im	Re
negative edge (ne)	Re	Im	-Re

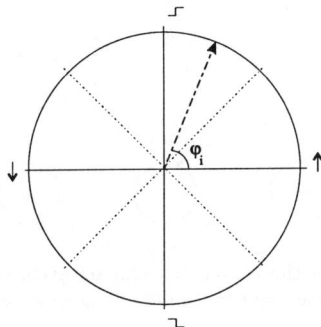

Figure 4. Phase gating for line and edge types.

cross sections. Mathematically, ideal and generalized (Gaussian-scaled) events are defined as follows (see also Fig. 3): ideal positive and negative lines are described by Dirac functions $\pm \delta(x)$, and generalized lines by Gaussian functions

$$\delta_g(x) = G(x|\sigma, 0, 1)/\sigma\sqrt{2\pi}. \tag{4}$$

Hence, a Dirac function follows from

$$\delta(x) = \lim_{\sigma \to 0} \delta_g(x). \tag{5}$$

In the same way, a generalized positive edge is defined by means of the error function

$$\phi_g(x) = \Phi(x/\sigma\sqrt{2}) \tag{6}$$

with

$$\Phi(x) = \frac{2}{\sqrt{\pi}} \int_0^x e^{-t^2} dt, \tag{7}$$

and an ideal positive edge is the step function

$$\phi(x) = \lim_{\sigma \to 0} \phi_g(x). \tag{8}$$

A characteristic diatom signal will never contain these ideal or generalized events. However, by introducing a multiscale Gabor filtering as well as a line/edge detection at each scale, the signal will be described by generalized events over multiple scales. This is how our visual system works and this will be exploited here in our contour feature extraction.

Having the complex Gabor filter response according to Eq. 3 at each scale, different line and edge detection schemes are possible. For in-depth discussions we refer to du Buf (1993, 1994) and du Buf and Fischer (1995). Basically, event positions can be detected by searching for local maxima of the modulus, and the phase at these positions determines the event type (phase gating, see Fig. 4). Because the modulus will be distorted when two events are too close, a better scheme is to detect a local maximum in one part of the response (either real or imaginary) and a corresponding zero crossing in the other part. Here we present a hybrid scheme based on the concept of grouping (Grossberg, 1999a,b).

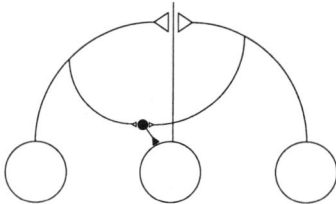

Figure 5. In the grouping model the center cell can be excitated by its two neighboring cells through the open triangles, whereas the small solid triangle causes a smaller inhibition to prevent oversaturation.

2.3 Line/edge detection and scale stability

This method starts by detecting local maxima of the modulus response, but then searches the neighborhood using maxima and zero crossings of the real and imaginary responses. To this end the grouping is applied to all positions in the scaled interval $[-3s\sigma, 3s\sigma]$ for each scale s, and the best event position is selected. At every position in the interval, the grouping is done according to the model shown in Fig. 5, combining the real and imaginary response parts according to Table 1, to detect the four event types. An event position is selected where the grouping has the strongest match with one of the event types. This grouping is described by

$$b_o = a_i + b_i + c_i - \max\{a_i, c_i\}. \qquad (9)$$

The cells a and c are located at positions $\pm\lambda/4$ from the center cell b, with $\lambda = s/\omega$ of the Gabor filter. The subscripts o and i stand for output and input respectively. Hence, each cell receives input from the real or imaginary part according to Table 1 and the cell b is also excitated and less inhibited (the max term in Eq. 9) by its neighbors. This is a very selective process because of the stronger excitation: for example, exactly *on* a positive line the real response is maximum and positive, whereas the imaginary responses at $\pm\lambda/4$ are also maximum but positive and negative, respectively. At each position the set of four grouping operators $S = \{G_{pl}, G_{nl}, G_{pe}, G_{ne}\}$ is applied, and the event type is determined by the largest response, i.e. $\max\{G_i\}$. Finally, this position is labeled with the event type as well as the amplitude. This process is repeated for all local modulus maxima and all filter scales s.

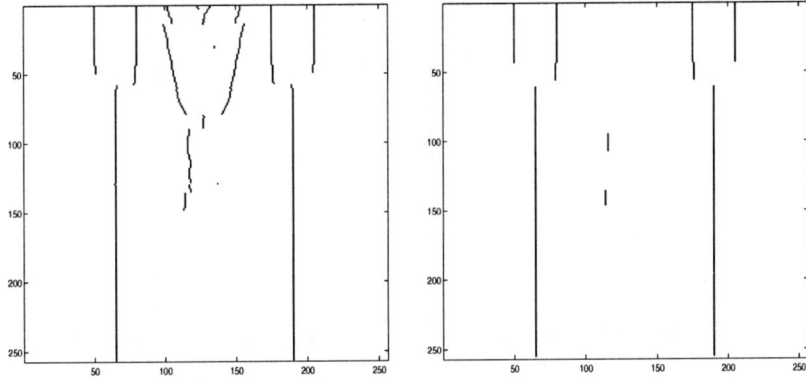

Figure 6. Lines and edges detected before (left) and after (right) the stability analysis in the case of a simple trapezoidal signal.

Figure 6 (left) shows the almost continuous scale space after the analysis of a simple trapezoidal signal with two plateaus connected by two ramped transitions. The smallest scale is at the top and we used 360 scales. At the smaller scales the 4 ramp edges (discontinuities) are detected and correctly labeled as positive and negative lines—the Gabor phase at a ramp edge corresponds to that of a line, which explains the Mach-band effect, see du Buf (1994). At larger scales the two ramped transitions are detected and labeled as edges (the figure does not show the event types, only the positions in black).

In a single-scale detection, but applied at different scales, there are also events detected that are not stable over neighboring scales due to Gabor response interference effects. Note that we do not apply a thresholding, i.e. noise can cause a detection of spurious events. To solve these problems, a scale stability analysis is performed such that only events that are stable over neighboring scales are preserved. In other words, an event detected at position x is stable if $\partial x/\partial s = 0$. In this stability analysis we apply the same type of grouping, in which the cells a, b and c are located at the same position x and the cells a and c are positioned at scales separated by Δs. The difference between the scales of the cells a and c and the center cell b allows to adjust the stability selectivity. In the case of a small difference (2 or 3 scales) only large instabilities are removed. Making the difference larger, e.g. to 5 scales, will also remove smaller instabilities as well as events that have been detected at less than 10 contiguous scales in this example. This grouping process cleans up the event map (Fig. 6, right), where a difference of ±5 scales has been applied. Figure 7 shows the results in the case of the characteristic signal from the *Navicula constans* specimen shown in Fig. 1. Figure 8 shows 3D plots of the event amplitudes (Gabor modulus) after the stability analysis for the trapezoidal signal (left) and the *Navicula constans* specimen (right). These amplitude variations over scales could be approximated by e.g. a second-order polynomial whose parameters complement the event position, type and scale information.

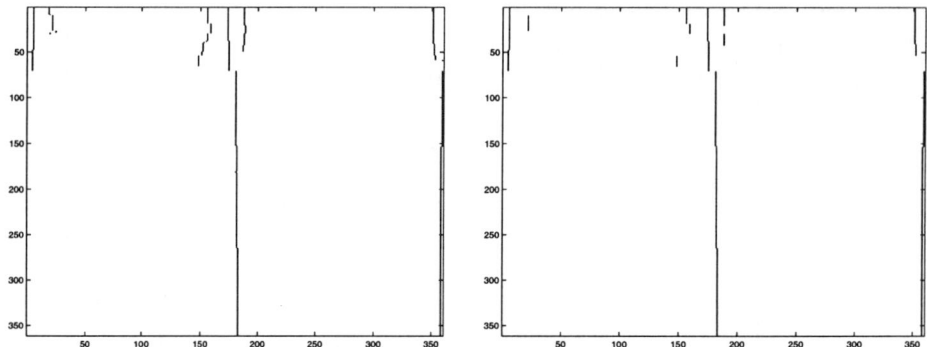

Figure 7. As Fig. 6 but applied to the characteristic profile of the *Navicula constans* specimen shown in Fig. 1 (right).

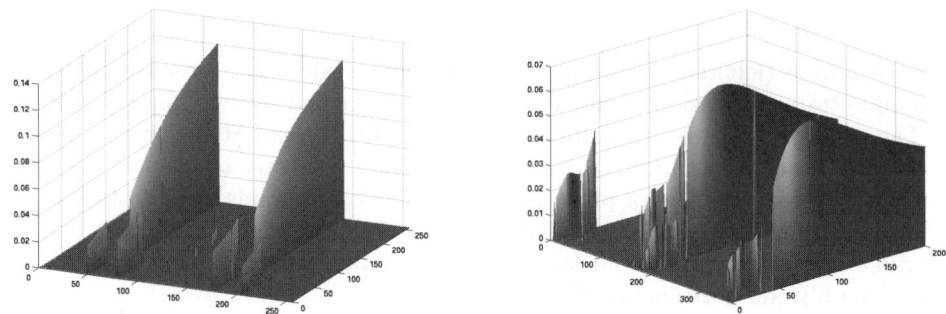

Figure 8. 3D plots of event amplitudes over scales for the trapezoidal signal (left) and the *Navicula constans* specimen (right).

Table 2. Features of the *Navicula constans* specimen sorted by the events' relevance. Only the 6 most relevant events from a total of 13 are listed.

no.	type τ_i	angle θ_i	I.Sc. s_i^I	F.Sc. s_i^F	I.Ampl Λ_i^I	F.Ampl Λ_i^F	MaxScale s_i^M	MaxValue Λ_i^M
1	PL	1	90	351	0.38	0.28	262	0.39
2	PL	181	121	351	0.39	0.40	198	0.43
3	PE	175	1	86	0.38	0.25	84	0.38
4	NE	6	1	81	0.30	0.20	80	0.31
5	PE	351	1	57	0.25	0.18	52	0.25
6	NE	188	1	38	0.10	0.12	5	0.14

Chapter 10: Gabor features

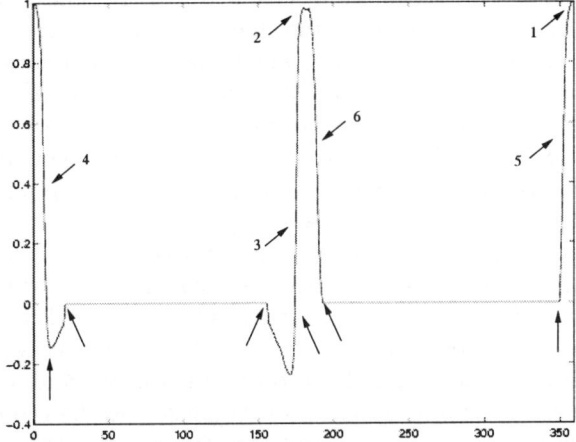

Figure 9. The characteristic signal and the positions which will generate significant events in the scale space, in the case of the *Navicula constans* specimen. The event numbers correspond to those in Table 2.

2.4 Syntactical features

The event map shown in Fig. 7 (right) is a very rich description of the characteristic signal because it includes the different event types at the different positions, the contiguous scale intervals as well as the event amplitudes at all positions and scales. Because of the almost continuous scaling, the individual line and edge features are too numerous to be used in a feature classification, i.e. diatom identification. In order to reduce the dimensionality we must select a few higher-level features that characterize the scale space as a whole, and therefore a contour. For each stable event (individual vertical line segment in Fig. 7, right), we select the event type $\tau_i \in \{1..4\}$, the angle θ_i of the event, its initial and final scales s_i^I and s_i^F, as well as the amplitudes at these two scales Λ_i^I and Λ_i^F. The event's amplitude is not constant over scales, usually it shows a maximum at a given scale. The scale at which this maximum is reached, as well as the amplitude's value at this scale, are represented by the features s_i^M and Λ_i^M respectively. The event type numbering corresponds to PL and NE etc, P/N meaning positive/negative and L/E meaning line/edge. Because there may be many events which are less relevant since they are stable over a relatively small scale interval, all events are sorted in relevance using the difference $s_i^F - s_i^I$. Insignificant events can be discarded and only the N most relevant ones used in the classification. In our experiments described below we use a threshold of $N = 6$. If the total number of events is smaller, then all are used of course. Table 2 presents the features of the most relevant events of the *Navicula constans* specimen used before.

It is important to explain the relevance of the small feature set for characterizing diatom contours, using the figures of the *Navicula constans* specimen. It was mentioned before that the characteristic signal is normalized, which means that the features are size invariant, although additional parameters from the first-fitted ellipse are available for e.g. size or ellipticity features. It is also important to keep in mind that the features are based on the characteristic signal: only the deviations from the best-fitted ellipse are reflected, which means for pennate diatoms the two nonelliptical valve endings (the signal will be zero in the case of perfectly elliptical contours). Hence, asymmetrical valve endings will lead to different events at angles around 0 (360) and 180°. Although we do not, at the moment, use a specific feature for the symmetry, any asymmetry will cause different amplitudes and scales. This can be seen in Figs. 1 and 9 as well as in Table 2. The *Navicula constans* specimen is *almost* symmetric about the vertical axis through the center, but not about the horizontal axis: the upper concave parts are more pronounced than the lower ones, and the left is larger than the right one. This yields a best-fitted ellipse that covers the lower contour part better than the upper. This small asymmetry is also reflected in Fig. 9 and Table 2: in the latter the 3rd and 4th events at 175° and 6° reflect the left flank of the right positive pulse and the right flank of the left positive pulse, respectively. The first and second events at 1° and 181° represent the two major line events representing the diatom poles. We can recognize the asymmetry referred to above. Looking at Fig. 9, we see that the negative minima are different. These smaller negative pulses are connected to the positive ones, they are in fact a continuation of these. The features in Table 2 do not include events directly related to the negative pulses, because they are not significant enough to pass the stability analysis and feature selection, but the 4th event (NE at 6°) is smaller in the range of scales (81 vs. 86) as well as in amplitudes (0.30 vs. 0.38; 0.20 vs. 0.25) if compared to the 3rd event (PE at 175°). Hence, this small asymmetry in the detail of the contour is reflected by the significant first PE and NE features. This is also the reason that we select six features for pennate shapes: the two main line events, each with their two flanking edges that can be connected to negative pulses (Fig. 9).

We would like to stress that the entire processing of a single contour, including all file I/O, takes less than 0.4 second on a 750 MHz Pentium III computer. This makes the method suitable for large batch jobs in which databases with hundreds or thousands of images must be processed in a relatively short time.

3 Ornamentation processing by grating and bar cell models

Diatoms are identified by taking into account both the valve outline and the ornamentation. The latter may possess different structures that can be used (Barber and Haworth, 1981), but we will only be concerned with the striation and the raphe of pennate diatoms. The striation density and orientation symmetry are the most important, and these features will be used in our identification tests. In a first pass the striated regions must be detected after which statistics concerning the density and orientations or symmetry type must be collected. Because the density as well as orientation can vary over the valve, a normal homogeneity-based segmentation (e.g. Bigun and du Buf, 1994) cannot be applied. One method to analyze the striation has been presented in Chapter 7, but here we will follow a different approach which, as for the contour processing, is inspired by the primate visual system.

Von der Heydt et al. (1992) have found bar and grating cells which, as opposed to

Chapter 10: Gabor features 197

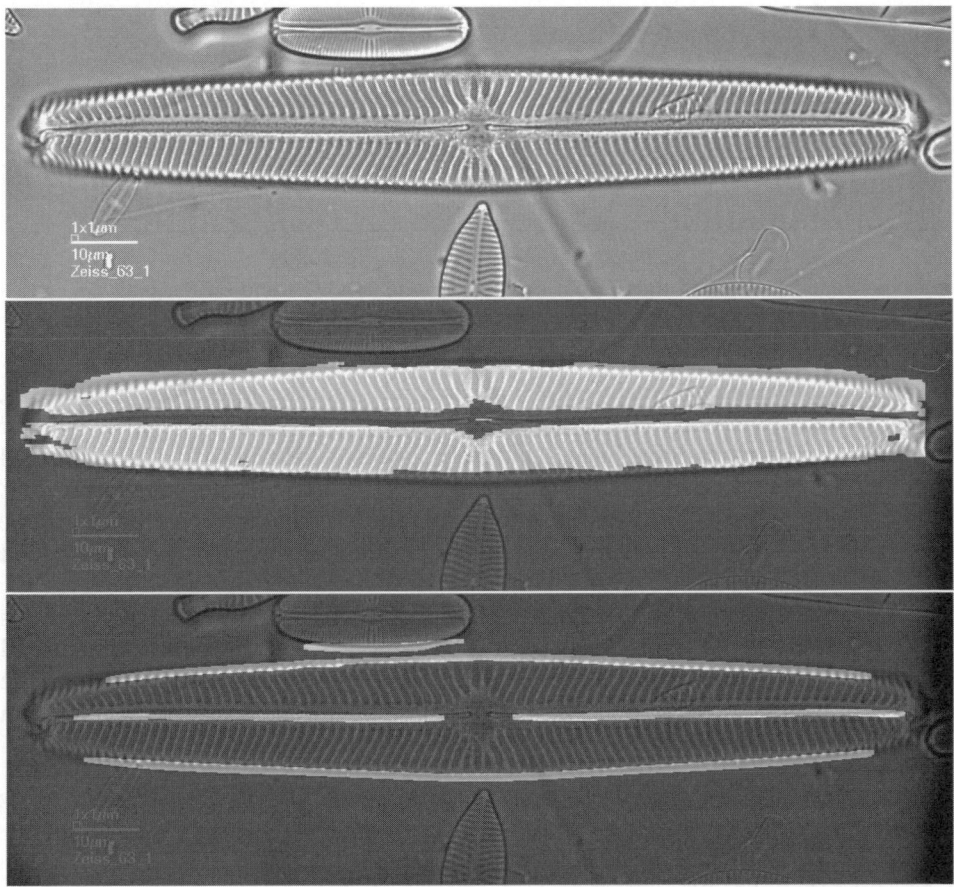

Figure 10. *Navicula oblonga* (top), grating cell outputs (middle) and bar cell outputs.

simple and complex cells, behave very nonlinearly: they respond either to isolated bars or to periodic gratings (linear textures), i.e. a bar cell does not respond to the individual bars in a grating and a grating cell does not respond to an isolated bar. Petkov and Kruizinga (1997) have published computational models, but these need improvement in order to be useful in our application. Firstly, their grating cell model detects the individual lines or bars in a linear texture, but then, in a heavy lowpass filtering, the boundary localization is lost. What we need is a binary detector with a precise boundary localization: at any position, for a given frequency and orientation, there either is a grating or there isn't. Secondly, their bar cell model is based on the detection of everything that is not detected by the grating cells, which includes blobs etc. Hence, this is not really a bar cell model. What we need is a model that responds only to isolated bars and nothing else. Our improved models also employ an anisotropic filtering scheme with Gabor wavelets that mimic cortical simple cells (Daugman, 1980; Marčelja, 1980), but a precise

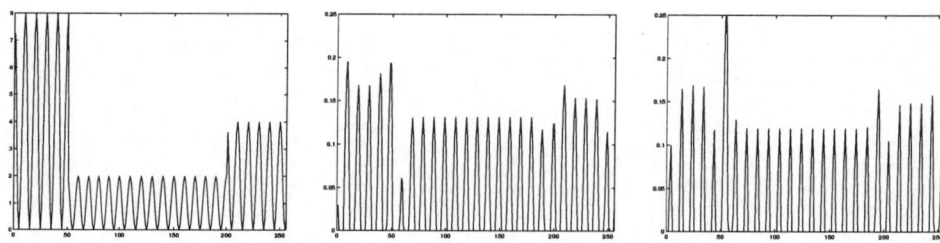

Figure 11. Left to right: input signal and normalized ON and OFF signals.

detection is obtained by again employing spatial grouping operators (Grossberg, 1999a,b). This precision is illustrated in Fig. 10: the two striated regions as well as the two raphe parts have been accurately detected. We note that the bar cell model also detected parts of the contour and even a part of another diatom, but these can be suppressed because the diatom contour is available for extracting shape descriptors like Legendre polynomials and the Gabor scale-space features described in the previous section.

Below we will assume the existence of filters and cells tuned to many frequencies and orientations. After explaining the improved models, i.e. the common front-end and the grouping operators for the grating and bar cells, the performance of the models will be illustrated by some more examples. Then the density and orientation features will be described.

3.1 The common front-end

Before applying the different grouping operators it is necessary to normalize the local contrast. This is done by using isotropic filters that mimic ON and OFF retinal ganglion cells in combination with shunting inhibition networks. This results in positive-only responses that are coded in the ON and OFF channels. After this pre-processing anisotropic Gabor filters with different orientations are applied together with a further sharpening of the ON and OFF responses, as well as a suppression of spurious responses.

3.1.1 Contrast normalization using ON and OFF isotropic channels

The normalization is obtained by two shunting center-surround networks based on ON and OFF retinal ganglion cells, i.e. isotropic DOG filters (Mingolla et al., 1999; Schiller et al., 1986): on-center with off-surround and off-center with on-surround channels. Let E and I be the excitatory and inhibitory responses of center and surround Gaussians (G_c, G_s). In the case of the ON channel G_c is positive and G_s is negative (the other way around for OFF). If L is the input image and $*$ denotes a convolution, then $E = G_c * L$ and $I = G_s * L$. The size of the Gaussian kernel G_s is larger than that of G_c, i.e. we use $\sigma_s = 3\sigma_c$. Both networks obey the shunting equation

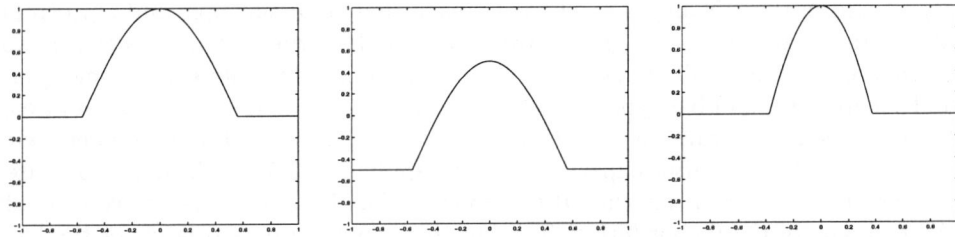

Figure 12. Sharpening, left to right: input signal, threshold subtraction and amplitude restoration. The amplitude is preserved and the width reduced.

$$\frac{dR}{dt} = -\alpha R + (\beta - R)E - (\gamma + R)I, \tag{10}$$

where R is either the ON response R^+ or the OFF response R^-. At equilibrium the ON response becomes

$$R^+ = \left[\frac{\beta E - \gamma I}{\alpha + E + I}\right]^+ \tag{11}$$

and the OFF response becomes

$$R^- = \left[\frac{\gamma I - \beta E}{\alpha + E + I}\right]^+, \tag{12}$$

where $[w]^+ = \max\{0, w\}$ is the halfwave rectification operator. Following Grossberg (1999a,b) we apply the values $\alpha = 1$, $\beta = 2$ and $\gamma = 1$. Hence, there are only two convolutions with the Gaussian kernels and at each image position Eqns. 11 and 12 need to be evaluated to obtain the ON (R^+) and OFF (R^-) responses. As a result, the ON channel will normalize local maxima (positive values relative to the local average), whereas the OFF channel will normalize local minima (negative values relative to the local average), and both ON and OFF responses will be positive or zero. This processing has the additional advantage that it assures a good response localization, i.e. an ON response in the case of a bright bar against a dark background will not be wider than the bar itself. This processing is illustrated in Fig. 11.

3.1.2 Simple cells and sharpening

Since linear gratings and bars are 1D events in a 2D image, we need anisotropic filtering to introduce an orientation selectivity. After both Marčelja and Daugman published their fundamental papers on simple cells in 1980, the Gabor (wavelet) model has been applied in many studies. A complex 2D Gabor kernel is defined by

$$G(x,y|s,\omega,\sigma_x,\sigma_y,\phi) = \exp\left[-\frac{1}{2s^2}\left(\frac{x^2}{\sigma_x^2} + \frac{y^2}{\sigma_y^2}\right) + i\left(\frac{\omega x}{s} + \phi\right)\right], \tag{13}$$

in which σ_x and σ_y define the size of the Gaussian envelope in x and y, and ω is the spatial frequency. The scaling factor s depends on the size of the Gaussian kernel

G_c as used in the ON and OFF channels, such that both isotropic and anisotropic filters are tuned to the same grating frequency and bar width. The frequency ω was optimized experimentally to achieve a good compromise for different gratings (sine- and squarewave) and bar types. The kernel can be rotated to cover all orientations. Defining the aspect ratio as $\gamma = \sigma_x/\sigma_y$, Jones and Palmer (1987) found cells with $0.23 < \gamma < 0.92$, and after experimenting we apply $\gamma = 0.8$. A phase $\phi = 0$ is used for the ON and $\phi = \pi$ for the OFF responses, but in this front-end we use only the real part of the Gabor filters. The imaginary parts are used later for the local orientation and frequency estimation (see below). We note that the actual filtering is done again in the frequency domain. For N orientations and M frequencies this requires only 1 forward FFT and $N \cdot M$ inverse FFTs.

Instead of only convolving the results $R^\pm$ of Eqns. 11 and 12 with all Gs, we keep separating the ON and OFF responses:

$$P^\pm = \left[R^\pm * \mathrm{Re}(G)\right]^+. \qquad (14)$$

Furthermore, in order to sharpen the results and to increase the robustness to noise, we apply the nonlinearity

$$Q^\pm = \frac{\max|P^\pm|}{\max|P^\pm| - T^\pm} \left[P^\pm - T^\pm\right]^+, \qquad (15)$$

where T are global threshold values (global instead of local because of the previous contrast normalization). The first term in Eq. 15 is necessary to preserve amplitudes. Taking $T^\pm = 0.5 \cdot \max|P^\pm|$ leads to a value of 2 and $\max Q = \max P$. This nonlinearity amplifies the local responses that are above the threshold level and attenuates those that are below it. In combination with the previous contrast normalization, this leads to an edge sharpening of the ON and OFF responses, whereas spurious responses due to less periodic gratings or ones that are slightly below or above the frequency tuning of the filters are suppressed. This improves the selectivity of the final grouping operators. This processing is illustrated in Fig. 12.

3.2 Grating and bar cell groupings

Up to this point we have positive or zero ON and OFF responses for different frequencies and orientations that are very selective. These responses are the bright and dark bars inside a linearly textured region, but we want to detect the striated regions. As in the case of contour processing, we apply a grouping following Grossberg (1999a,b). This grouping is done in two steps: (1) the detection of events along a line and (2) an amplitude completion or filling-in process. A grouping cell c that is connected with two cells a and b on the left and two cells d and e on the right, see Fig. 13, is defined by again using

$$c_o = a_i + b_i + c_i + d_i + e_i - \max\{a_i, b_i, d_i, e_i\}. \qquad (16)$$

The nonlinearity introduced by the max term prevents the cell c from becoming oversaturated. Since we don't know *a priori* the saturation level of a single cell, we model this by considering the maximum of the cell's neighbors.

The five cells receive input from equidistantly sampled ON and OFF responses, the $Q^\pm$ signals. For a grating grouping, the $\{a, c, e\}$ cells should be located in

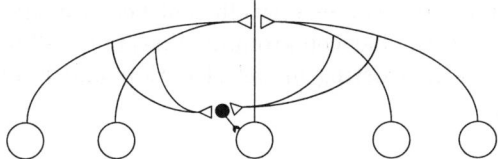

Figure 13. In the grouping model the center cell can be excited by its four neighboring cells. See caption Fig. 5.

another type of response than the $\{b, d\}$ cells, i.e. both ON,OFF,ON,OFF,ON and OFF,ON,OFF,ON,OFF are valid neighborhoods, since a grating does not have polarity. Some of the cells can be activated because they are positioned on a bar, whereas others cannot because they lie between two bars. If the cell c lies between two bars, it will be activated by some of its neighbors, i.e. the bars are nevertheless grouped together. In addition, this grouping only takes place between events, not beyond a region boundary. It is even capable of completing a grating locally where there is some irregularity in the striation.

Since the grouping is supposed to take place only in the preferred orientation of a rotated Gabor filter (e.g. in x as used in Eq. 13), taking into account also the periodicity $1/\omega$, the grouping operator is rotated in the same sense as the filter and the spacing between the five cells is scaled using the same scale factor s. This requires a bilinear interpolation of the ON or OFF responses if a cell's position differs from a pixel position. Because of the contrast normalization and the sharpening in the front-end, the responses at the four closest pixel positions can be averaged, instead of using a bilinear interpolation. This also works well and mimics the summation in a cell's dendritic field. Finally, after the grouping the outcome should be binary: linear grating (texture) or not. This requires another thresholding, in which the value c_o is compared with an "activation threshold." To allow for the detection of less ideal events, the grouping takes place if all cells but one are activated (4 out of 5). This follows from Johnson and Dudgeon's (1993) detection formula (page 208)

$$\gamma = \sqrt{L}\sigma Q^{-1}(\alpha) \tag{17}$$

with $L = 5$ and other suitable parameters measured and estimated in our experiments ($\alpha = 10^{-3}$, $Q^{-1}(\alpha) = 3$ and $\sigma = 0.13$ yields $\gamma = 0.2$).

The bar cell model employs the same front-end but a different grouping scheme. In the case of a bright bar against a dark background the output of the front-end will be a center ON response flanked by two OFF responses. For a dark bar against a bright background this will be a center OFF flanked by two ON responses. For detecting a bright bar, the three grouping cells b, c and d must all be inside the bar edges and activated by an ON response. The outer two cells a and e must be outside the bar edges and activated by neighboring OFF responses. The ON and OFF responses must be exchanged for detecting a dark bar. Hence, there are two groupings with OFF,ON,ON,ON,OFF and ON,OFF,OFF,OFF,ON, and we apply Eq. 16 and the same thresholding (4 cells activated). In this way we achieve two goals: (1) the bar detection becomes very efficient because only a different grouping needs to be done, and (2) the bar-cell grouping does not lead to a detection of other patterns, in contrast to the existing model of Petkov and Kruizinga (1997).

3.3 Results

Figure 14 shows Linköping's "ploop" image (Granlund and Knutsson, 1995) with the output of Petkov and Kruizinga's model (one frequency and one orientation) and outputs of our model, i.e. two different frequencies and orientations and the combination of a set of 16 grating cells with 2 frequencies and 8 orientations. In the latter the logical OR function was used and small gaps in the outer ring are visible. These can be avoided by using more orientations. The figure clearly demonstrates the good localization and selectivity properties of our model as well as the possibility to detect entire regions. The input image has a size of 256×256 pixels and contains frequencies ranging from π (the Nyquist frequency if the distance between pixels is 1) to $\pi/10$. Our models were tuned to 8 and 10 pixels/cycle. Other results, obtained with non-diatom images, are presented elsewhere (Santos and du Buf, 2001).

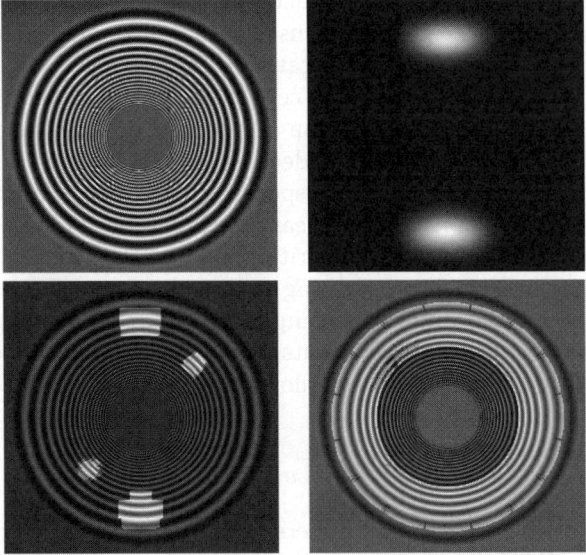

Figure 14. Input image (top left) and outputs of Petkov and Kruizinga's grating-cell model (one cell, top right) and our improved model (two and sixteen cells, bottom left and right, respectively).

Concerning the diatom images shown in Figs. 10 as well as 15 to 22, in most cases we applied 8 orientations and experimented with one or two frequencies between 8 and 15 pixels/cycle. The outputs of the grating cells are always combined using the logical OR operator. In the case of applying bar cells, the tuning frequency normally differs from that or those of the grating cell(s). In most cases the results are very good and sufficient to extract statistics of the striation, i.e. density and

Chapter 10: Gabor features

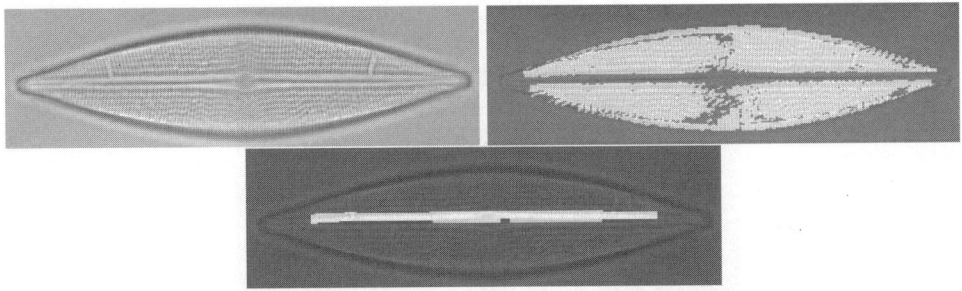

Figure 15. *Brachysira serians* with grating and bar cell outputs.

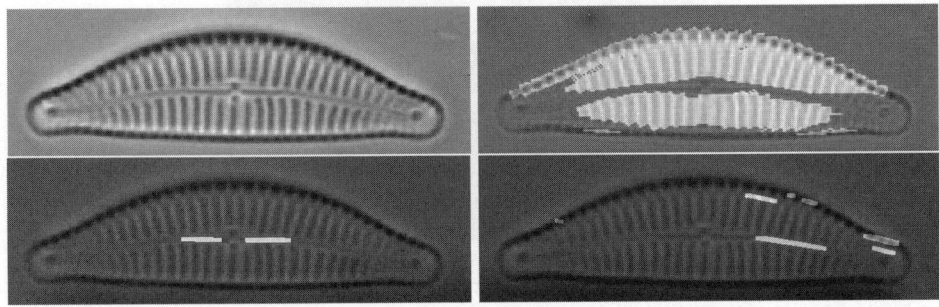

Figure 16. *Cymbella affinis* with two bar cell outputs for orientations 0 and $\pi/16$.

orientation, and the raphe (e.g. Figs. 10, 15, 16 and 18). The biggest problem is the raphe detection. In Fig. 16 we applied two bar cells with orientations of 0 and $\pi/16$ (orientation $-\pi/16$ gives the left part of the raphe and the three results must be combined). In Figs. 17 and 19 the raphe is not complete due to low contrast and small size. Actually, the missing output of grating cells along the raphe could be used to complement the bar cell information. Figures 21 and 22 show diatoms with debris which may seriously affect the quality of a boundary extraction. This concerns especially one application in which diatoms from human organs are analyzed in forensic research (drowning cases, Fig. 22). Here we can see that the outputs of the grating cells could be used to guide the boundary extraction for a shape or contour analysis.

The missing regions are due to missing filter frequencies, i.e. increasing the number of filters and grating cells will improve the results, although a complete detection up to the valve endings is questionable because of the lower contrast and the thresholding (this is not crucial for diatom identification).

3.4 Striation density and orientation

The detection of the raphe proved to be a problem, although in most examples parts have been detected and during postprocessing the bar and grating results could be combined to complement the information. The identification process benefits from,

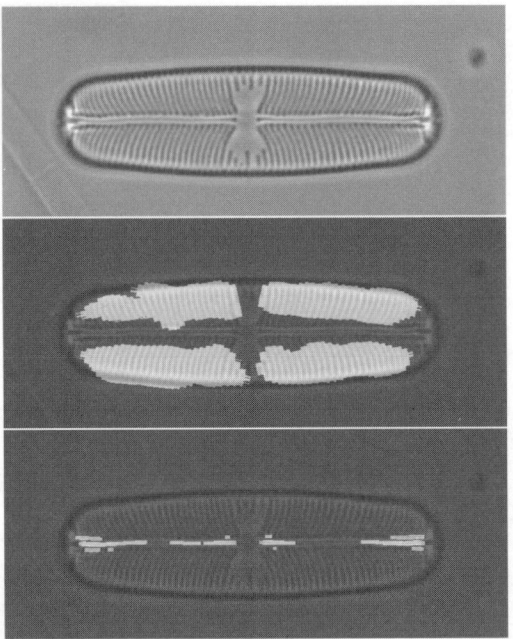

Figure 17. *Sellaphora pupula*. Here the dark raphe is partially detected and flanked by outputs of a bright-bar detector that could be suppressed.

Figure 18. *Navicula angusta*.

apart from contour features, the striation density and orientation plus, if there is a raphe, its shape (linear, curved, sigmoidal, its symmetrical or asymmetrical position, etc). Due to insufficient time, we do not use raphe features here, and will keep this problem for future research.

The striation results discussed above were obtained by experimenting with se-

Chapter 10: Gabor features

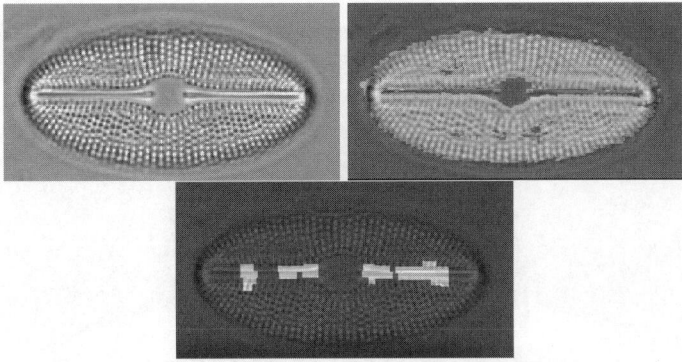

Figure 19. *Diploneis ovalis*. The striae here consist of very coarse individual pores.

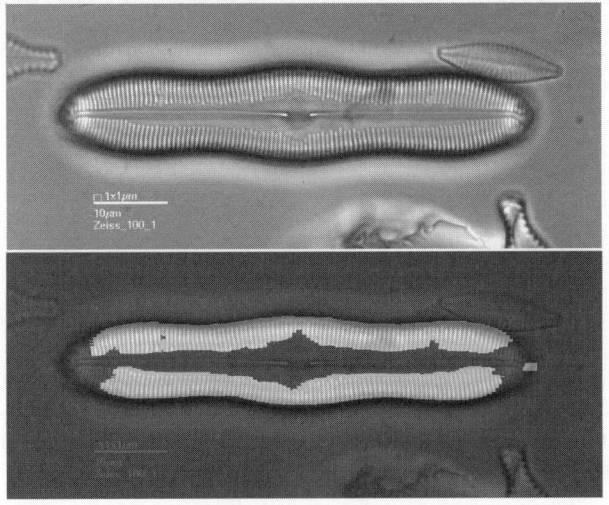

Figure 20. *Caloneis silicula*.

veral frequencies and eight orientations. Two frequencies proved to work well for the different test sets, i.e. most images in the *Sellaphora pupula* and the mixed genera set yielded most parts of the striated regions, enough for extracting statistics. Only in a few cases it was necessary to apply three frequencies. A fixed set of three frequencies could have been used for the entire sets, but then the computations would require significantly more time.

For a future *unsupervised* processing it may be better to apply a *complete* set of filters and cells that cover sufficiently all possible orientations as well as frequencies, for example eight orientations and five frequencies. This is how information is processed in our visual cortex, but it can also be implemented if the computational cost is acceptable or parallel processing can be applied. The CPU time (without disk access) of applying all filtering once (once for center-surround and once for Gabor)

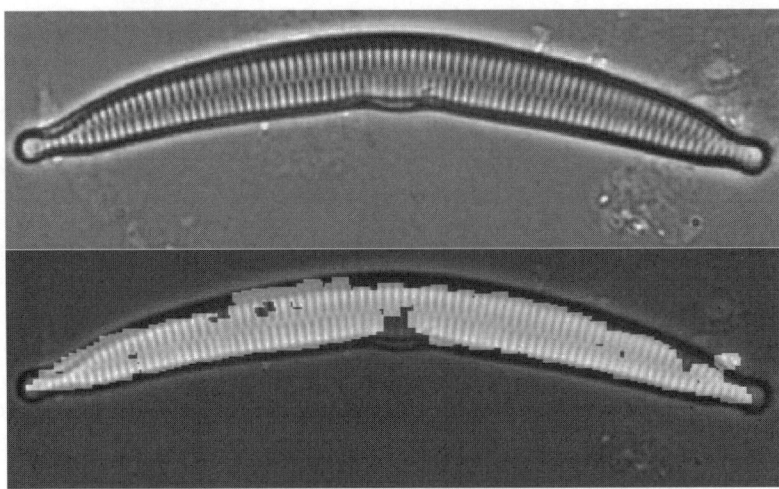

Figure 21. *Hannaea arcus*, a diatom without raphe, surrounded by debris.

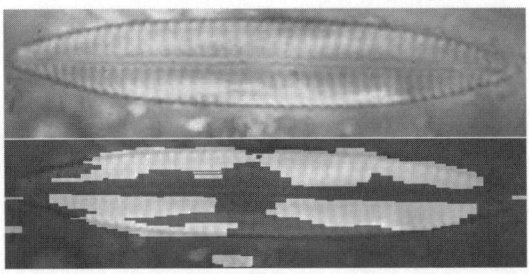

Figure 22. *Navicula tripunctata* with low contrast and a lot of debris, which is characteristic for forensics (drowning cases, see Chapter 3).

in the frequency domain, plus the contrast normalization, sharpening and grouping, takes about one second on a 750 MHz Pentium III computer. The application of a set of e.g. 5 frequencies and 8 orientations will take much less than $5 \times 8 = 40$ seconds because there are only two forward FFTs and 41 inverse FFTs. In addition, the sine and cosine tables for the inverse FFTs need to be computed only once before a big loop over the frequencies and orientations. Such a big loop is ideal for a parallelization, because only the grouping results need to be assembled on one CPU, and a speedup close to the number of CPUs can be expected (for comparable code we measured a factor of 3.4 on an older 4-CPU SMP system).

Applying a complete set of filters and cells has the additional advantage that the "complete" local Gabor power spectrum is available for measuring the local frequency and orientation inside the striated regions. This is illustrated in Fig. 23. The estimated local frequency corresponds to the center frequency of the Gabor filter with the maximum modulus, whereas the local orientation is obtained by a

Chapter 10: Gabor features

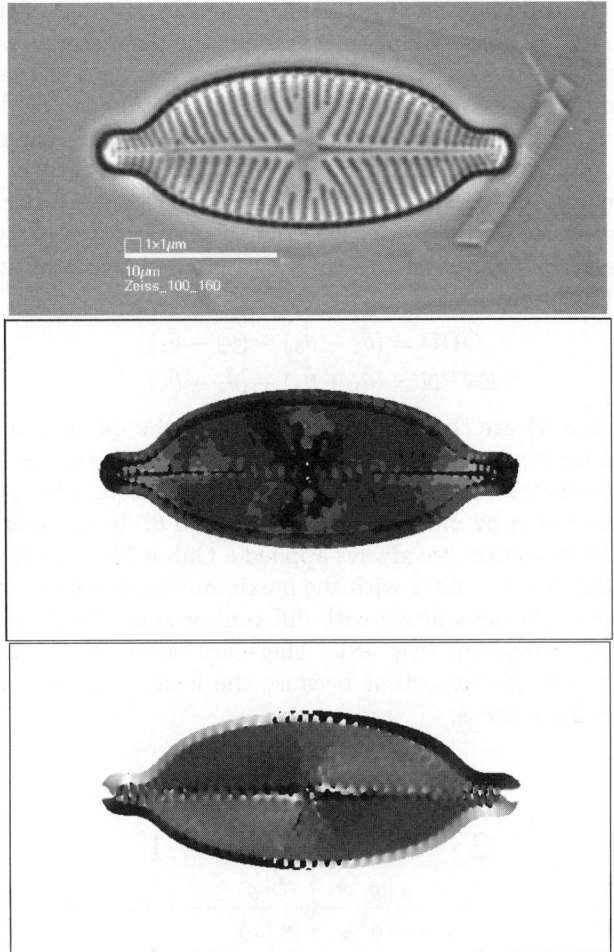

Figure 23. *Navicula menisculus* with local frequency (middle) and orientation (bottom) estimates shown as gray levels. The orientation wraps from black to white when crossing the horizontal axis.

complex double-angle summation (Granlund and Knutsson, 1995) using

$$Z = \sum_{j=0}^{N-1} |Z_j| \exp\left(i\frac{2\pi}{N}j\right), \tag{18}$$

in which Z_j represents the amplitude responses of the input image convolved with N Gabor kernels (Eq. 13), separated by π/N degrees, from which the angle on the interval $[-\pi, \pi]$ is computed, i.e. $\theta = \tan^{-1}(Z)$. The local frequency estimation can be improved by applying a continuous Gaussian model to the responses of the Gabor filters (du Buf, 1992) or by other methods (Cumming and Perea-Vega, 1999). The Gaussian model also allows for the estimation of the variance (bandwidth), which is a measure of the certainty of the frequency.

Our striation features consist of (a) the average and variance of the frequency in the detected striated regions, taking into account the pixel size of each image (this number is available in the databases), and (b) four averages of the orientation in the four quadrants. This is important, since the orientation can be different in each quadrant, although using four averages is an approximation—the orientation can vary within each quadrant—that needs to be improved in the future. A combination of these four orientations (e.g. by adding them together) is not advisable due to the symmetry properties of diatoms. Because the orientation symmetry types around the central axes and/or raphe of the valve, we created two additional symmetry features, called ODD and EVEN:

$$\text{ODD} = (\bar{\theta}_1 - \bar{\theta}_3) - (\bar{\theta}_2 - \bar{\theta}_4) \tag{19}$$

$$\text{EVEN} = (\bar{\theta}_1 + \bar{\theta}_4) + (\bar{\theta}_2 + \bar{\theta}_3) \tag{20}$$

where $\bar{\theta}_i, i \in \{1, 2, 3, 4\}$ are the mean orientations in the quadrants. EVEN equals 2π in the case of perfectly symmetric striae, whereas ODD equals -2π. Figure 24 illustrates the symmetry relations for the case of a diatom with perfect striation symmetry, characterized by one mirrored orientation θ. In the identification tests described in the next section, we always applied a Gabor filter bank with 7 frequencies and 4 orientations. The filter with the maximum modulus response determines the local frequency, and the 4 filters with different orientation at this frequency are used for the local orientation (Eq. 18). There are other solutions, but the exact method proved to be less important because the local frequency and orientation are averaged over large areas.

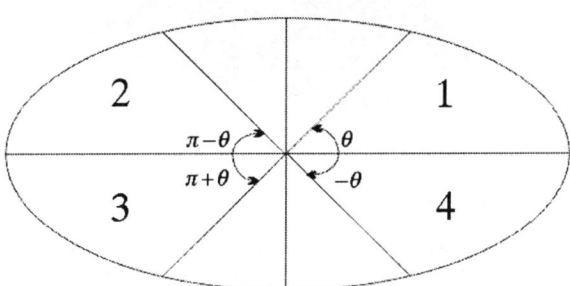

Figure 24. A perfect striation symmetry characterized by one angle θ that is mirrored in the four quadrants.

4 Identification tests

In this section we describe the feature sets used, the classifiers applied, and the experimental results obtained with the *Sellaphora pupula* and mixed genera image test sets (Chapter 4). In the case of the *Sellaphora pupula* both "classical" and Gabor striation features were used. Because of the good results obtained with the Gabor striation features, we restricted ourselves to using these in the mixed genera tests.

4.1 Feature sets

We carried out tests using only contour features, only striation features, and mixtures of contour and striation features.

4.1.1. Contour features

- In the case of the Gabor contour features, as described in Section 2 of this chapter, the *Sellaphora pupula* set was tested by selecting the six most significant events, each having 11 features, with a total of 66. For the mixed genera set, a total of nine events were used, again each with 11 features. The reason for this is that this set contains more complex shapes, like the one shown in Fig. 25 (left), with events at 90°.
- For comparison purposes we also included classical morphometric features (see Chapter 7): rectangularity, triangularity, circularity, ellipticity and compactness, although some of these may be highly correlated in the case of *Sellaphora pupula* diatoms. Because diatomists use the absolute length for separating some demes, we also included this feature.
- We also included Legendre polynomials, with a total of 34 features. Half of these represent the upper part of a diatom's contour, and the other half the lower part. See Chapter 9 for details.

4.1.2. Striation features

- The Gabor striation features (Section 3 of this chapter) comprised both density and orientation. For the density we used both the average and the variance of the local frequency estimate at each position in the detected striated areas. For each quadrant of the detected areas, the local orientation was averaged, yielding 4 features, supplemented by the two symmetry features ODD and EVEN.
- "Classical" striation features were extracted from the gradient vector direction histogram (Sun, 1995). The first four moments of the histogram were used as orientation features. The density is extracted from the edges in the striated region. See Chapter 7 for details.

4.2 Classification methods

We applied the following classifiers and methods:
- A fully-connected neural-network (NN) classifier with one hidden layer. The number of input units equals the number of features. The hidden layer has the same number of units as the input layer, and the output layer has as many units as there are classes (taxa). After doing some experiments, a back-propagation learning method called Quickprop (Fahlman, 1988) was chosen because it is much faster than normal learning and the results were as good. This method uses information about the curvature of the error surface. It requires the computation of the second-order derivatives of the error function. Quickprop assumes the error surface to be locally quadratic and attempts to jump in one step from the current position directly into the minimum of the parabola. The methodology adopted for training was the

following: the NN was trained until the error rate measured when classifying the validation set reached a local minimum (Bishop, 1997).

- A naive Bayes, also called maximum-likelihood classifier (Kendall, 1975). In using this classifier, we estimated the probability density function of the features of each class in the training set. When a new object needs to be classified, we compute the conditional probabilities of each class, given its feature vector, and assign the object to the class with the highest probability.

- Decision trees (DT) using Quinlan's (1996a,b) software packages C4.5 and C5.0, see also Chapter 7. Different cross-validation tests have been done using leave-1-out or leave-n-out, with and without windowing (C4.5) or boosting (C5.0). Although these operations require more time while training the classifiers, the time to classify the test set remains approximately the same. In the first tests we used the cross-validation method with a total of 7 trials, first leaving out 60 samples (a typical 50% split), then 24, 15, 12, 4, 2 and 1 (leave-1-out) samples. We also performed additional tests, using only two distinct training and test sets.

In the leave-1-out methodology, one sample (image) is excluded from the data, a DT is constructed using the remaining $n - 1$ samples, and the excluded sample is classified using the constructed tree. This procedure is repeated n times. The leave-1-out cross-validation was found to be a nearly unbiased estimator of the expected error for nonparametric classifiers (Fukunaga and Kessel, 1971).

In other tests, we applied the following strategy: (1) the entire set is divided into two equally-sized sets, a training set and a test set; (2) the classifier is first trained on the training set, and then applied to the test set; (3) then the two sets are reversed, i.e. the *test* set is used to train, and the *training* set is used to test. This provides a test on the symmetry (equality) of the sets.

The boosting operation implemented in C5.0 (Freund and Schapire, 1996; Quinlan, 1996a,b) generates several classifiers rather than just one. When a new sample is to be classified, each classifier votes in order to determine the final class (taxon). The classifier that was trained last has the highest weight in the final decision. To generate several classifiers from a single data set, the following procedure is used: (1) A list of weights is established for each sample in the training set. A single decision tree is constructed from the training data. This classifier is likely to make mistakes on some samples in the data. (2) When the second classifier is constructed, more attention is paid to these samples, increasing their weight in an attempt to classify them correctly. Hence, the second classifier will generally be different from the first one, but it also will make errors on some samples, but these become the focus of attention during the construction of the third classifier. This process continues for a pre-determined number of trees. In our tests we used 5 trees, and the last one is applied to the test set.

The windowing operation (C4.5) uses a subset of all available training images to build the classifier, and then tests it on the samples that were not used. New samples are added to the initial subset in order to train a new classifier with a better performance. This process is repeated several times (in practice, we noticed that a number of iterations between six and nine were enough), until all the samples which were not used could be assigned with an identification rate of 100%.

The NN and Bayes classifiers were applied using 50% training and 50% test

Chapter 10: Gabor features   211

sets, but only for the *Sellaphora pupula* set (120 images split into two groups of 60). These classifiers have not been used for the mixed genera set.

We selected these classifiers because there is no *ideal* classifier, i.e. each has advantages and disadvantages. In some cases, a NN can perform very well, but it is almost impossible to explain the results. A Bayesian classifier is often used to obtain a baseline for a comparison with other methods. It provides some information of the distribution of the features in the feature space.

4.3 Results obtained with the Sellaphora pupula *set*

This set contains 120 images of 6 demes, each deme being represented by 20 images. Training and test sets, each with 10 images of all demes, were created randomly at Royal Botanic Garden Edinburgh. The deme names are: "cf. rectangular," "elliptical," "pseudoblunt," "pseudocapitate," "tidy" and "large." For an in-depth discussion of this set and representative images we refer to Chapter 4.

4.3.1. First tests – cross-validations

The set of 120 images was used to create 120 decision trees, each time leaving out one different image, forming a training set of 119 images and a test set of only one. Each tree classified the image that was left out and the accuracy of the classifier is given by the total fraction of the 120 tests in which these images were correctly classified.

Table 3 shows the typical convergence in the creation of a decision tree using windowing, in which the tree size denotes the number of decision criteria in the tree. The error rate decreases from 68.7 to 0%. After each iteration the error rate is obtained by applying the tree to the images not used. It shows that a DT with 50/50 training/testing (in this case leave-60-out because of the 120 images) will have problems when using small sets.

Table 4 shows the results obtained with cross-validation tests using only Gabor contour features. The table shows the number of images left out, the average ID

Table 3. Example of the formation of a decision tree using the C4.5 method. The windowing is clearly visible in the number of images used.

iteration	tree size	images used	not used	wrong	error rate (%)
1	20	24	96	66	68.7
2	49	57	63	38	60.3
3	68	76	44	22	50.0
4	68	87	33	13	39.4
5	80	93	27	9	33.3
6	75	97	23	3	13.0
7	75	100	20	1	5.0
8	75	101	19	0	0.0

rate obtained when classifying the images left out, as well as the standard error (S.E.). The right column shows the results obtained by the best decision tree when classifying the same left-out images. We can see that up to the leave-4-out, the best trees were able to classify the remaining images with 100% accuracy. We can also see that the mean error rate in these three cases is about 37%. The small standard error is an indication that the large mean errors are not caused by outliers, i.e. most trees performed about equally. In the leave-one-out case, the average result converged to an ID rate of 64.2%. In the case of C4.5, the result of leave-one-out was only 44.2%, even using windowing. This reflects one general trend that we found for the *Sellaphora pupula* set: C5.0 performed better than C4.5.

Table 5 shows the results of a special experiment in which we used cross-validations by means of the Bayes maximum likelihood classifier, i.e. the procedure and features were equal to those used to produce Table 4, but no boosting was applied. This simple procedure yields about the same results when compared to C5.0 with boosting: only a few percent are lost when the training sets are large enough (up to leave-4-out). A future experiment could be to study the effect of boosting in combination with the Bayes classifier. If boosting was to increase the ID rate by several percent, the performance of the Bayes classifier may equal that of decision trees.

4.3.2. Classifier comparison using distinct training and test sets

Table 6 presents a comparison of the different classifiers when applied to 60 training and 60 test images, using only Gabor contour features, as well as the reverse test after swapping the training and test sets. The neural network performed worst. Quinlan's improved decision tree version C5.0 performed significantly better than the C4.5 version. It also outperformed the naive Bayes classifier, but not by much. Both C5.0 and Bayes were able to achieve a maximum score on the training and reverse training sets, but their performance on the test and reverse test sets dropped by about 35%. We might conclude that the two subsets created randomly are well balanced, because the test and reverse test results are quite similar. It can also

Table 4. Cross-validation tests (C5.0 with boosting) for the *Sellaphora pupula* set of 120 images, using only Gabor contour features.

leave out	number of trees	mean ID rate (%)	S.E.	best tree (%)
1	120	64.2	4.4	100
2	60	62.5	4.7	100
4	30	63.3	4.3	100
12	10	67.5	5.8	91.7
15	8	61.7	4.7	86.7
24	5	64.2	2.8	70.8
60	2	62.5	2.5	65.0

Chapter 10: Gabor features

Table 5. Cross-validation tests (using Bayes classifier) for the *Sellaphora pupula* set of 120 images, using only Gabor contour features.

leave out	number of classifications	mean ID rate (%)	S.E.	best (%)
1	120	60.8	4.5	100
2	60	61.2	4.7	100
4	30	60.8	4.8	100
12	10	59.2	5.3	83.3
15	8	63.3	3.0	73.4
24	5	59.1	4.0	62.5
60	2	52.1	1.24	53.3

Table 6. Comparative results (ID rates in %) of several classifiers for the *Sellaphora pupula* set. The first column shows the classifier used. The entire set of 120 images was split into a training and a test set of 60 images each. After the normal training and testing the two sets were reversed. The last column shows the average ID rate obtained on the test set and reversed test set.

	Gabor contour only				
classifier	train	test	reverse train	reverse test	aver
C4.5	96.7	58.3	95.0	58.3	58.3
C5.0	100	65.0	100	71.7	68.3
Bayes	100	60.0	100	70.0	65.0
NN	75.4	42.5	73.2	41.5	42.0

be observed that, although the Bayes classifier learned the training set very well, its generalization capabilities are inferior to those of the C5.0 classifier. Finally, the C5.0 result of 68.3% is better than leave-one-out (64.2%) and comparable with leave-12-out (67.5%, see Table 4).

Table 7 shows a comparison of results when using only striation features in combination with C5.0 and boosting. O/E refers to the orientation symmetry features ODD and EVEN. In the case of Gabor features, the average stria orientations in the four quadrants were used, the average stria orientation complemented with the symmetry features, the striation density (mean and variance), and the combination of all features. In the latter case a performance of 27.5% was achieved. This is significantly better than the results obtained with the "classical" striation features as described in Chapter 7. We do not show the results obtained with the C4.5 method. Occasionally C4.5 gives better results for some features, but on average the results are worse compared to C5.0 (by about 2%). The result of 27.5% was a surprise because the taxonomic experts emphasized that they use mainly the contour in identifying *Sellaphora pupula* demes.

Table 8 shows the best results obtained for the individual feature sets, i.e. Gabor contour, Legendre contour, morphometric and Gabor striation. In all cases the C5.0 method gave the best results. Despite the fact that Legendre polynomials

Table 7. Comparative results for the *Sellaphora pupula* set using only striation features. The columns show the results obtained on the normal training and test sets, plus the reversed sets. The last column shows the average ID rate (%) of the test set and reversed test set. O/E refers to ODD and EVEN symmetry features.

	C5.0 with boosting				
features	train	test	reverse train	reverse test	aver
Gabor:					
orientation	85.0	20.0	80.0	20.0	20.0
orientation + O/E	71.7	26.7	61.7	18.3	22.5
density	71.7	23.3	65.0	25.0	24.2
density + orientation + O/E	85.0	30.0	77.0	25.0	27.5
classical:					
orientation	58.3	26.7	60.0	21.7	24.2
density	45.0	23.3	31.7	15.0	19.2
orientation + density	63.3	20.0	43.3	18.3	19.2

can approximate a contour very precisely, the results obtained with these were clearly inferior relative to using Gabor contour features. The morphometric contour features performed worse.

Table 9 shows the results for feature combinations that all include the Gabor contour features. The best result (85.0%) was obtained with the C4.5 method using almost all feature sets. This best result is only 3% better than using only Gabor contour and morphometric features in combination with the C5.0 method.

4.3.3. Final observations for Sellaphora pupula

Above we have presented a small selection of all experimental results. In order to be able to compare the performance of different classifiers in relation to different feature sets, the entire set of 120 images was split into two sets of 60 each. As shown in Table 6 this procedure leads to a lower performance of the C5.0 classifier (also C4.5), but it is the most "honest" comparison. A general conclusion is that we can attain a performance of 85% with the methods as described, although there is still room for improvements (see Discussion). This result is rather encouraging, given the difficulty that this data set poses even for expert diatomists. The experts that participated in the "groundtruthing," as described in Chapter 5, obtained an average of 82%, although the best experts were able to do significantly better, up to 98.3%. Hence, our methods must still be improved. The best results were obtained by combining Gabor contour and Legendre contour features (plus others). This is not a complete surprise because the two contour feature sets complement each other. The Legendre polynomials can model the upper and lower contour parts with very high precision, but are less accurate in describing the valve endings (poles). The Gabor contour features have been optimized to model the valve endings.

Table 8. Summary of identification results for the *Sellaphora pupula* set, using individual feature sets.

features	best classifier	best test	best r.test	aver
Gabor contour	C5.0	65.0	71.7	68.3
Legendre	C5.0	60.0	48.3	54.2
morphometric	C5.0	21.7	25.0	23.4
Gabor striation	C5.0	30.0	25.0	27.5

Table 9. Summary of identification results for the *Sellaphora pupula* set. Shown are the results obtained with the Gabor contour features, in combination with the ones listed in each row.

features: Gabor contour +	best classifier	best test	best r.test	aver
Legendre + striae + length	C4.5	81.7	88.3	85.0
morphometric	C5.0	78.2	86.5	82.4
Legendre + length	C5.0	80.0	80.0	80.0
striae + length	C5.0	75.0	71.7	73.4
Legendre	C5.0	68.3	75.0	71.6
striae	C5.0	63.3	66.7	65.0

4.4 Results obtained with the mixed genera set

In this section we present the results obtained with the larger test set of 37 different taxa and a total of 781 images, as presented in Chapter 4, with at least 20 images per taxon. This set includes many different shapes, some of which are more complex than *Sellaphora pupula*. The striation patterns also show a more pronounced variation. Because there were no pre-defined training and test sets, we only applied the decision trees C4.5 and C5.0 with the cross-validation strategy leave-one-out.

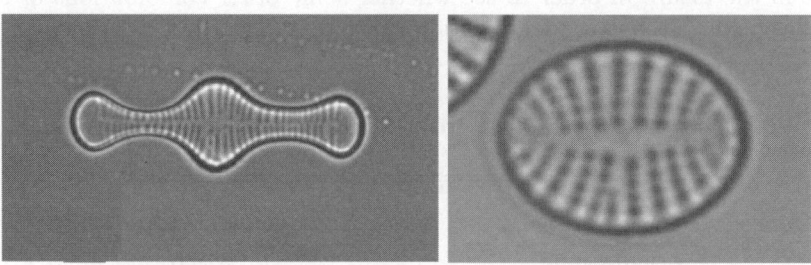

Figure 25. Two non-pennate valve shapes in the mixed genera set. The example on the left also produces events around 90 and 270°. On the right we can see an almost elliptical shape. The characteristic profile will be almost zero, or will have very small amplitude deviations. Since both quasi-circular and quasi-elliptical shapes yield very similar characteristic profiles, it is difficult to distinguish such shapes. The mixed genera set contains almost 29% of such cases.

Since the striation is more diverse in this set, we might expect a better performance of the Gabor striation features compared to the results obtained with the *Sellaphora pupula* set. Table 10 shows that the C4.5 classifier with all the Gabor striation features achieved an identification rate of 52%. The same test on the *Sellaphora pupula* set yielded only 27.5%. This huge difference can only be explained by the greater diversity of striation in the mixed genera set.

It is impossible to predict the performance of the Gabor contour features because these have been optimized for valve endings of pennate shapes, and the characteristic profile based on the best-fitted ellipse can have problems with more complex or elliptical shapes such as those shown in Fig. 25. Instead of selecting the six most significant events it was necessary to include more events, so we experimented with nine, each one having 11 features, giving a total of 99.

Table 11 shows results obtained with individual contour feature sets. C5.0 gave better results than C4.5. As expected, Legendre polynomials directly computed from the two contour parts (Chapter 9) performed much better than the Gabor contour features (84.9% vs. 37%).

Table 12 shows the results obtained with different combinations of contour and striation features. Since a decision tree will select the features that produce the best decision criteria, the increased performance of 87.9% with combined Legendre and Gabor features shows that the latter still contribute, i.e. they provide additional information at the poles where Legendre polynomials are less accurate.

4.4.1. Final observations for the mixed genera set

Using only Gabor striation features, an ID rate of 52% was obtained. The Legendre contour features were clearly better than the Gabor contour features (84.9 vs. 37%, Table 11). The best result (87.9%) was obtained by combining both contour feature sets, without the striation features, using the C5.0 decision trees (Table 12). The number of images and features, the latter combining the two contour sets and the striation set, unfortunately exceeds the limitation of the C5.0 version that was available to us, and the performance of the C4.5 version without limitations was clearly worse. Hence, additional tests with a C5.0 version without limitations must be done in the future in order to see whether some other combinations of feature sets could yield even better identification rates.

Table 10. Performance of different Gabor striation features for the mixed genera set, using C4.5 and leave-one-out.

features	ID rate (%)
orientation	19.6
orientation + symmetry	22.0
density	38.6
density + orientation + symmetry	52.0

Table 11. Summary of identification results for the mixed genera set, using only contour features, with and without boosting or windowing.

classifier features	ID rate
C5.0:	
Gabor contour (boosting)	37.0
Legendre	81.1
Legendre (boosting)	84.9
C4.5:	
Gabor contour	30.8
Gabor contour (windowing)	32.1
Legendre	80.7
Legendre (windowing)	80.2

Table 12. Summary of identification results for the mixed genera set, using contour and striation features.

classifier features	ID rate
C5.0:	
Gabor contour + Legendre	82.2
Gabor contour + Legendre (boosting)	87.9
C4.5:	
Gabor contour + Legendre	80.4
Gabor contour + Legendre (windowing)	81.1
Gabor contour + striae (windowing)	57.2
Legendre + striae	81.0
Gabor contour + Legendre + striae	81.2

5 Discussion

In this Chapter we have introduced new Gabor feature sets for the contour as well as the striation. The latter can be detected quite accurately by an improved model of grating cells. The best results obtained, i.e. 85% in the case of the *Sellaphora pupula* and 88% in the case of the mixed genera sets, are promising but need improvement. The Gabor contour features are based on the characteristic profile, which is extracted after dynamic ellipse fitting. This procedure is not appropriate when a contour does not have a central elliptical part. One possible improvement could be to apply the Gabor contour features directly to a signal that represents a contour, without ellipse fitting.

In general, better results are obtained by combining different feature sets, which may lead to hundreds of features, but often identification rates are unpredictable. In other chapters (7, 8, 12), very good results are obtained by carefully selecting several good features from the entire feature sets. Due to time constraints, only a preliminary test on this was carried out. Using only Gabor contour features with the *Sellaphora pupula* test set, the C4.5 classifier with the leave-one-out method achieved an ID rate of only 41.7% when all 66 features were considered. However, a feature reduction based on a linear search improved the ID rate to 78.7% when using only 6 features. Similar improvements by linear search or random subspace methods have been reported elsewhere (Devijever and Kittler, 1982; Fukunaga, 1990; Ho, 1998a,b; Uger and Gauch, 2000). Hence, we can expect a significant improvement when selecting the best features from the different sets.

The Gabor contour features include amplitudes at an event's initial and final scales, plus the maximum amplitude and its associated scale. As mentioned before, the amplitudes at all scales can be approximated by a low-order polynomial, up to the third or fifth order, say, and the polynomial coefficients can be used as contour features. Because such a polynomial approximation will describe the amplitudes more precisely, and thereby an event's distribution over scale, we can expect a further improvement.

Regarding the ornamentation, it will be difficult, if not impossible to improve the striation density and orientation features if only a few features (e.g. averages in four quadrants) are considered. The grating cell model can accurately detect the striated regions, but the bar cell model requires further processing to detect reliably raphes where they are present. Only when additional analyses could extract a central axis, the stria density and orientation can be sampled along this axis, and polynomial approximations can be applied. The polynomials' coefficients will provide better striation features, but will result in more features.

References

Abbasi, S., Mokhtarian F. and Kittler J. (2000) Enhancing curvature scale space based shape retrieval for objects with shallow concavities. Image and Vision Computing, Vol. 18, pp. 199-211.

Barber, H.G. and Haworth, E.Y. (1981) A guide to the morphology of the diatom frustule. Freshwater Biological Association, The Ferry House, Far Sawrey, Ambleside, Cumbria LA22 0LP, UK.

Bigun, J. and du Buf, J.M.H. (1994) N-folded symmetries by complex moments in Gabor space. IEEE Trans. Pattern Analysis and Machine Intelligence, Vol. 16, pp. 80-87.

Bishop, C.M. (1997) Neural network for pattern recognition. Claredon Press, Oxford, UK.

Ciobanu, A., Shahbazkia, H. and du Buf, J.M.H. (2000) Contour profiling by dynamic ellipse fitting. Proc. Int. Conf. Pattern Recognition, Barcelona (Spain), Vol. III, pp. 758-761.

Cumming, I.G. and Perea-Vega, D. (1999) Local frequency estimation in interferograms using a multiband pre-filtering approach. FRINGE'99, Liège (Belgium), November 10-12. See http://www.ece.ubc.ca/sar/FRINGE96/FRINGE99_diego.pdf

Daugman, J. (1980) Two-dimensional spectral analysis of cortical receptive field profiles. Vision Research, Vol. 20, pp. 847-856.

Devijver, P.A. and Kittler, J. (1982) Pattern recognition: a statistical approach. Prentice-Hall, Englewood Cliffs (NJ), USA.

du Buf, J.M.H. (1992) Abstract processes in texture discrimination. Spatial Vision, Vol. 6, pp. 221-242.

du Buf, J.M.H. (1993) Responses of simple cells: events, interferences, and ambiguities. Biological Cybernetics, Vol. 68, pp. 321-333.

du Buf, J.M.H. (1994) Ramp edges, Mach bands, and the functional significance of the simple cell assembly. Biological Cybernetics, Vol. 70, pp. 449-461.

du Buf, J.M.H. and Fischer, S. (1995) Modeling brightness perception and syntactical image coding. Optical Engineering, Vol. 34, pp. 1900-1911.

Fahlman, S.E. (1988) Faster-learning variations on back-propagation: An empirical study. In: Connectionist models summer school, T.J. Sejnowski, G.E. Hinton and D.S. Touretzky (eds), Morgan Kaufmann.

Fischer, S., Binkert, M. and Bunke, H. (2000) Feature based retrieval of diatoms in an image database using decision trees. Proc. Advanced concepts for intelligent vision systems 2000, Baden-Baden, Germany, August 1-4, pp. 67-72.

Freund, Y. and Schapire, R.E. (1996) Experiments with a new boosting algorithm. Proc. 13th Int. Conf. on Machine Learning, Morgan Kaufmann, pp. 146-148.

Frigo, M. (1999) A fast Fourier transform compiler. Proc. ACM SIGPLAN Conf. on Programming Language Design and Implementation (PLDI'99), Atlanta, Georgia. http://www.fftw.org/

Fukunaga, K. and Kessel, D. (1971) Estimation of classification error. IEEE Trans. Computers, Vol. 20, pp. 1521-1527.

Fukunaga, K. (1990) Introduction to statistical pattern recognition. Academic Press, San Diego (CA), USA (2nd Edition).

Granlund, G.H. and Knutsson, H. (1995) Signal processing for computer vision. Kluwer Academic, The Netherlands.

Grossberg, S. (1999a) How does the cerebral cortex work? Learning, attention and grouping by the laminar circuits of visual cortex. Spatial Vision, Vol. 12, pp. 163-186.

Grossberg, S. (1999b) The laminar architecture of visual cortex and image processing technology. Proc. 10th Int. Conf. Image Analysis and Processing, Venice (Italy), pp. 2-9.

Ho, T.K. (1998a) C4.5 decision forests. Proc. 14th Int. Conf. Pattern Recognition, Brisbane (Australia), pp. 545-549.

Ho, T.K. (1998b) The random subspace method for constructing decision forests. IEEE Trans. Pattern Analysis and Machine Intelligence, Vol. 20, pp. 832-844.

Jones, J. and Palmer, A. (1987) An evaluation of the two-dimensional Gabor filter model of simple receptive fields in cat striate cortex. J. Neurophysiology, Vol. 58, pp. 1233-1258.

Johnson, D.H. and Dudgeon, D.E. (1993) Array signal processing – concepts and techniques. Prentice Hall, Englewood CLiffs.

Kendall, M. (1975) Multivariate analysis. Charles Griffin and Co. London.

Lindeberg, T. (1994) Scale-space theory in computer vision. Kluwer Academic. Dordrecht, Netherlands.

Marčelja, S. (1980) Mathematical description of the responses of simple cortical cells. J. Optical Society of America, Vol. 70, pp. 1297-1300.

Mingolla, E., Ross, W. and Grossberg, S. (1999) A neural network for enhancing boundaries and surfaces in synthetic aperture radar images. Neural Networks, Vol. 12, pp. 499-511.

Petkov, N. and Kruizinga, P. (1997) Computational models of visual neurons specialised in the detection of periodic and aperiodic oriented visual stimuli: bar and grating cells. Biological Cybernetics, Vol. 76, pp. 83-96.

Pessoa, L. (1996) Mach bands: how many models are possible? Recent experimental findings and modeling attempts. Vision Research, Vol. 36, pp. 3205-3227.

Quinlan, J.R. (1996a) Improved use of continuous attributes in C4.5. J. Artificial Intelligence Research, Vol. 4, pp. 77-90.

Quinlan, J.R. (1996b) Bagging, boosting and C4.5. Proc. 13th National Conf. Artificial Intelligence, AAAI/MIT Press, pp. 725-730.

Santos, L. and du Buf, J.M.H. (2001) Improved grating and bar cell models in diatom analysis. Submitted for publication in Image and Vision Computing.

Schiller, P.H., Sandell, J.H. and Maunsell, J.H.R. (1986) The functions of the ON and OFF channels of the visual system. Nature, Vol. 322, pp. 824-825.

Sun, C. (1995) Symmetry detection using gradient information. Pattern Recognition Letters, Vol. 16, pp. 987-996.

Uger, S. and Gauch, S. (2000) Feature reduction for document clustering and classification. Technical report, Computing Dept., Imperial College, London (UK).

von der Heydt, R., Peterhans, E. and Dürstele, M.R. (1992) Periodic-pattern-selective cells in monkey visual cortex. J. Neuroscience, Vol. 12, pp. 1416-1414.

CHAPTER 11

IDENTIFICATION BY MATHEMATICAL MORPHOLOGY

MICHAEL H.F. WILKINSON, ANDREI C. JALBA, ERIK R. URBACH
AND JOS B.T.M. ROERDINK

This chapter describes diatom identification methods based on algorithms from mathematical morphology. Two types of feature extraction method are described: a contour-based method which analyzes features of the diatom outline, and another method which computes features of the valve ornamentation, i.e. the striation pattern. The first method uses morphological curvature scale spaces, in combination with unsupervised cluster analysis, in order to select the best scale-space features of contours. The second method computes multi-scale ornamentation features on the basis of size and shape distributions. The sets of feature vectors were used as input for decision-tree classifiers with bagging. An identification rate of 83.7 ± 11.7% was obtained in the case of the *Sellaphora pupula* data set, whereas the mixed genera data set was identified with a rate of 89.6 ± 1.9%). The latter result increased to 97.5% when the rules were relaxed to allow the occurrence of the correct taxon within the top-five matches.

1 Introduction

This chapter concentrates on methods based on mathematical morphology for automatic diatom identification. Two types of feature extraction method are described, a contour-based method which analyzes features of the diatom *outline*, and another which computes features of the *ornamentation*, that is, the striation pattern. The resulting feature vectors provide the input for automatic identification by a pattern classifier. Identification results based on the separate as well as the combined feature sets are presented.

Mathematical morphology in its original form is a set-theoretical approach to image analysis (Serra, 1982). It studies image transformations with a simple geometrical interpretation and their algebraic decomposition and synthesis in terms of elementary set operations. Such a decomposition enables fast and efficient implementations on digital computers, which explains their practical importance, see e.g. Serra (1986), Giardina and Dougherty (1988) and Haralick et al. (1987). In order to reveal the structure of binary images, small subsets, called structuring elements, of various forms and sizes are translated over the image to perform shape extraction. In the case of gray-level images one works with structuring functions, see Sternberg (1986) and Serra (1988). The method can easily be extended to signals of any dimension. Nowadays, mathematical morphology is widely used for image filtering, segmentation and analysis (Serra and Vincent, 1992; Roerdink and Meijster, 2000; Heijmans, 1994).

The method for diatom outline analysis described below is based on extraction of contour features by multi-scale mathematical morphology applied to one-dimensional curvature functions (Jalba et al., 2001). After extracting the contour of a valve, it is smoothed adaptively, encoded using Freeman's chain code, and converted into a curvature representation which is invariant under translation and scale change. A curvature scale space is built from these data, and the most important features are extracted from it by unsupervised cluster analysis. The resulting pattern vectors, which are translation, rotation and scale invariant, provide the input for an automatic pattern classifier.

To analyze the valve ornamentations, again a multi-scale method from mathematical morphology is used (Wilkinson et al., 2001). Two techniques are combined: (1) size distributions and (2) shape distributions. Size distributions, or granulometries, form an important class of multiscale tools in mathematical morphology. They were initially introduced by Matheron (1975), and have found many applications. For a recent review of granulometries see Vincent (2000).

The organization of this chapter is as follows. Section 2 describes diatom contour analysis based on curvature scale spaces. We focus on the curvature measure extraction with adaptive smoothing, which is used to cope with noise-related problems, and describe the clustering method used for the extraction of the most important features from the curvature scale space data. Section 3 describes the analysis of diatom ornamentation. We describe texture feature vectors based on different pattern spectra. Section 4 presents the classification technique (decision tree classifier) used for diatom identification. Experimental results are reported in Section 4, and conclusions are drawn in Section 5.

2 Diatom contour analysis

In both human and computer vision, curvature extrema of the contour of an object are thought to contain important information about the shape (Leyton, 1987). Here we present a technique to extract this information by multi-scale mathematical morphology, and use this for automatic diatom identification by a decision tree classifier. The methods used are applicable to other shapes besides diatoms.

Several techniques for multi-scale shape analysis exist, such as size distributions, or granulometries, which are used to quantify the amount of detail in an image at different scales (Breen and Jones, 1996; Nacken 1994). A similar method, based on sequential alternating filters, has been proposed by Bangham et al. (1996a,b). Their method is used on 1-D signals, although they do discuss extensions to higher dimensions. We have developed a different multi-scale approach to the analysis of 1-D signals, based on work by Leymarie and Levine (1988). They developed a morphological curvature scale space for shape analysis, based on sequences of morphological top-hat or bottom-hat filters with increasing size of the structuring element. A problem not addressed by Leymarie and Levine is that of extracting the most important features from the scale space. We will present some modifications of their technique, and include a method by which the features in the scale space may be clustered in an unsupervised way. The aim is to obtain a small set of rotation-, translation- and scale-invariant shape parameters, which contain as much information as possible about the shapes of interest.

2.1 Curvature based shape recognition

A common approach to shape recognition is by focusing on the curvature of the contour (Pavlidis, 1980). Since the curvature of the contour defines a shape completely, it should in theory be possible to classify diatom shapes using the curvature. An obvious way is to use Fourier descriptors (Gonzalez and Woods, 1992, Sect. 8.1.1) of the curvature function, but these have a number of drawbacks. The most notable one is that each Fourier descriptor contains information about all parts of the contour, i.e., localization of features is lost. Even global properties of the contour, such as symmetries, can be difficult to obtain from the descriptors in the presence of noise. Methods based on mathematical morphology (Bangham et al., 1996a/b; Leymarie and Levine, 1988) do not suffer from the delocalization problem, and for this reason they will be explored here.

Whatever the method used, all approaches start by computing a contour representation of the object. Various methods for computing the curvature along the contour of an object use Freeman's chain-code representation as a starting point. This code gives the direction of travel as the points on the contour are followed. Differences between adjacent contour codes can in principle yield curvatures, but this direct approach results in a highly noisy curvature measure due to the small number of possible directions of travel (four or eight). Despite its good compactness, the chain code representation has serious drawbacks when used as contour descriptor. In order to circumvent these problems we follow the approach by Leymarie and Levine (1988).

In order to obtain reliable estimates of image measurements (e.g. perimeter, area, moments, curvature, etc.) based on contour information, one must take into account the noisy nature of discrete contours, which is due to many factors, ranging from discrete sampling and quantization errors to segmentation errors. A frequently used method for dealing with these types of errors, before extracting some useful information, is to smooth the discrete contours first. In general, the purpose of smoothing is twofold: noise is eliminated to facilitate further processing, and features irrelevant to a given problem are ruled out to reduce complexity.

One common method of smoothing is to filter the contour using a Gaussian kernel $G_\sigma(x)$ with width σ, defined as

$$G_\sigma(x) = \frac{1}{\sigma\sqrt{2\pi}} \exp\left(\frac{-x^2}{2\sigma^2}\right). \tag{1}$$

If $\theta(x)$ denotes the chain code, then the curvature $k(x)$ is given by

$$k(x) = (\theta * G_\sigma)'(x) = (\theta * G'_\sigma)(x). \tag{2}$$

Here $*$ denotes convolution and the prime indicates differentiation. This simple smoothing method did not prove to be suitable when applied to diatom contours (Wilkinson et al., 2000), because in straight regions it undersmoothed, whereas in strongly curved regions oversmoothing occurred. Since our purpose is to extract the most important curvature features of the contour, and the curvature signal is very noisy, we need to apply *adaptive* smoothing according to a suitable criterion in order to obtain a tradeoff between the loss of information due to smoothing and the noise reduction. We choose to obtain maximal information in the strongly curved regions, where the mean curvature in a given window has a large value, and to increase the degree of smoothing as the mean curvature decreases, i.e. in weakly curved regions. Therefore, we need to build a function for varying the width of the Gaussian derivative kernel at every point along the contour, with the property that it decreases when the mean curvature increases and vice-versa. A simple example of such a function is given by

$$\sigma(\mu) = \sigma_{min} + (\sigma_{max} - \sigma_{min})/(1 + \frac{\sigma_{max} - 2 \cdot \sigma_{min}}{\mu_0 \cdot \sigma_{min}} \cdot \mu), \qquad (3)$$

where σ_{min} and σ_{max} are the minimum and maximum width of the smoothing kernel and μ is the mean curvature computed in a fixed window of width W obtained after a "pilot" smoothing of the contour with a fixed width σ_p of the kernel. In all experiments below, a value of $\mu_0 = 0.015$ is used. The values of the other three constants ($\sigma_{min} = 0.02$, $\sigma_{max} = 0.08$ and $\sigma_p = 0.035$) were obtained empirically such that the best identification performance was obtained. The steps of the adaptive smoothing method can be summarized as follows:

1. Smooth the contour with a derivative Gaussian kernel of fixed width $\sigma_p = 0.035$.

2. Compute the mean curvature in every point along the diatom contour, in a window of width W. The window width (W), also obtained empirically, is a fraction (5%) of the length of the diatom contour.

3. Evaluate the width of the kernel using the function given in eq. (3) in every point along the contour.

4. Smooth the contour a second time, but now adaptively, using the newly computed values $\sigma(\mu)$ for the width of the kernel.

Figure 1 shows some diatom outline images and their curvature functions after ordinary and adaptive Gaussian smoothing. First of all, it is clear that the different diatoms shown here have very distinct curvature patterns. To demonstrate the advantage of adaptive Gaussian smoothing we shown in Fig. 2 an enlargement of two regions in the curvature function of the left diatom image in Fig. 1, i.e., the intervals [0.2, 0.3] and [0.3, 0.6], respectively. The curvature function on the interval [0.2, 0.3], which corresponds to the leftmost part of the diatom image, shows two distinct peaks in the adaptively smoothed signal, whereas ordinary smoothing has removed one of the peaks. In contrast, the curvature function on the interval [0.3, 0.6], which corresponds to the slowly varying middle lower part of the diatom image, shows a nice smooth curve in the adaptively smoothed signal, whereas ordinary smoothing has failed to remove some of the small noise peaks.

2.2 From curvature to scale-space features

The multi-scale representation is a very useful tool for handling image structures at different scales in a consistent manner (Lindeberg, 1994), and there has been an increasing trend to use multiple scales in image analysis and computer vision.

Chapter 11: Mathematical morphology

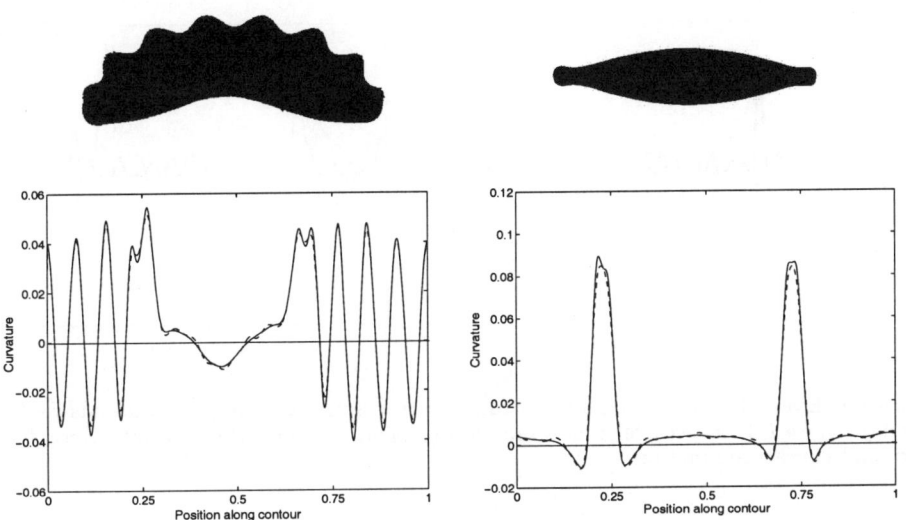

Figure 1. Diatom contours and their corresponding curvatures as a function of position along the contour; *dashed:* Gaussian smoothing, *continuous:* adaptive Gaussian smoothing.

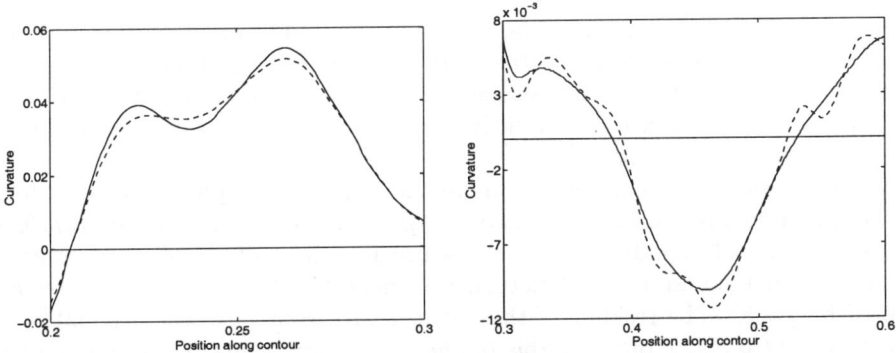

Figure 2. Two enlargements of the curvature function of the left diatom image in Fig. 2.1 corresponding to the intervals [0.2, 0.3] and [0.3, 0.6], respectively; *dashed:* Gaussian smoothing, *continuous:* adaptive Gaussian smoothing. Note the scale difference with Fig. 1.

The basic idea is to embed the original signal into a stack of gradually smoothed signals, in which the fine scale details are successively suppressed.

Consider a signal $f : \mathbb{R}^n \to \mathbb{R}$ and a family of smoothing kernels $g_\sigma : \mathbb{R}^n \times \mathbb{R}$, where σ is the scale parameter. The signal smoothed at scale σ is $F : \mathbb{R}^n \times \mathbb{R} \to \mathbb{R}$:

$$F(x, \sigma) = (f * g_\sigma)(x), \qquad (4)$$

where $*$ again denotes convolution. F is then a function in an $(n + 1)$-dimensional space, called *scale space*, and is known as the *scale-space image* of the signal (Witkin

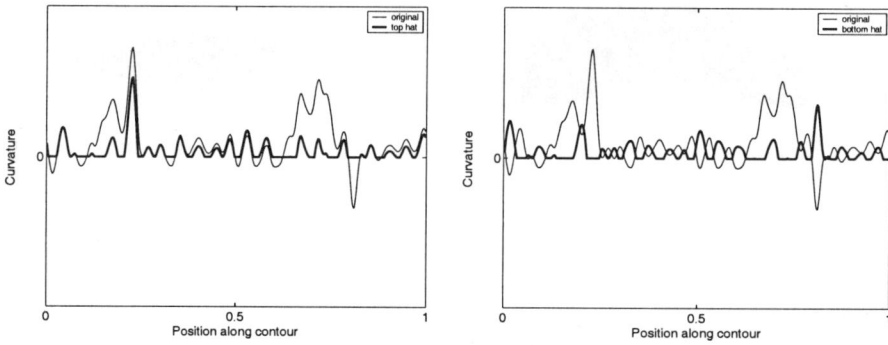

Figure 3. Example of a top-hat and bottom-hat transform, showing the residuals $f - (f \circ K)$ and $f - (f \bullet K)$, respectively, as a function of position along the contour. Here f is the original curvature function.

et al., 1987).

Morphological operators (Serra, 1982; Heijmans, 1994) can remove structure from a signal and therefore they were found suitable for scale-space smoothing (Jackway and Deriche 1996; Chen and Yan, 1989; Park and Lee, 1996). In order to extract the convex and concave contour parts, which correspond to the *peaks* and *valleys* of the curvature signal, we use a method adapted from Leymarie and Levine (1988). This represents an alternative approach for the construction of curvature scale space, in which the position and height of extrema in the signal are preserved with increasing scale until they vanish, rather than obtaining an "evolution" of zero-crossings of the curvature.

The *hat-transforms* represent an important class of morphological transforms used for detail extraction from signals or images. Here we apply the hat-transforms to one-dimensional signals, i.e. the curvature functions of the diatom contours. Assume a signal f and a 1-D structuring element K. The opening of f by K is denoted by $f \circ K$. The residual of the opening with respect to the original signal, that is, $f - (f \circ K)$, represents the *top-hat* transformation. Those parts of the signal where the structuring element K does not fit are removed by the opening. Thus, after subtraction of the opened signal from the original one obtains a signal which contains the desired detail (see Fig. 3). Its dual, the *bottom-hat* transform, is defined as the residual of a closing $f \bullet K$ with respect to the original signal f. Therefore, one can use hat-transforms with increasing size of the structuring element to extract details of increasing size. By performing repeated hat-transforms with increasing size of the structuring element on the curvature signal, we can build the morphological curvature scale spaces. A top curvature scale space consists of a number of levels $\ell = 1, 2, \ldots, L$, where each level ℓ corresponds to a top-hat transform with size δ_ℓ of the structuring element, where δ_ℓ increases with ℓ. Let the smoothed curvature signal be stored in an array C, K_ℓ denote the structuring element used at level ℓ and T_ℓ denote the top-hat $C - (C \circ K_\ell)$. All nonzero elements of T_ℓ are parts of features at scales δ_ℓ or smaller. Starting at level 1, we apply top-hat transforms to extract peaks of maximum curvature. At each level ℓ, T_ℓ is compared to $T_{\ell-1}$. If a peak which is present at level $\ell - 1$ stops increasing at level ℓ, i.e. $T_\ell = T_{\ell-1}$, it is removed by subtraction from the original curvature signal C, and its extremal curvature value, mean curvature, extent and location are stored in a node of a doubly-linked list. This process ends when either all elements of array C are zero, or the largest scale L is reached. This yields the top scale space, in which every peak is precisely localized along the contour. A similar approach is used to obtain the bottom scale space (by performing bottom-hat transforms), in which all valleys are described. At the end of this we have obtained two curvature scale spaces, a top scale space of peaks and a bottom scale space of valleys.

Both scale spaces can contain spurious detail due to noise and therefore a filtering step is necessary to remove those features with a curvature magnitude smaller than a given threshold (0.005 gave us the best results in the final clustering). Also, another approach we found useful for noise removal is to increase the width of the structuring element at each level ℓ not by one, but by a fraction of the contour length. Note that because of the noise filtering it is not possible to reconstruct the original curvature signal from the scale-space data.

Another issue that we need to deal with is nested features. Nested features occur frequently in the data, e.g. when the ends (poles) of diatoms are more or less square. This is the case in the left-hand diatom in Fig. 1. Its ends are indicated by two broad peaks at positions 0.3 resp. 0.7 along the contour, each with smaller peaks superimposed. The smaller peaks correspond to the sharper corners of either end of the diatom. Leymarie and Levine (1988) do not deal with this case, stating that only one curvature feature may be associated with a given section of the contour. The key difference between our approach and that of Leymarie and Levine is that we do allow multiple features to be associated with a given section of the contour. We do this simply by storing all features extracted by the algorithm in the linked list, sorted by position in ascending order. Using position and width information it is possible to determine whether peaks are nested, although we do not yet use this possibility explicitly in the further analysis yet.

A further important difference between our method and that of Leymarie and Levine is that we do not split the boundary into convex and concave parts; instead, we run the top- and bottom-hat transforms on the entire signal.

The scale spaces can be visualized by plotting each feature as a box of either the maximum or the average height at the appropriate point in the curvature graph. If nested features are present, we can simply stack the features in the plot, as has been done in Fig. 4. The current method is sensitive to differences in the relative locations of the curvature features. This means that, for example, an elongated rectangle and a square are separable, because the widths of the major valleys in the curvature correspond to the distances between major peaks, so it is possible to discriminate between these. We gain scale invariance by computing the ratio between the extent of a peak (valley) and the contour length. Obviously, the extracted features are also translation invariant.

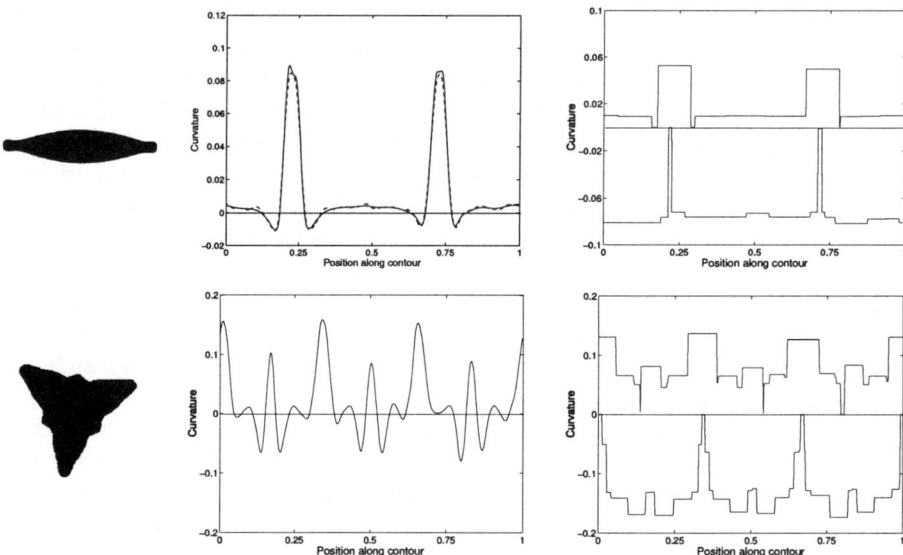

Figure 4. Building curvature spaces. Left: binary images of diatoms; center: curvature plots; right: top and bottom scale spaces represented as curves, showing scale-space features as blocks of the correct width and average height.

2.3 Cluster analysis

In the following we will describe an unsupervised, distribution-free method of cluster analysis, based on kernel density estimation. It should be stressed that any other unsupervised, distribution-free method could be used as well. The result of the cluster analysis is a set of feature vectors to be used for identification (discussed in Section 4).

One approach would be to use the scale-space data directly as pattern vectors for identification, but this causes several problems. The scale space may still contain spurious detail caused by noise. To remove this noise, one may set a limit to the length of the feature vectors, but this introduces the problem as to how this length should be determined. The simplest approach is to use some threshold and count the number of peaks above the threshold. However, finding a single threshold suitable for all diatom shapes is by no means trivial. By selecting a threshold we are in effect creating a very coarse histogram of scale-space features with just two bins. One may use more bins, but here we encounter another dimensionality problem: choosing too few bins results in a poor separation of features, whereas choosing too many implies that most bins will be empty. A better approach would be to set the boundaries between classes of scale-space features from the data themselves, which can be done by cluster analysis. However, no assumptions about the number of clusters or the shape of the distribution should be made *a priori*. Therefore, an unsupervised clustering method should be used.

We decided to use kernel density estimation as the unsupervised clustering method (Silverman, 1986). Kernel density estimates approximate an unknown probability density distribution $p(\mathbf{x})$ from N data elements $\mathbf{x}_i$ by $\hat{p}(\mathbf{x})$, given by

$$\hat{p}(\mathbf{x}) = h^{-d} N^{-1} \sum_{i=1}^{N} K((\mathbf{x}_i - \mathbf{x})/h), \qquad (5)$$

where $K(\mathbf{u})$ is called the kernel, d is the number of dimensions of $\mathbf{x}$, and h is the *window* or smoothing width. Fukunaga and Hostetler (1975) used kernel density estimation to derive the following clustering algorithm, also known as the "mean-shift" method:

1. For each data item $\mathbf{x}_i$: compute the center of gravity (centroid) $\bar{\mathbf{x}}_i$ of all data items which lie within h of $\mathbf{x}_i$ (including $\mathbf{x}_i$ itself!);

2. If, for any i, $\bar{\mathbf{x}}_i \neq \mathbf{x}_i$: replace all $\mathbf{x}_i$ by $\bar{\mathbf{x}}_i$, and repeat the first step;

3. Else: end procedure.

Note that this is different from k-nearest-neighbor (k-NN) methods, since we only ever use those neighbors within radius h. Therefore, the number of neighbors used to assign a new value to each $\mathbf{x}_i$ may vary per iteration.

The method can be made unsupervised by using the optimal, or automatic, choice h_{opt} for the window width which minimizes the mean square integrated error (MSIE):

$$h_{\text{opt}} = \sqrt[d+4]{\frac{8(d+4)2^d \pi^{d/2}}{c_d N}} \sigma, \qquad (6)$$

in which c_d is the volume of a d-dimensional unit sphere, and σ is the square root of the average of the variances of the data in each of the d dimensions. As a final step in the curvature feature extraction, we construct two types of feature vectors:

Type 1 : for both top and bottom curvature scale spaces, select the first two clusters containing the scale-space features with the largest absolute curvatures, and for each cluster compute the number of peaks, mean curvature and variance.

Type 2 : for both top and bottom curvature scale spaces, select the first two clusters containing the scale-space features with the largest absolute curvatures, and for each cluster compute the mean curvature, and the extent as well as variance of the points with the highest curvature.

Hence, in both cases the size of the feature vector is 12 ($2 \times 2 \times 3$). Since we no longer concern ourselves with the positions of points on the contour, the curvature measure is automatically rotation invariant. At the end of this a small set of rotation, translation and scale-invariant shape descriptors is obtained.

2.4 Preliminary identification results

In order to decide which feature set is to be chosen, we performed a number of identification experiments using a decision tree classifier (Jalba et al., 2001). We

Table 1. Identification performance using the decision tree classifier (without bagging), with 10-fold cross-validation for error estimation. The column "$\bar{x}$" shows the average number of errors; the column "σ" lists the standard deviation of the number of errors; the columns "min." and "max." give the minimum and maximum number of errors; the column "perf. (%)" lists the average and standard deviation of samples identified correctly.

feature set	$\bar{x}$	σ	min.	max.	perf. (%)
Sellaphora pupula data set:					
type-1	2.7	0.8	1	4	76.7 ± 2.7
type-2	2.5	1.45	0	4	79.2 ± 4.8
mixed genera data set:					
type-1	34.5	2.5	32	39	56.0 ± 8.3
type-2	21.2	3.5	16	26	73.0 ± 11.7

applied our method to two sets of diatom contour files, the *Sellaphora pupula* data set and the mixed genera data set, as described in Chapter 4.

Table 1 shows the identification performance computed by 10-fold cross-validation of the decision tree based classifier, for both types of feature vectors. The results show that the feature vectors of type 2 give better results than those of type 1, especially for the mixed genera data set. The same observations were found when using other classifiers, such as k-nearest neighbor classification. On the basis of these results we decided to use type-2 feature vectors in the final identification experiments.

In order to further improve robustness of the curvature scale-space features we added to the type-2 pattern vector two other curvature descriptors (not related to curvature scale spaces). The first descriptor is *bending energy*, computed as the integral along the contour of the squared curvature, and the second is *boundary straightness*, i.e. the ratio between the total number of boundary points and the number of boundary points where the contour direction changes significantly. This results in a feature vector of size 14 to be used in the final identification experiments, see Section 4.

3 Analysis of diatom ornamentation

In this section we will discuss the use of a multi-scale method from mathematical morphology for the analysis of diatom ornamentations, using a combination of size and shape distributions (Matheron, 1975; Vincent, 2000). Intuitively, a size distribution can be seen as a set of sieves of different grades, each allowing details of certain sizes to pass. They can be used to classify or extract image details of different size classes (scales). Usually, the width of each detail is the relevant size criterion. Apart from their use in image filtering, size distributions can be used to generate morphological pattern spectra, which summarize the action of a size distribution on a particular image in a single, 1-D array (Nacken, 1994, 1996).

More formally, a size distribution consists of an ordered set of operators, each of which converts an image to a new image in which features smaller than a particular size are absent. These filters must be idempotent, anti-extensive, and increasing, which means they must be *openings*. Many, though not all (Nacken, 1994, 1996), types of openings can be used as granulometries.

Fairly recently, Breen and Jones (1996) proposed a new type of size distribution based on attribute openings, which belong to the class of connected filters. These allow the use of many size criteria other than width, such as area, length of the diagonal of the minimum enclosing rectangle, moment of inertia, etc., to define the "grades" of the morphological sieves. They also put forward the idea of attribute thinnings, which allow image filtering based on shape, rather than size criteria.

Also *shape distributions* will be used, which are ordered sets of scale-invariant thinnings, which are idempotent, anti-extensive, but not increasing (Urbach and Wilkinson, 2001). These allow extraction of pattern spectra based on shapes, rather than sizes of details.

We will first discuss connected-set filters which form the basis of our methods, followed by a theoretical comparison of size and shape distributions. Then we will show how 2-D pattern spectra computed from combined size and shape distributions are obtained. Finally, we will discuss several methods for extracting a small set of rotation-, translation- and scale-invariant texture features, and compare the performance of the corresponding features for diatom identification.

3.1 Theory

The theory of size distributions and connected-set filters is presented only very briefly here. For more detail the reader is referred to Matheron (1975), Vincent (2000), Breen and Jones (1996) as well as Salembier et al. (1998). In the following discussion, binary images X and Y are defined as subsets of the image domain $\mathbf{M} \subset \mathbb{Z}^n$ (usually $n = 2$), and gray-scale images are mappings from $\mathbf{M}$ to $\mathbb{Z}$.

3.1.1 Size distributions

Definition 1 *A binary size distribution or granulometry is a set of operators $\{\alpha_r\}$ with r belonging to some totally ordered set Λ (usually $\Lambda \subset \mathbb{R}$ or $\mathbb{Z}$), with the following three properties*

$$\alpha_r(X) \subset X \tag{7}$$

$$X \subset Y \Rightarrow \alpha_r(X) \subset \alpha_r(Y) \tag{8}$$

$$\alpha_r(\alpha_s(X)) = \alpha_{\max(r,s)}(X), \tag{9}$$

for all $r, s \in \Lambda$.

Since equations (7) and (8) define α_r as anti-extensive and increasing, respectively, and eq. (9) implies idempotence, it can be seen that size distributions are openings. Generalization to the gray-scale case is straightforward (Nacken, 1994, 1996; Vincent, 2000).

3.1.2 Shape Distributions

Let us define a scaling X_λ of set X by a scalar factor $\lambda \in \mathbb{R}$ as

$$X_\lambda = \{x \in \mathbb{Z}^n | \lambda^{-1} x \in X\}. \tag{10}$$

Likewise, a scaling f_λ of a gray-scale image f is defined as

$$f_\lambda(x) = f(\lambda^{-1}x) \quad \forall \lambda^{-1}x \in \mathbf{M}. \tag{11}$$

An operator ϕ is said to be *scale invariant* if

$$\phi(X_\lambda) = (\phi(X))_\lambda \quad \text{or} \quad \phi(f_\lambda) = (\phi(f))_\lambda \tag{12}$$

for all $\lambda > 0$. A scale-invariant operator is therefore sensitive to shape rather than to size. If an operator is scale, rotation and translation invariant, we call it a *shape operator*.

Definition 2 *A binary shape distribution is a set of operators $\{\beta_r\}$ with r belonging to some totally ordered set Λ, with the following three properties*

$$\beta_r(X) \subset X \tag{13}$$
$$\beta_r(X_\lambda) = (\beta_r(X))_\lambda \tag{14}$$
$$\beta_r(\beta_s(X)) = \beta_{\max(r,s)}(X), \tag{15}$$

for all $r, s \in \Lambda$ and $\lambda > 0$.

Thus, a shape distribution consists of operators which are anti-extensive, and idempotent, but not necessarily increasing. Therefore, the operators must be thinnings, rather than openings. To exclude any sensitivity to size, we add property (14), which is just scale invariance for all β_r. Extension to the gray level case is straightforward:

Definition 3 *A gray-scale shape distribution is a set of operators $\{\beta_r\}$ with r from some totally ordered set Λ, with the following three properties*

$$(\beta_r(f))(x) \leq f(x) \tag{16}$$
$$\beta_r(f_\lambda) = (\beta_r(f))_\lambda \tag{17}$$
$$\beta_r(\beta_s(f)) = \beta_{\max(r,s)}(f), \tag{18}$$

for all $r, s \in \Lambda$ and $\lambda > 0$.

We will later show that certain attribute thinnings are shape filters with the desired properties for shape distributions.

3.1.3 Shape and size pattern spectra

The pattern spectra $s_\alpha(X)$ and $s_\beta(X)$ obtained by applying size and shape distributions $\{\alpha_r\}$ and $\{\beta_r\}$ to a binary image X are defined as

$$(s_\alpha(X))(u) = -\left.\frac{\mathrm{d}A(\alpha_r(X))}{\mathrm{d}r}\right|_{r=u} \tag{19}$$

and

$$(s_\beta(X))(u) = -\left.\frac{\mathrm{d}A(\beta_r(X))}{\mathrm{d}r}\right|_{r=u}, \tag{20}$$

in which $A(X)$ is a function denoting the Lebesgue measure in $\mathbb{R}^n$. In the gray-scale case, the pattern spectrum is usually defined in terms of the sum or integral of the gray levels over the image domain as a function of r:

$$(s_\alpha(f))(u) = \left.\frac{d \int_M (\alpha_r(f))(x)dx}{dr}\right|_{r=u}, \qquad (21)$$

and likewise for $s_\beta(f)$.

In the discrete case, we can compute a pattern spectrum by repeatedly filtering an image by each α_r or β_r, in ascending order of r. After each filtering step, the sum S_r of the gray levels of the resulting image $\alpha_r(f)$ is computed. The pattern spectrum value at r is the difference between S_{r^-} and S_r, with r^- the scale (or shape class) immediately preceding r. In practice, faster methods for computing pattern spectra can be used (Meijster and Wilkinson, 2001; Nacken; 1994, 1996; Breen and Jones, 1996), see also Section 3.3.

3.1.4 Shape decomposition

If we wish to decompose an image into its constituent components based on shape rather than size, we would intuitively require that after filtering with a gray-scale attribute thinning ϕ_r^T derived from a binary attribute thinning Φ_r^T, no structures which do not meet the criterion T are present in the resulting image. Moreover, the difference image $f - \phi_r^T(f)$ should only contain structures which do not meet criterion T. Let the *peak component* P_h^k at gray level h be the kth connected component of the threshold set $X_h(f)$ of the image f:

$$X_h(f) = \{x \in M | f(x) \geq h\}. \qquad (22)$$

The intuitive requirements now mean that all peak components of $\phi_r^T(f)$ meet T, and all peak components of $f - \phi_r^T(f)$ do not. More formally,

$$\Phi_r^T(X_h(\phi_r^T(f))) = X_h(\phi_r^T(f)) \qquad (23)$$

and

$$\Phi_r^T(X_h(f - \phi_r^T(f))) = \emptyset \qquad (24)$$

for all h.

3.1.5 Attribute operators

Binary attribute openings and thinnings are based on binary connected openings. The binary connected opening $\Gamma_x(X)$ of X at point $x \in M$ yields the connected component of X containing x if $x \in X$, and $\emptyset$ otherwise. Thus Γ_x extracts the connected component to which x belongs, discarding all others. Breen and Jones then use the concept of trivial thinnings Φ_T, which use a non-increasing criterion T to accept or reject connected sets. A criterion T is increasing if the fact that C satisfies T implies that D satisfies T for all $D \supseteq C$. The trivial thinning Φ_T with criterion T of a connected set C is just the set C if C satisfies T, and is empty otherwise. Furthermore, $\Phi_T(\emptyset) = \emptyset$. The binary attribute thinning is defined as follows:

Definition 4 *The binary attribute thinning Φ^T of set X with criterion T is given by*

$$\Phi^T(X) = \bigcup_{x \in X} \Phi_T(\Gamma_x(X)). \tag{25}$$

It can be shown that this is a thinning because it is idempotent and anti-extensive (Breen and Jones, 1996). The attribute thinning is equivalent to performing a trivial thinning on all connected components in the image, i.e., removing all connected components which do not meet the criterion. It is trivial to show that if the criterion T is scale invariant, i.e.

$$T(C) = T(C_\lambda) \quad \forall \lambda > 0 \wedge C \subseteq \mathbf{M}, \tag{26}$$

so are Φ_T and Φ^T. Assume $T(C)$ can be written as $\tau(C) > r$, $r \in \Lambda$, with τ some scale-invariant attribute of the connected set C. Let the attribute thinnings formed by this T be denoted as Φ_r^τ. It can readily be shown that

$$\Phi_r^\tau(\Phi_s^\tau(X)) = \Phi_{\max(r,s)}^\tau(X). \tag{27}$$

Therefore, $\{\Phi_r^\tau\}$ is a shape distribution, since attribute thinnings are anti-extensive, and scale invariance is provided by the scale invariance of $\tau(C)$.

Examples of scale-invariant shape attributes abound. Obvious choices are P^2/A, with P the perimeter and A the area of a component, or I/A^2, which is the ratio of the moment of inertia I to the square of the area. These attributes attain minimal values for circular discs, and increase as the objects become more elongate. For computational reasons, the latter shape attribute is used, because both moment of inertia and area can easily and accurately be computed incrementally, whereas this is more difficult in the case of perimeter. A fuller discussion of shape attributes is given by Breen and Jones (1996) and Salembier et al. (1998).

Various gray-scale generalizations of these filters have been proposed and are compared in Salembier et al. (1998). An efficient algorithm for computing these filters based on a tree of *flat zones* in the image was also put forward in the same paper. The filtering is reduced to different methods of "pruning" this tree structure. In the following subsection this algorithm and various thinnings based on different pruning strategies are discussed.

3.2 The Max-Tree

The *Max-Tree* representation was introduced by Salembier et al. (1998) as a more versatile structure to separate the filtering process in three steps: construction, filtering and restitution. It is a tree where the nodes represent sets of flat zones, i.e. connected components of pixels of constant gray value. The Max-Tree node C_h^k consists of the subset of a peak component P_h^k with gray level h. The root node represents the set of pixels belonging to the background, that is the set of pixels with the lowest intensity in the image. The Max-Tree is a rooted tree: each node has a pointer to its parent, i.e., the nodes corresponding to the components with the highest intensity are the leaves (see Fig. 5), hence the name Max-Tree: the leaves correspond to the regional maxima. This means that the Max-Tree can be used for attribute openings or thinnings. Conversely, a tree in which the leaves

Chapter 11: Mathematical morphology 235

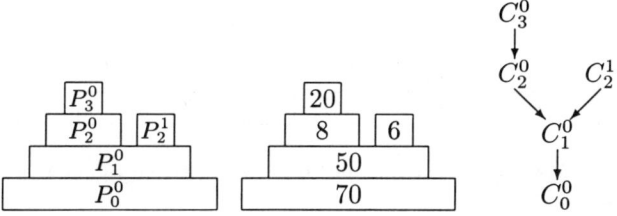

Figure 5. The peak components of a gray level image X (left), the corresponding attributes (middle) and the resulting Max-Tree (right).

correspond to the minima is called a Min-Tree and can be used for attribute closings or thickenings. During the construction phase, the Max-Tree is built from the flat zones of the image.

After this, the tree is processed during the filtering phase. Based on the criterion value $T(P_h^k)$ of a node C_h^k, the algorithm takes a decision to preserve or to remove it. Two classes of strategies exist: (i) pruning strategies, which remove all descendants of C_h^k, if C_h^k is removed, and (ii) non-pruning strategies in which the parent pointers of children of C_h^k are updated to point at the oldest "surviving" ancestor of C_h^k. If T is non-increasing, such as perimeter or moment of inertia divided by the square of the area, pruning strategies must either remove nodes which meet T, or leave nodes which do not. It is trivial to see that no pruning strategy can meet both decomposition properties of equations (23) and (24). Salembier et al. (1998) describe four different rules for the algorithm to filter the tree: the *Min*, the *Max*, the *Viterbi* and the *Direct* decision. The first three are pruning strategies. In addition to these, we introduce a new, non-pruning strategy: the *Subtractive* decision. Of these five, the subtractive method alone satisfies both equations (23) and (24). The decisions of these rules are as follows:

Min : A node C_h^k is removed if $T(P_h^k) < r$ or if one of its ancestors is removed.

Max : A node C_h^k is removed if $T(P_h^k) < r$ and for all its $C_{h'}^j$, $T(P_{h'}^j) < r$.

Viterbi : The removal and preservation of nodes is considered as an optimization problem. For each leaf node, the path with the lowest cost to the root node is taken, where a cost is assigned to each transition. For details see Salembier et al. (1998).

Direct : A node C_h^k is removed if $T(P_h^k) < r$; its pixels are lowered in gray level to the highest ancestor which meets T, but its descendents are unaffected.

Subtractive : As above, but the descendents are lowered by the same amount as C_h^k itself.

Figure 5 shows the peak components of a 1-D signal, their attribute values, and the corresponding Max-Tree. The results of applying the Min, Max, Direct and Subtractive methods to this example with $r = 10$ are shown in Fig. 6. Which of these rules is the most appropriate one depends mainly on the application, e.g.

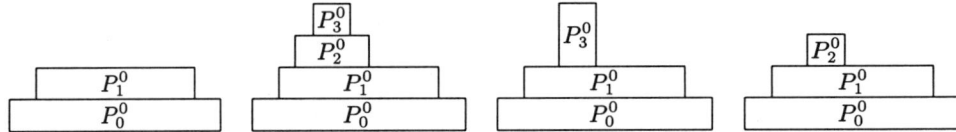

Figure 6. Result after filtering the example in Fig. 5 with (left to right) Min, Max, Direct and Subtractive decisions.

image filtering or decomposition. From experiments with the Min, Max, Direct and Subtractive methods, we concluded that the subtractive rule is one which works best for decomposition (Wilkinson et al., 2001). For this reason, we have chosen to use it in the computation of shape pattern spectra.

3.3 Algorithms

To compute the shape-size distribution of the diatom ornamentation we require two images: one gray-scale image and one binary image of the valve, both at the same scale. The latter image is used as a mask to indicate which pixels in the gray-scale image belong to the ornamentation. All image processing operations described below are restricted to the ornamentation pixels.

3.3.1 2-D Shape-size pattern spectra

Suppose we have a size distribution $\{\alpha_r\}$ with r from some index set Λ_r, and a shape distribution $\{\beta_s\}$ with s from some index set Λ_s. With these distributions, we can compute a 2-D pattern spectrum, similar to that of Ghosh and Chanda (1998). We can then store a 2-D pattern spectrum, which contains both shape and scale information in a 2-D array S, in which $S(r,s)$ contains the sum of gray levels of those features in the image which fall in the size class between r^- and r, and shape class between s^- and s. Note that r^- and s^- are the members of Λ_r and Λ_s immediately preceding r and s, respectively. The computation of a 2-D pattern spectrum from a diatom image is performed as follows:

- Set all elements of the array S to zero;

- Compute a Max-tree according to the algorithm in Salembier et al. (1998);

- As the Max-tree is built, compute the area $A(P_h^k)$ and moment of inertia $I(P_h^k)$ of each node;

- For each node C_h^k:
  - Compute the size class r from the area $A(P_h^k)$;
  - Compute the shape class s from the shape parameter $I(P_h^k)/A^2(P_h^k)$;
  - Compute the gray level difference δ_h between the current node and its parent;
  - Add the product of δ_h and $A(P_h^k)$ to $S(r,s)$.

In our case $r,s \in \{1,2,\ldots,32\}$.

3.3.2 Extracting feature vectors from pattern spectra

The mapping of a 2-D pattern spectrum to a 1-D feature vector can be done in various ways. Three methods were implemented: (i) down-sizing the 2-D pattern spectrum into a 4×4 matrix, which is then mapped onto a 1-D vector (in lexicographic order), (ii) computing pattern spectrum moments, which are well-studied for computing a compact feature vector from a pattern spectrum (Sand and Dougherty, 1992), and (iii) selection of the 16 "most important" elements from the pattern spectrum.

The first method is the simplest, and reduces to computing a pattern spectrum with a greatly reduced set of classes, each of which contains a wider range of scales and shapes.

The second method uses moments of the pattern spectrum to compute the feature vector $\vec{v} = \{x_0, x_1 ... x_{15}\}$ from the pattern spectrum. Instead of using the discrete pattern spectrum array, this method uses the criterion values of the Max-tree directly for more precise results. This way, the pattern spectrum can be considered as being continuous. Four kinds of moments (Gonzalez and Wintz, 1987) are computed:

Moments: $$m_{pq} = \int_{-\infty}^{\infty} \int_{-\infty}^{\infty} x^p y^q f(x,y)\, dx\, dy \qquad (28)$$

Central moments: $$\mu_{pq} = \int_{-\infty}^{\infty} \int_{-\infty}^{\infty} (x-\bar{x})^p (y-\bar{y})^q f(x,y)\, dx\, dy \qquad (29)$$

Normalized moments: $$n_{pq} = m_{pq}/m_{00}^{\gamma} \qquad (30)$$

Normalized central moments: $$\eta_{pq} = \mu_{pq}/\mu_{00}^{\gamma} \qquad (31)$$

where

$$\bar{x} = m_{10}/m_{00}, \quad \bar{y} = m_{01}/m_{00} \quad \text{and} \quad \gamma = 1 + (p+q)/2.$$

In the implementation the integrations $\int_{-\infty}^{\infty}$ in these formulae are replaced by $\sum_{D_0}^{D_1}$, where D_0 and D_1 specify the range of the criterion as discussed earlier. The p and q specify the order of the moment about respectively the horizontal and the vertical axis of the pattern spectrum.

The *Moments* method creates a two-dimensional array A of 4×4 elements using these moment functions. The normalized central moments are the most appropriate to use, but $\eta_{00} = 1$ and $\eta_{10} = \eta_{01} = 0$. Therefore, this method computes A as follows:

$$A[p,q] = \begin{cases} m_{00} & \text{if } p=0 \wedge q=0, \\ n_{01} & \text{if } p=0 \wedge q=1, \\ n_{10} & \text{if } p=1 \wedge q=0, \\ \eta_{pq} & \text{otherwise.} \end{cases}$$

The mapping of A to $\vec{v}$ is done lexicographically, as in the down-sizing method.

The third method has proven to be rather arbitrary, and lacks robustness, unless reasonable objective criteria can be put forward on how to select good features.

We have done some experiments using population variance or entropy of features, but the results were far from satisfactory, hence further refinements were not attempted. The feature vector described above is not necessarily the final one, as will be explained below.

3.3.3 Combination

Until now the assumption was made that pattern spectra are based on 2-D size-shape distributions. However, many images are identified using both the bright and the dark patterns. To obtain information about the dark patterns anti-size and anti-shape distributions are required, which are based on closings and thickenings, respectively. Instead of designing new algorithms for the computation of pattern spectra based on these anti-size and anti-shape distributions, these were implemented by computing pattern spectra of the inverted image.

The question now arises whether the pattern spectrum based on the normal or the inverted image should be used, or maybe even a combination of these two. Therefore the following combinations were implemented:

- *Normal*: The pattern spectrum based on the normal image is used. This means that only information about the bright structures in the image is known.

- *Inverted*: This one uses the pattern spectrum of the inverted image, which means that the dark structures in the image can be investigated.

- *Add*: Two temporary feature vectors are computed according to the methods *Normal* and *Inverted* described above: $\vec{v_n}$ and $\vec{v_i}$ respectively. The final feature vector $\vec{v_r}$ is computed as $\vec{v_r} = \vec{v_n} + \vec{v_i}$. An advantage of this method is that both bright and dark image details are represented in this vector. Furthermore, dark and bright details with the same criterion values are represented by the same class. This means that this feature vector is invariant to inversion, i.e., inverting an image does not change the result, which is useful for diatom identification because choosing a different focus depth can make bright structures turn dark, or vice-versa.

- *Join*: Like in the previous method, two vectors $\vec{v_n}$ and $\vec{v_i}$ are computed. Now the final feature vector $\vec{v_r}$ is the concatenation of $\vec{v_n}$ and $\vec{v_i}$. Note that this vector is twice as large as with the other methods. Again bright and dark image details are present in the result, but now bright and dark details with the same criterion values are represented by different classes. The vector $\vec{v_r}$ is not invariant to inversion, which can be a disadvantage, but on the other hand this also means that $\vec{v_r}$ now contains more information about the image than with the previous method.

3.4 Preliminary identification tests

A preliminary identification test was first tried on the mixed genera data set (see Chapter 4). This test was performed using all possible configurations of the mapping and the combination methods in order to find the best configuration. The

Table 2. Identification results using the decision tree classifier with bagging for different configurations of the mapping and combination methods. For an explanation of the columns, see Table 1.

feature set	$\bar{x}$	σ	min	max	perf. (%)
matrix invert	49.9	5.3	43	57	73.0 ± 2.84
matrix normal	48.9	5.7	42	59	73.6 ± 3.06
matrix join	32	6.5	23	46	82.7 ± 3.51
matrix add	40.8	7.4	27	47	77.9 ± 4.01
moments invert	79.8	6.7	70	92	56.9 ± 3.63
moments normal	82.7	5.1	76	91	55.3 ± 2.73
moments join	55.8	4.8	49	63	69.8 ± 2.58
moments add	94.7	5.1	86	101	48.8 ± 2.78

identification results using decision trees with bagging (see Section 4 for a more detailed description of the procedure) are shown in Table 2.

The first column shows the different configurations of the mapping and combination methods. The next four columns show the the average, standard deviation, minimum and maximum number of errors, respectively. The last column lists the average and standard deviation of the percentage of diatoms identified correctly. Based on these results we decided to use the *Matrix* mapping method and the *Join* combination method in the final identification experiments, see Section 4.

4 Identification results

In this section we describe the final identification results based on the feature sets chosen on the basis of the preliminary identification tests described above. We present here the results when using decision trees with bagging, as described in detail in Chapter 7. In the experiments, the C4.5 algorithm is used for constructing the decision trees (Quinlan, 1993). We give results both for separate as well as combined feature sets.

Again, the experiments were conducted on the *Sellaphora pupula* and the mixed genera data sets (see Chapter 4). In both cases, each taxon is represented by at least 20 samples. The *Sellaphora pupula* set has six demes and a total of 120 samples, whereas the subset of the mixed genera set used here counts 37 taxa with a total of 781 samples.

The procedure employed for constructing the bagging predictors resembles that in Breiman (1996) and is as follows:

1. Make a random division of the data into a training set and a test set, such that the test set contains exactly five samples of each taxon. The size of the test set then becomes about 25% of the total data set.

2. Construct 25 new training sets from the initial training set using bootstrapping, and build a decision tree for each of them.

3. Evaluate the 25 decision tree classifiers on the test set, and take a majority vote on the outputs of all trees.

Table 3. Identification performance for the *Sellaphora pupula* data set, based on three morphological feature sets: contour features (curvature scale space), ornamentation features (size/shape distribution), and a combination of these. The numbers between parentheses refer to the number of features. For an explanation of the columns, see Table 1. The sizes of the training and test sets were 90 resp. 30 samples.

feature set	$\bar{x}$	σ	min.	max.	perf. (%)
contour features (14)	7.6	5.8	2	10	75.1 ± 19.5
ornamentation features (32)	9.0	4.4	5	12	70.0 ± 14.6
combined features (46)	4.9	3.5	2	8	83.7 ± 11.7

Table 4. As Table 3, but for the mixed genera data set. The sizes of the training and test sets were 596 and 185 samples, respectively.

feature set	$\bar{x}$	σ	min.	max.	perf. (%)
contour features (14)	38.9	4.4	32	44	79.0 ± 2.4
ornamentation features (32)	34.4	4.1	28	40	81.4 ± 2.2
combined features (46)	19.2	3.5	15	25	89.6 ± 1.9

4. Repeat the above steps 10 times and average the results.

4.1 Identification performance

Tables 3 and 4 show the identification performance when using (a) contour features (curvature scale space, 14 features), (b) ornamentation features (size/shape distribution, 32 features), and (c) the combination of these. Each table lists the minimum and maximum errors for each feature set, the average error, and the standard deviation. The performance is given as the percentage (average over all runs with standard deviation) of samples identified correctly. In the case of the *Sellaphora pupula* data set, the performance based on contour features is higher than that based on ornamentation features. This is not surprising, since the demes of the *Sellaphora pupula* set have very similar ornamentations. The combined feature sets yielded a performance of about 84% correct identification. However, we note the large relative variances for this data set (total size of the training and test sets was 90 and 30 samples, respectively). In the case of the mixed genera data set, it is possible to correctly identify about 80% of the diatoms using only the contour or ornamentation features. By combining contour and ornamentation features, the ID performance increases to 90%.

4.2 Identification performance with ranking

Instead of using majority voting of the bagged predictor, we can also rank the outputs of the individual decision trees with respect to the frequency by which a certain class label is chosen. We did not treat ties in a special way. For example, if the rank-1 and rank-2 frequencies were equal, the order was kept as produced by the sorting algorithm. Also, in the bagging procedure, each individual classifier

Table 5. Identification performance (%) for the mixed genera data set using bagging with ranking. The numbers between parentheses refer to the number of features. The columns labelled "rank 1" to "rank 5" correspond to the rank-r classifier, with r ranging from 1 to 5. The rank-1 classification is identical to majority voting.

feature set	rank 1	rank 2	rank 3	rank 4	rank 5
contour features (14)	79.0	87.9	91.5	93.5	94.4
ornamentation features (32)	81.4	90.8	94.0	95.4	96.3
combined features (46)	89.6	94.9	96.6	97.2	97.5

had an equal weight. Table 5 shows the results when we extend the search to the first five ranked matches. The column labelled "rank 1" represents the result of majority voting, whereas the next columns present the 2nd, 3rd ... 5th best matches. We can observe that, as the rank is increased, the performance steadily increases, approaching almost 98% for the combined feature set.

5 Conclusions

In this chapter we have reported methods based on mathematical morphology for automatic diatom identification. Two types of feature extraction methods were described. The first method uses morphological curvature scale spaces, combined with unsupervised cluster analysis, to analyze features of the diatom outline for contour feature extraction. We have used a multi-scale approach to the analysis of 1-D signals based on the work by Leymarie and Levine (1988), proposed some modifications of their method to include nested features, and developed a method for salient feature extraction. The unsupervised clustering of raw scale-space data turned out to be a useful method for this purpose. A second method was used to compute features of the ornamentation, i.e., the valve striation. Again, a multi-scale method from mathematical morphology was used, combining size and shape distributions.

On the basis of preliminary identification experiments, feature sets of size 14 and size 32 were chosen to represent the contour and the ornamentation, respectively. These sets of feature vectors were used as input for decision trees with bagging as pattern classifier. Identification results based on separate as well as combined feature sets were presented. In the case of the mixed genera data set, we found identification rates of 79 ± 2.4% for the contour-based features, 81.4 ± 2.2% for the ornamentation features, and 89.6 ± 1.9% when combining both feature sets.

The identification performance of the ornamentation feature set is among the best of all texture operators tested so far within ADIAC (see Chapters 7, 10 and 12). Only the "classical" texture features from the Bern group perform better (84.3 ± 1.9%), but this difference may not be significant taking the standard deviations into account. Besides, the latter method uses some 80 features, compared to only 32 features of the pattern-spectrum method.

In the case of the *Sellaphora pupula* data set, we found identification rates of 75.1 ± 19.5% for the contour-based features, 70 ± 14.6% for the ornamentation features, and 83.7 ± 11.7% when combining both feature sets. These results are worse than those obtained for the mixed genera data set, but the difficulty of this set is reflected in the large standard deviations of the identification rates.

Many improvements of the methods as presented here are possible. For the contour analysis, more research is needed on how to choose various parameters, which have here been determined empirically. For the ornamentation features, a study of robust criteria for the selection of pattern-spectrum features is needed. Also, a comparison with diatom identification based on other methods is needed as well, including computational efficiency as one of the relevant criteria.

References

Bangham, J.A., Chardaire, P., Pye, C.J. and Ling, P.D. (1996a) Multiscale nonlinear decomposition: the sieve decomposition theorem. IEEE Trans. Pattern Analysis and Machine Intelligence, Vol. 18, pp. 529-538.

Bangham, J.A., Ling, P.D. and Harvey, R. (1996b) Scale-space from nonlinear filters. IEEE Trans. Pattern Analysis and Machine Intelligence, Vol. 18, pp. 520-528.

Breen, E.J. and Jones, R. (1996) Attribute openings, thinnings and granulometries. Computer Vision and Image Understanding, Vol. 64, pp. 377-389.

Breiman, L. (1996) Bagging predictors. Machine Learning, Vol. 24, pp. 123-140.

Chen, M.H. and Yan, P.F. (1989) A multiscale approach based on morphological filtering. IEEE Trans. Pattern Analysis and Machine Intelligence, Vol. 11, pp. 694-700.

Fukunaga, K. and Hostetler, L.D. (1975) Estimation of the gradient of a density function with applications in pattern recognition. IEEE Trans. Information Theory, Vol. 21, pp. 32-40.

Ghosh, P. and Chanda, B. (1998) Bi-variate pattern spectrum. Proc. SIBGRAPI'98, IEEE Computer Society, Rio de Janeiro, pp. 476-483.

Giardina, C.R. and Dougherty, E.R. (1988) Morphological methods in image and signal processing. Prentice Hall, Englewood Cliffs, NJ.

Gonzalez, R.C. and Wintz, P. (1987) Digital image processing. Addison-Wesley, Reading, MA, 2nd Edition.

Gonzalez, R.C. and Woods, R.E. (1992) Digital image processing. Addison-Wesley, Reading, MA.

Haralick, R.M., Sternberg, S.R. and Zhuang, X. (1987) Image analysis using mathematical morphology. IEEE Trans. Pattern Analysis and Machine Intelligence, Vol. 9, pp. 532-550.

Heijmans, H.J.A.M. (1994) Morphological image operators. Vol. 25 of: Advances in electronics and electron physics, Supplement. Academic Press, New York.

Jackway, P.T. and Deriche, M. (1996) Scale-space properties of the multiscale morphological dilation-erosion. IEEE Trans. Pattern Analysis and Machine Intelligence, Vol. 18, pp. 38-51.

Jalba, A.C., Wilkinson, M.H.F., Roerdink, J.B.T.M., Bayer, M.M. and Juggins, S. (2001) Automatic diatom identification using contour analysis by morphological curvature scale spaces. Technical report 2001-9-05, Inst. for Mathematics and Computing Science, Univ. of Groningen, The Netherlands.

Leymarie, F. and Levine, M.D. (1988) Curvature morphology. Technical Report TR-CIM-88-26, Computer Vision and Robotics Laboratory, McGill Univ., Montreal, Quebec, Canada.

Leyton, M. (1987) Symmetry-curvature duality. Computer Vision Graphics and Image Processing, Vol. 38, pp. 327-341.

Lindeberg, T. (1994) Scale-space theory: a basic tool for analysing structures at different scales. J. of Applied Statistics, Vol. 21, pp. 225-270.

Matheron, G. (1975) Random sets and integral geometry. John Wiley.

Meijster, A. and Wilkinson, M.H.F. (2001) Fast computation of morphological area pattern spectra. Proc. Int. Conf. on Image Processing, Oct. 7-10, Thessaloniki, Greece, pp. 668-671.

Nacken, P.F.M. (1994) Chamfer metrics in mathematical morphology. J. Mathematical Imaging and Vision, Vol. 4, pp. 233-253.

Nacken, P.F.M. (1996) Chamfer metrics, the medial axis and mathematical morphology. J. Mathematical Imaging and Vision, Vol. 6, pp. 235-248.

Park, K.-R. and Lee, C.-N. (1996) Scale-space using mathematical morphology. IEEE Trans. Pattern Analysis and Machine Intelligence, Vol. 18, pp. 1121-1126.

Pavlidis, T. (1980) Algorithms for shape analysis of contours and waveforms. IEEE Trans. Pattern Analysis and Machine Intelligence, Vol. 2, pp. 301-312.

Quinlan, J.R. (1993) C4.5: programs for machine learning. Morgan Kaufmann Publishers, San Francisco, CA.

Roerdink, J.B.T.M. and Meijster, A. (2000) The watershed transform: definitions, algorithms, and parallelization strategies. Fundamenta Informaticae, Vol. 41, pp. 187-228.

Salembier, P., Oliveras, A. and Garrido, L. (1998) Anti-extensive connected operators for image and sequence processing. IEEE Trans. Image Processing, Vol. 7, pp. 555-570.

Sand, F. and Dougherty, E.R. (1992) Statistics of the morphological pattern spectrum moments for a random grain model. J. of Mathematical Imaging and Vision, Vol. 1, pp. 121-135.

Serra, J. (1982) Image analysis and mathematical morphology. Academic Press, New York.

Serra, J. (1986) Introduction to mathematical morphology. Computer Vision Graphics and Image Processing, Vol. 35, pp. 283-305.

Serra, J., (1988) Image analysis and mathematical morphology. II: Theoretical advances. Academic Press, New York.

Serra, J. and Vincent, L. (1992) An overview of morphological filtering. Circuits, Systems and Signal Processing, Vol. 11, pp. 47-108.

Silverman, B.W. (1986) Density estimation for statistics and data analysis. Chapman and Hall, London.

Sternberg, S.R. (1986) Grayscale morphology. Computer Vision Graphics and Image Processing, Vol. 35, pp. 333-355.

Urbach, E.R. and Wilkinson, M.H.F. (2001) Shape distributions and decomposition of grey scale images. IWI-report 2000-9-15, Inst. for Mathematics and Computing Science, Univ. of Groningen, The Netherlands.

Vincent, L. (2000) Granulometries and opening trees. Fundamenta Informaticae, Vol. 41, pp. 57-90.

Wilkinson, M.H.F., Roerdink, J.B.T.M., Droop, S. and Bayer, M. (2000) Diatom contour analysis using morphological curvature scale spaces. Proc. 15th Int. Conf. on Pattern Recognition (ICPR'2000), Barcelona, Spain, Sept. 3-7, pp. 656-659.

Wilkinson, M.H.F., Urbach, E.R. and Roerdink, J.B.T.M. (2001) Connected size-shape distributions for texture-based identification of diatoms. IEEE Trans. on Image Processing, submitted.

Witkin, A.P., Terzopoulos, D. and Kass, M. (1987) Signal matching through scale space. Int. J. Computer Vision, Vol. 1, no. 2, pp. 133-144.

CHAPTER 12

MIXED-METHOD IDENTIFICATIONS

MICHEL A. WESTENBERG AND JOS B.T.M. ROERDINK

In this chapter we report on identification experiments based on combinations of feature sets, as described in the previous chapters, using the mixed genera data set with 37 taxa. We developed an application framework that integrates the contributions of the project partners to make these mixed-method identifications possible. Identification performance is measured by bootstrap aggregating (bagging) C4.5 decision trees. Combinations of contour-based features show that over 90% of the diatoms can be identified correctly, and a similar result is obtained using ornamentation features. If all features are combined, the identification rate increases to almost 97%. From the analysis of a collection of 25 decision trees, a set of 17 robust features, that were used by at least 12 trees, is selected. This small feature set yields an identification rate of almost 96%. Because a few feature sets were still under development at the time of writing (convex/concave curvature, Legendre polynomials), it is expected that the same experiments on the basis of the final feature sets will result in even better ID rates. This chapter also describes a web-based application, called *ADIACweb*, that allows users to interact with the automatic identification system. Currently, it can identify 37 different taxa.

1 Introduction

This chapter presents the identification performance obtained through various combinations of the feature sets which have been described in Chapters 7 to 11, as available about half a year before the end of the ADIAC project. Using the bagged C4.5 decision tree classifier, combined contour-based features result in correct identification of over 90% of the diatoms, and a similar result is obtained using ornamentation features. If all features are combined, identification performance increases to almost 97%. Unfortunately, the convex/concave curvature features (Chapter 8) and Legendre polynomials (Chapter 9) were still under development. However, this is not a problem because similar experiments based on combinations of feature sets that include the two sets mentioned, can only yield even better results. To make these mixed-method identifications possible, an application framework has been developed that integrates the contributions of the project partners.

Furthermore, this chapter describes a web-based application called *ADIACweb*, which allows users to interact with the automatic identification system. Currently, *ADIACweb* can identify 37 different taxa. Usage is restricted to a set of preselected images that can be submitted to the system, but it can be easily extended such that users are allowed to submit their own images for identification.

The organization of this chapter is as follows. Section 2 describes the framework used for the integration. This framework is then applied to combine feature sets, and the identification results are reported in Section 3. Section 4 discusses *ADIACweb*.

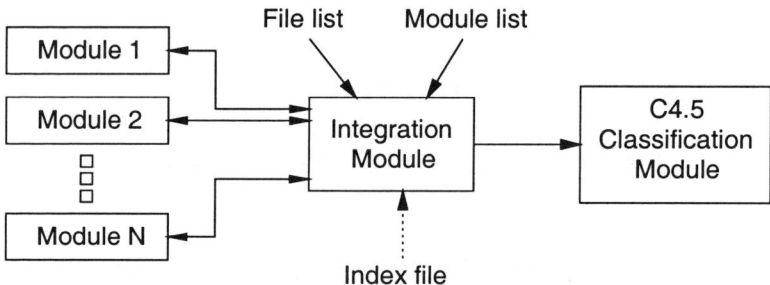

Figure 1. Module integration framework.

2 Integration module framework

The development of software with a team that is distributed geographically and organizationally is a difficult problem. When software is developed in the context of a pilot project like ADIAC, it is even more difficult, since such projects entail much more than just an implementation of existing methods. In such cases, making a formal specification of the system architecture is not advantageous for the development cycle. To bypass these problems, it is better to let all partners develop their own software independently, and to pose only high level requirements. This section describes such an approach.

We opted for integration at the level of executable programs and not at the level of source code. These executable programs are called *feature extraction modules*. This approach allows for freedom in the choice of programming language and paradigm for each module, and makes testing much simpler, since no dependencies exist between the different modules. We selected the decision tree algorithm C4.5 (Quinlan, 1993) for all identifications using separate and/or combined feature extraction modules.

The modules have to comply only with the following requirements:

1. Uniformity of command line interface.

2. Each module accepts two inputs: an image file and a contour file. The image file is an 8-bit, one sample per grey level TIFF file. The contour file is a proprietary format ASCII file which contains the (x,y) coordinates of the outline of the diatom.

3. The output of a module is conform to the C4.5 specification.

The bridge between the modules and the decision tree algorithm C4.5 is formed by the *integration module*, which takes as its input a list of modules and a list of diatom images with corresponding contour files. In the training phase, that is, when C4.5 is building a classifier from input data, the integration module also uses an index file to associate the diatom images with their respective classes (taxa). The integration module runs all modules on all images, collects the outputs, and prepares these for C4.5. Since a large amount of ASCII file processing is involved,

Table 1. Identification performance of each feature set separately, based on results obtained using the mixed genera data set. (O) denotes an ornamentation feature, and (C) denotes a contour feature. The column "$\bar{x}$" contains the average number of errors; the column "σ" contains the standard deviation of the number of errors; the columns "min." and "max." contain the minimum and maximum number of errors, respectively; the column "perf. (%)" contains the percentage (average with standard deviation) of samples identified correctly, cf. Eq. 2. Sizes of the training and test sets were 596 and 185 samples, respectively.

feature set	$\bar{x}$	σ	min.	max.	perf. (%)
Classical features (Univ. of Bern, Chapter 7)					
raphe (O)	180	0.0	180	180	2.7 ± 0.0
symmetry class (C)	171.6	1.0	170	173	7.2 ± 0.5
undulation (O)	167.1	1.6	165	169	9.7 ± 0.9
striae (O)	113	3.1	108	118	38.9 ± 1.7
frequencies (O)	110.5	5.9	100	120	40.3 ± 3.2
moment invariants (C)	59.4	4.9	50	66	67.9 ± 2.6
geometric (C)	51.2	4.0	46	58	72.3 ± 2.2
simple shape (C)	46.3	3.8	40	53	75.0 ± 2.1
form (C)	44.8	4.4	38	51	75.8 ± 2.4
Fourier descriptors (C)	29.5	4.1	23	34	84.1 ± 2.2
texture (O)	29.1	3.4	22	33	84.3 ± 1.9
Contour features (Univ. of Faro, Chapters 8 to 10)					
curvature (C)	104.4	4.2	98	110	43.6 ± 2.3
Gabor (C)	84.2	5.8	77	94	54.5 ± 3.2
characteristic profile (C)	46.2	4.6	41	57	75.0 ± 2.5
Morphological features (Univ. of Groningen, Chapter 11)					
curvature scale space (C)	38.9	4.4	32	44	79.0 ± 2.4
size/shape distribution (O)	34.4	4.1	28	40	81.4 ± 2.2

the integration module was implemented in the programming language Perl (Wall et al., 2000). Figure 1 shows the module integration framework.

3 Identification results

This section describes the identification results obtained by combining the feature sets produced by the various modules. For a detailed description of these feature sets, the reader is referred to the respective chapters in this book.

3.1 The set of diatom images

The experiments described in this chapter were conducted on a set of 37 common diatom taxa – mostly pennate freshwater taxa – which are listed in Chapter 4 (mixed genera data set). Each taxon is represented by at least 20 images, and the total number of images is 781. The species *Tabularia* sp. 1 and *Gomphonema* sp. 1 are new, undescribed, species and they do not yet have a full binomial name.

3.2 Bagging

Bootstrap aggregation (bagging) is a method to obtain an aggregated predictor by building multiple predictors, and combining the outcomes in some way. The procedure starts by making a random division of the data into a training set and a test set. Multiple training sets are constructed by bootstrapping, i.e., by drawing samples at random, but *with* replacement, from the initial training set. Next, a predictor is built for each of the new training sets. To obtain a single predicted class from the multiple predictors, a majority vote is taken. It has been shown (Breiman, 1996) that bagging helps a good but unstable classifier, such as a decision tree, towards optimality.

The procedure used to obtain the identification performance in this chapter is as follows:

1. Make a random division of the data into a training set and a test set, making sure that the test set contains exactly five samples of each taxon/class. The size of the test set then becomes 185 samples, which is about 25% of the data (total number of samples is 781).

2. Construct 25 new training sets (each containing 596 samples) from the initial training set using bootstrapping, and build a decision tree for each of them.

3. Evaluate the 25 decision tree classifiers on the test set, and take a majority vote on the outcomes of each tree.

4. Repeat the above steps 10 times, and average the results.

3.3 Identification performance

The identification performance is defined as follows. In each run i, $i = 1, 2, \ldots, N_{\text{runs}}$ of the procedure, the data set D is divided into a training set $D_{\text{tr}}^{(i)}$ and a test set $D_{\text{test}}^{(i)} = D \setminus D_{\text{tr}}^{(i)}$, which depend on the run. The sizes of training and test sets are constant in all runs. Let $X = V \times Y$ be the space of labelled instances, i.e., diatom images with class names. Here V denotes the space of unlabelled instances and Y the set of distinct class labels. In our case, we have a finite number of classes indexed by $j = 1, 2, \ldots, N_{\text{classes}}$, where N_{classes} is the number of classes (37 in our case), and the label of class j is denoted by y_j. Let $D_{\text{tr}}^{(i)} = \{x_1, x_2, \ldots, x_n\}$ be the training set consisting of n labelled instances, where $x_k = \langle v_k \in V, y_k \in Y \rangle$. The decision tree algorithm C4.5 with bagging and majority voting, denoted by $I_{\text{bag}}^{(\text{mv})}$, is an *inducer* which maps the training set $D_{\text{tr}}^{(i)}$ into a classifier $C^{(i)} = I_{\text{bag}}^{(\text{mv})}(D_{\text{tr}}^{(i)})$, which depends on the run i. The classifier $C^{(i)}$ maps an unlabelled instance $v \in V$ to a label $y \in Y$. Then $I_{\text{bag}}^{(\text{mv})}(D_{\text{tr}}^{(i)}, v) = (I_{\text{bag}}^{(\text{mv})}(D_{\text{tr}}^{(i)}))(v)$ denotes the label assigned to v by the classifier $I_{\text{bag}}^{(\text{mv})}(D_{\text{tr}}^{(i)})$ built from the training set $D_{\text{tr}}^{(i)}$.

The identification accuracy in the i-th run is given by

$$acc^{(i)} = \frac{1}{N} \sum_{\langle v_k, y_k \rangle \in D_{\text{test}}^{(i)}} \text{Match}(I_{\text{bag}}^{(\text{mv})}(D_{\text{tr}}^{(i)}, v_k), y_k), \qquad (1)$$

Chapter 12: Mixed-method identifications

where $Match(y_k, y_l) = 1$ if $y_k = y_l$ and 0 otherwise. Here N denotes the size of the test set $D_{\text{test}}^{(i)}$ (in our case $N = 185$). The *identification performance* is defined as the average of the accuracies over all runs:

$$Performance = \frac{1}{N_{\text{runs}}} \sum_{i=1}^{N_{\text{runs}}} acc^{(i)}, \qquad (2)$$

where in our case $N_{\text{runs}} = 10$.

3.4 Separate feature sets

We tested the decision tree classifier with bagging on each of the separate feature sets produced by the different project partners involved in pattern recognition (Bern, Faro, Groningen).

Table 1 shows the identification performance. Due to different testing conditions, it is possible that the numbers presented in this table differ from those given in the chapters devoted to the individual feature sets. The table lists the minimum and maximum errors for each feature, the average error, and the standard deviation. The performance is given as the percentage of samples identified correctly, averaged over all runs, cf. Eq. 2. The Fourier descriptors and texture features showed the best performance for the contour and ornamentation sets, respectively. It is interesting to see that it is possible to correctly identify over 80% of the diatoms using only the contour or ornamentation structure. For a detailed evaluation of the performance of each of the separate feature sets, we refer to the respective chapters in this book.

3.5 Combined feature sets

Next, we present the results of the decision tree classifier with bagging on various combinations of feature sets. Table 2 shows the identification performance for the following combinations:

1. Combinations of feature sets of individual partners (Bern, Faro, Groningen).

2. Combination of all contour-based features (labelled *contour*).

3. Combination of all ornamentation features (labelled *ornamentation*).

4. Combination of all features (labelled *all*).

5. The table entry labelled *good* corresponds to the combination of features that each have a performance of at least 75%. This combination consists of the following features: simple shape, form, texture, Fourier descriptors, characteristic profile, curvature scale space, and size/shape distribution.

6. Combination of all features that are used by at least 12 decision trees (labelled *robust*). This entry will be discussed in detail later on in this section.

The table shows that by combining either all contour-based features or all ornamentation features, the performance already exceeds 90%. The performance becomes slightly better by combining the good features (94%), and reaches a maximum for the combination of all features, resulting in a performance of almost 97%. The classical features consist of a good and extensive set of contour-based and ornamentation features, which explains the good performance.

Table 2. As Table 1 for combined feature sets. The numbers between parentheses refer to the number of features present in the corresponding feature set.

feature set	$\bar{x}$	σ	min	max	perf. (%)
Ch. 8 to 10 contour features (123)	29.8	4.8	25	40	83.9 ± 2.6
Ch. 11 morphological features (46)	19.2	3.5	15	25	89.6 ± 1.9
ornamentation (125)	15.6	2.1	13	20	91.6 ± 1.2
contour (204)	14.4	3.2	11	20	92.2 ± 1.8
good (221)	10.4	3.8	6	19	94.4 ± 2.1
robust (17)	8.3	2.7	5	14	95.5 ± 1.4
Ch. 7 classical features (160)	6.6	3.2	2	13	96.4 ± 1.7
all (329)	5.7	2.3	2	9	96.9 ± 1.2

3.6 Identification performance with ranking

Instead of using majority voting to get a bagged predictor, we can also rank the outcomes of the individual decision trees with respect to the frequency by which a certain class label is chosen. To make this precise, we introduce an index ℓ for each of the trees generated by C4.5, and denote the corresponding inducer by $I^{(\ell)}$, $\ell = 1, 2, \ldots, N_{\text{trees}}$ (in our case $N_{\text{trees}} = 25$). The frequency F_j of class j for a given unlabelled instance v is the number of times label y_j is chosen:

$$F_j = \#\{\ell : I^{(\ell)}(D_{\text{tr}}^{(i)}, v) = y_j\}, \quad j = 1, 2, \ldots, N_{\text{classes}}. \tag{3}$$

Recall that $D_{\text{tr}}^{(i)}$ is the training set in run i, and N_{classes} is the number of classes (diatom taxa, 37 in our case). Now we sort the frequencies $\{F_j\}$ in decreasing order, producing a permutation $R(1), R(2), \ldots, R(N_{\text{classes}})$ of class indices with a corresponding sorted sequence of frequencies:

$$F_{R(1)} \geq F_{R(2)} \geq \ldots \geq F_{R(N_{\text{classes}})}.$$

For example, if $R(1) = y_M$, this means that M is the class most often assigned to instance v among the 25 trees. We call $R(1), R(2), \ldots$ the class of rank 1, rank 2, etc. Note that the distribution of ranks depends on the input sample v. To each rank r corresponds a bagged rank-r predictor $I_{\text{bag}}^{(r)}$, with an accuracy defined by

$$acc^{(i)} = \frac{1}{N} \sum_{\langle v_k, y_k \rangle \in D_{\text{test}}^{(i)}} Match(I_{\text{bag}}^{(r)}(D_{\text{tr}}^{(i)}, v_k), y_k), \tag{4}$$

Table 3. Identification performance (%) of combined feature sets using bagging with ranking. The numbers between parentheses refer to the number of features present in the corresponding feature set. The columns labelled "rank 1" to "rank 5" correspond to the rank-r classifier defined in Eq. 5, with r ranging from 1 to 5. The rank-1 classifier is identical to the majority vote classifier, cf. Table 2. Performance was computed according to Eq. 2, with $acc^{(i)}$ as defined in Eq. 4.

feature set	rank 1	rank 2	rank 3	rank 4	rank 5
Ch. 8 to 10 contour features (123)	83.9	90.9	92.9	94.1	94.8
Ch. 11 morphological features (46)	89.6	94.9	96.6	97.2	97.5
ornamentation (125)	91.6	95.7	97.6	98.4	98.9
contour (204)	92.2	96.5	97.9	98.6	98.8
good (221)	94.4	97.3	98.1	98.5	98.8
robust (17)	95.5	97.9	98.6	98.9	99.2
Ch. 7 classical features (160)	96.4	98.2	98.8	99.0	99.2
all (329)	96.9	98.6	99.1	99.4	99.6

where

$$Match(I_{\text{bag}}^{(r)}(D_{\text{tr}}^{(i)}, v_k), y_k) = \max_{\ell \in \{1,2,\ldots,r\}} Match(I^{R(\ell)}(D_{\text{tr}}^{(i)}, v_k), y_k). \quad (5)$$

That is, $I_{\text{bag}}^{(r)}$ counts the number of matches when looking for the correct result in the first r classes after the sorting procedure. Note that the rank-1 predictor equals the majority vote classifier, i.e., $I_{\text{bag}}^{(1)} = I_{\text{bag}}^{(\text{mv})}$.

Table 3 shows the identification performance computed according to Eq. 2, but with $acc^{(i)}$ as defined in Eq. 4. The columns labelled "rank 1" to "rank 5" correspond to the rank-r classifier, with $r = 1, 2, 3, 4, 5$. The rank-1 classifier is identical to the majority vote classifier, cf. Table 2. We can observe that, as the rank is increased, the performance steadily increases, approaching a perfect identification for several of the combined feature sets.

As already mentioned in Chapter 11, ranking is useful in order to give a user a number of options that must be checked, where the user can assume that the right taxon is one of the listed taxa (ID rate very close to 100%). Such a procedure can be seen as a semi-automatic identification, which is perhaps preferable, rather than getting one taxon knowing that it could be wrong.

A few remarks are in order. First, we did not treat ties (*ex-aequo* matches) in a special way. For example, if the rank-1 and rank-2 frequencies were equal, the order was kept as produced by the sorting algorithm. Second, no weight was placed on the contribution of each inducer $I^{R(\ell)}$ to the combined classifier $I_{\text{bag}}^{(r)}$. E.g., if for all samples of the test set one tree gives the correct result, whereas all other trees give the same but wrong answer, then the rank-2 match is still 100%, because the correct answer is found among the first two ranks. It is possible to develop corrections for dealing with these effects, but we have not yet done so.

3.7 Robust features

A better insight into the features chosen by C4.5 is obtained by studying the decision trees it generates. A set of 25 trees was selected from one of the runs of the bagging

Table 4. Analysis of 25 decision trees obtained in one run of the bagging procedure on the *all* feature set. The column "first node" shows the feature that is used to make a decision in the first node of the decision tree. P_{tr} and P_{test} are the performances (%) on the training set and test set, respectively. The average performance on the test set is 83.5%. The features of trees 1–8 and 10–25 are described in Chapter 7, and the feature of tree 9 is described in Chapter 9.

T	first node	P_{tr}	P_{test}	T	first node	P_{tr}	P_{test}
1	Fourier desc. 3	99.3	79.5	14	mom. inv. 6 (Hu)	99.3	88.1
2	width	99.5	80.0	15	Fourier desc. 3	99.5	82.2
3	width	99.5	89.2	16	width	99.7	84.3
4	width	98.8	78.4	17	Fourier desc. 3	99.7	81.6
5	striae direction	99.0	85.9	18	mom. inv. 6 (Hu)	99.2	82.7
6	width	99.8	82.2	19	width	99.0	87.6
7	width	99.2	84.3	20	triangularity	99.3	89.7
8	Fourier desc. 3	99.5	85.9	21	Fourier desc. 3	99.2	82.2
9	AVang338-346	99.2	82.2	22	Fourier desc. 3	99.7	77.8
10	width	99.0	82.7	23	width	99.7	82.2
11	form left direction	99.3	81.6	24	width	99.3	87.0
12	mom. inv. 6 (Hu)	99.3	84.9	25	width	99.7	83.2
13	width	99.0	81.1				

procedure on the combination of all features. Table 4 shows the data collected on the performance and the first node of each of the trees. The performance on the test set ranges from 77.8 to 89.7%, with an average of 83.5%, indicating that the decision tree classifier is unstable. A comparison of these performances of individual trees with that of the bagged classifier (the entry "all" in Table 2) shows that the bagging procedure indeed considerably improves performance.

An interesting observation can be made by looking at the first node of each decision tree. The width of the diatom valve is the first node for 12 trees, which, incidentally, is also a strong character for diatomists. Fourier descriptor 3 (Chapter 7) is the first node in 6 decision trees. The first node of tree 9 ("AVang338-346") is from the characteristic profile feature set (Chapter 9), and the first node of tree 20 ("triangularity") from the simple shape descriptors feature set (Chapter 7).

We continued the analysis by investigating which features are actually used by the decision trees. As a result of this investigation, we selected a set of "robust" features, i.e. features that are used by at least 12 decision trees. From the full set of features (329 in total), 17 remain, and they are listed in Table 5. The features named "morph. matfeat" are from the set of morphological size/shape distribution features, and those named "morph. pvar 1" and "morph. vmean2" are from the set of morphological curvature scale space features, see Chapter 11. The table shows that some features that do not perform well when used separately (cf. Table 1), turn out to be very robust when used in combination with other features. For instance, undulation is used by 23 trees, and striae density by all trees.

Another performance evaluation was carried out, using only the 17 robust features. The results show that these features form a good combination, yielding a performance of 95.5% (see Table 2). The performances of the last three entries in Table 2 are almost equal, and taking the standard deviation of the number of errors

Chapter 12: Mixed-method identifications

Table 5. The collection of "robust" features, i.e. features that are used by at least 12 decision trees in Table 4. Width and striae density are used by all decision trees. The first 12 features in this table are described in Chapter 7 and the remaining features in Chapter 11.

feature	1	2	3	4	5	6	7	8	9	10	11	12	13	14	15	16	17	18	19	20	21	22	23	24	25
undulation	•	•	•	•	•	•	•	•	•	•	•	•	•	•	•	•	•		•		•	•	•	•	•
striae density	•	•	•	•	•	•	•	•	•	•	•	•	•	•	•	•	•	•	•	•	•	•	•	•	•
striae direction	•	•	•		•	•		•			•	•	•	•		•	•	•		•	•	•	•	•	•
width	•	•	•	•	•	•	•	•	•	•	•	•	•	•	•	•	•	•	•	•	•	•	•	•	•
Fourier desc. 6	•		•			•	•		•				•	•		•	•		•		•		•	•	
mom. inv. 5 Hu		•	•	•	•		•		•		•	•	•	•		•	•		•			•	•	•	•
mom. inv. 1 Fusser	•	•	•	•	•		•	•			•	•	•	•		•	•		•			•	•	•	•
form left			•			•		•	•		•	•		•	•	•	•		•			•	•	•	•
form right	•	•	•			•	•	•	•		•		•		•	•	•		•			•	•	•	
form top	•	•	•				•	•			•		•			•	•	•	•			•	•	•	
form bottom				•		•		•	•	•	•	•		•		•	•		•			•	•	•	•
form property 1	•	•	•	•		•		•	•			•		•	•	•	•		•			•	•	•	
morph. matfeat10	•		•	•	•	•			•			•	•	•		•	•		•		•		•	•	
morph. matfeat13		•	•		•				•				•			•	•		•			•	•	•	•
morph. matfeat30	•	•		•	•			•	•				•	•		•	•					•	•	•	•
morph. pvar1	•	•	•			•	•	•		•	•	•	•	•	•	•	•	•	•		•		•	•	•
morph. vmean2	•	•			•	•	•				•		•	•	•	•	•	•			•		•	•	•

into account, it is clear that they do not differ significantly from each other.

3.8 Example decision tree

To gain some insight into the kind of decisions taken by the C4.5 algorithm when constructing the decision trees, this section provides a drawing of a typical decision tree. From the trees listed in Table 4, we selected tree 3 because it has the best performance of the trees that use the width of the diatom valve in their first node. The features used by this tree are listed in Table 6 (non-bold entries). Entries in boldface correspond to names of feature sets, with the numbers between parentheses indicating the total number of features in that feature set. The tree consists of 89 nodes, and a drawing is shown in Fig. 2. For display purposes the tree was split into a main tree and four subtrees, denoted by S1, S2, S3 and S4, respectively. The nodes of the tree are formed by the features, and along the edges the decision criterion can be seen. The boxed nodes are the leaves of the tree, which correspond to the taxa.

3.9 Conclusions

We evaluated the identification performance of separate feature sets of diatoms and various combinations thereof, using the bagged C4.5 decision tree classifier with majority voting. Combinations of contour-based features showed that over 90% of the diatoms can be identified correctly, and a similar result was obtained using ornamentation features. If all features are combined, identification performance increases to almost 97%. From the analysis of a collection of 25 decision trees, a set of 17 robust features was derived by selecting those features which were used

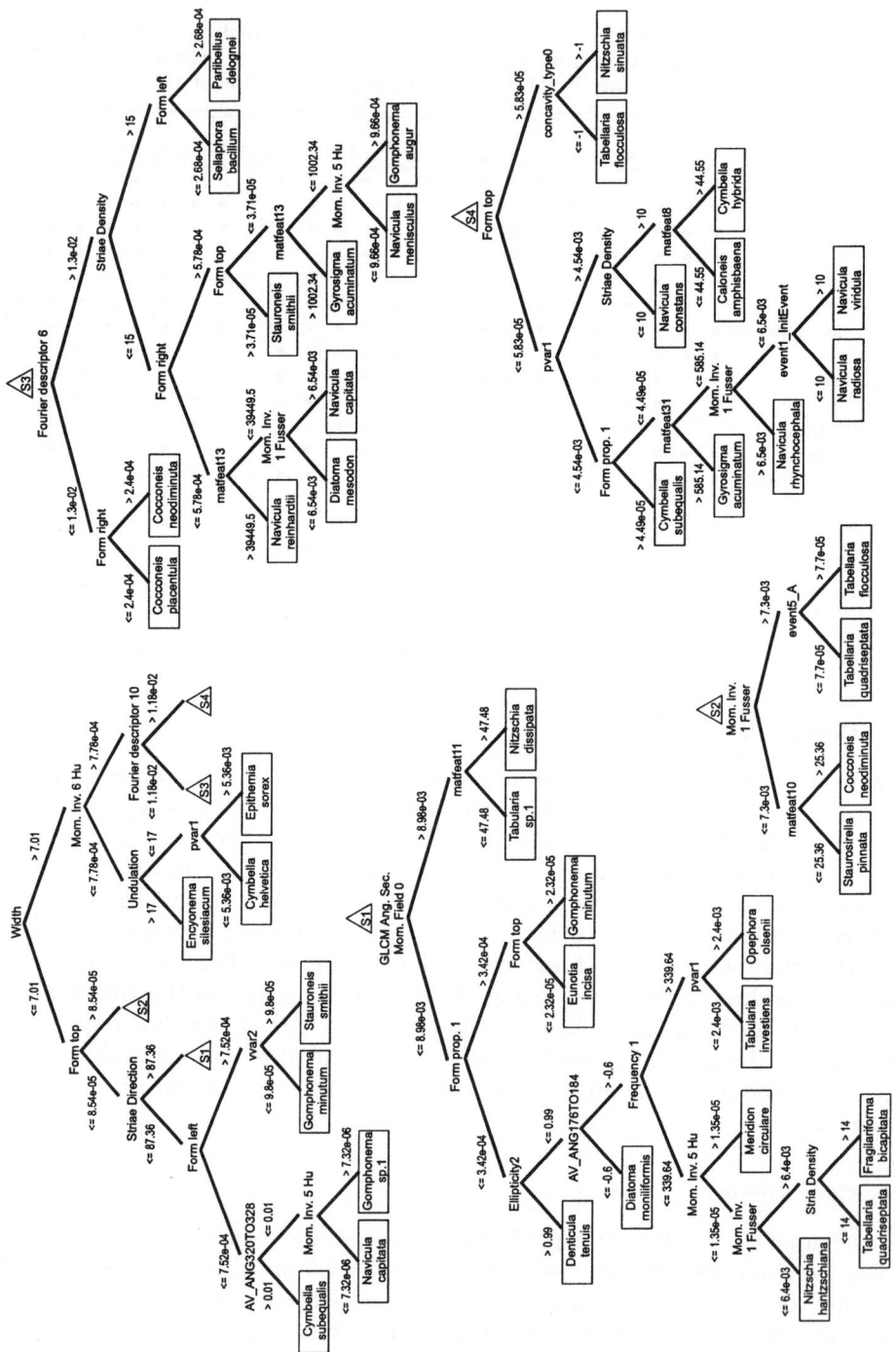

Figure 2. Decision tree 3 of Table 4. For explanation, see section 3.8.

Table 6. List of features (non-bold entries) used by decision tree 3 in Table 4. Entries in boldface correspond to names of feature sets, with the numbers between parentheses indicating the total number of features in that feature set.

Ch. 7 geometric properties (4)	**Ch. 7 frequencies (4)**
width	frequency 1
Ch. 7 undulation (1)	**Ch. 7 simple shape descriptors (7)**
undulation	ellipticity2
Ch. 7 Fourier descriptors (30)	**Ch. 11 size/shape distribution (32)**
Fourier descriptor 6	matfeat_8
Fourier descriptor 10	matfeat_10
Ch. 7 moment invariants (11)	matfeat_11
moment invariant 5 (Hu)	matfeat_13
moment invariant 6 (Hu)	matfeat_31
moment invariant 1 (Fusser)	**Ch. 11 curvature scale space (14)**
Ch. 7 striae features (3)	pvar1
striae density	vvar2
striae direction	**Ch. 9 characteristic profile (40)**
Ch. 7 texture features (84)	AV_ANG176TO184
(field 0) GLCM angular second moment	AV_ANG320TO328
Ch. 7 form features (14)	**Ch. 10 Gabor (81)**
form left	event1_Initevent
form right	event5_A
form top	concavity_type0
form property 1	

by at least 12 trees. The performance of this feature set showed an identification performance of almost 96%. Instead of using majority voting to get a bagged predictor, we also ranked the outcomes of the decision trees with respect to the frequency by which a certain class label is chosen by the individual decision trees. A rank-r classifier was defined which counts the number of matches when looking for the correct result in the first r classes after the ranking procedure. As the rank was increased, the performance was found to approach perfect identification for several of the combined feature sets.

This section has given a first impulse to the analysis of the feature sets. A more elaborate study is needed, for instance by performing a principal component analysis. Furthermore, no comparison of computational cost of the feature sets has been made. One can imagine a system that automatically counts diatoms on a slide, in which case computational costs become important. When two feature sets show similar identification performance, the one that has the lowest computational cost is favorable over the more costly one.

4   ADIACweb – an online identification system

The Internet plays an increasingly important role in scientific computing, and enables researchers to share their methods and results with a large community. To make the software that was produced as part of the ADIAC project accessible

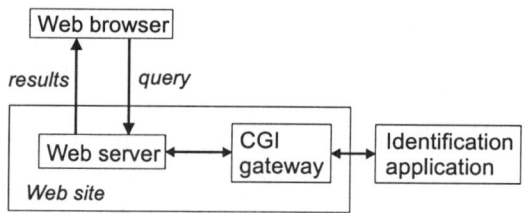

Figure 3. *ADIACweb* software architecture.

to other researchers, we have developed a web-based demonstration system called *ADIACweb*, which is available online.[a]

ADIACweb has as a three-tier client/server architecture, as shown in Fig. 3. Tier 1 is the user interface, that is provided by a standard web browser, from which users can send queries. Tier 2 is a server-side CGI program that dispatches the queries to tier 3, which is the part that performs diatom identification. For performance reasons, tier 3 runs on a dedicated server. The approach is general and platform independent. Any machine connected to the Internet can use the system, and users have no concerns regarding machine compatibility. Furthermore, there are no code portability problems, since the software operates on the server's host machines.

The core of *ADIACweb* is formed by the identification application. The input to this application is a set of features extracted from a diatom image. These features are passed through the same 25 decision trees as listed in Table 4, and an identification is obtained by taking a majority vote on the outcomes. A confidence value is computed as the percentage of trees that contribute to the final identification.

Using *ADIACweb* is straightforward. The user is offered a selection of diatom query images, and, by clicking on an image, a request to identify that sample is sent to the web server. The query images are taken from the test set and have not been used for training. To keep waiting time to a minimum, the features of all query images have been precomputed. The identification application then identifies the sample, and returns the full taxon name of the sample including the confidence value. The user can then click on the taxon name to submit a query to the ADIAC Diatom Image Database[b], which returns a collection of reference images for that taxon. Figure 4 shows *ADIACweb* running in a web browser.

The current setup does not yet allow users to submit their own images. The main reason for this is that fully automatic contour extraction is not feasible at this moment. A full working system is intended as part of a future project. This would allow users to submit their own images and have identifications returned to them.

References

Breiman, L. (1996) Bagging predictors. Machine Learning, Vol. 24, pp. 123-140.

---

[a] http://www.cs.rug.nl/hpci/demos/adiac/demo.html
[b] http://www.rbge.org.uk/ADIAC/db/adiacdb.htm

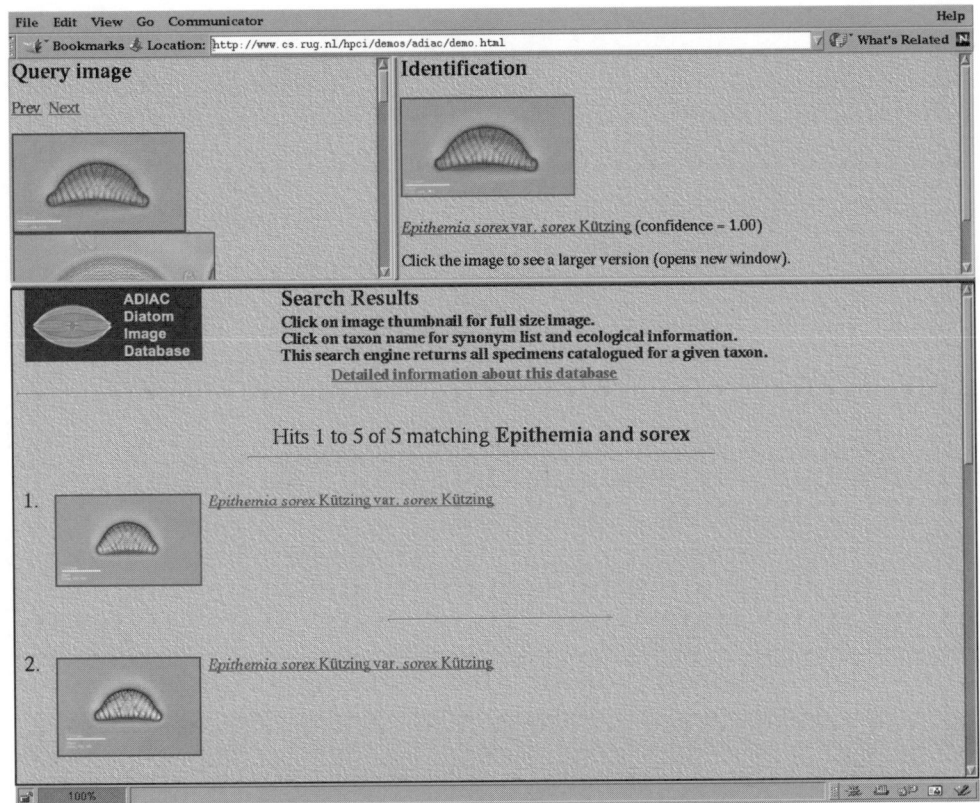

Figure 4. *ADIACweb* demo running in a web browser. The top left frame shows the list of query images. Identification results are reported in the top right frame. The bottom frame shows the search results of the ADIAC Diatom Image Database.

Quinlan, R. (1993) C4.5: Programs for machine learning. Morgan Kaufmann, San Mateo, USA.

Wall, L., Christiansen, T. and Orwant, J. (2000) Programming Perl. O'Reilly and Associates, Beijing (etc.), 3rd Edition.

CHAPTER 13

AUTOMATIC SLIDE SCANNING

JOSÉ L. PECH-PACHECO AND GABRIEL CRISTÓBAL

In this chapter we describe an automatic slide scanning and autofocusing system that consists of three components. First, images are captured at low magnification. This yields a panoramic view of an entire slide which allows the extraction of the positions and sizes of all particles. Then, a particle screening is carried out at medium magnification, in order to eliminate particles that are not diatom-like, i.e. debris that does not comply with pre-defined selection criteria. This method selects about 80% of all usable valves. In a third step at high magnification, images of particles, mostly diatoms, are captured using autofocusing and multifocus fusion. The output of the system is an annotated gallery of images that can be used for diatom identification. The main limitation of the automatic slide scanning system consists of the mechanical parts. The average time to position the motor-controlled microscope stage and to acquire an image is about 3 seconds.

1 Introduction

Although some studies have explored the analysis of images at medium magnification, they have not considered the problem at the next level down, i.e. starting the automation process from a mosaic of low magnification images. This is a very challenging problem, because a diatom sample is normally contaminated with debris, broken diatom valves and valves which lie at an angle. We distinguish between two broad classes of diatom identification systems: semi-automatic and automatic. Semi-automatic methods require human interaction to select specimens from a low resolution image. Obviously, this provides good results but it is still time consuming, tedious and demands a lot of concentration. Methods for diatom detection and identification have been studied by Cairns et al. (1972, 1979) and Pech-Pacheco and Alvarez-Borrego (1998). Cairns et al. (1972) have proposed diatom identification methods based on coherent optics and holography. However, they did not consider the problem of automation with low resolution images. Culverhouse et al. (1996) described methods for identifying dinoflagellates, using neural networks, but again they did not provide a fully automatic method. Pech-Pacheco and Alvarez-Borrego (1998) proposed a hybrid optical-digital method for the identification of five different species of phytoplankton, using operators invariant to translation, rotation and scale. In Pech-Pacheco et al. (1999) correlation techniques are used for identifying phytoplankton specimens.

At low magnification it is difficult, if not impossible, to discriminate between diatom and non-diatom particles. Therefore, it is necessary to detect all possible "candidate" particles through image binarization on the basis of thresholding. Several methods have been proposed in the literature for threshold determination.

Here, we will propose a simple but efficient method based on a modified histogram thresholding. Another key aspect in the automation process is a reliable and fast autofocusing method. Groen et al. (1985) have identified and compared 11 different autofocus algorithms. Different focusing criteria have been proposed (Krotkov, 1987; Vollath, 1987; Subbarao et al., 1993; Yeo et al., 1993; Nayar and Nakagawa, 1994; Bocker et al., 1996; Neveu, 1999). Most of them extract a focus measure that is maximum for the best-focused image. Autofocus algorithms can be classified into two categories: those based on the statistical variance of pixel values, and others based on the spatial frequency content of the image. Here, we propose an adaptive thresholding method that provides a good compromise between detectability and noise rejection. Also, a new autofocusing algorithm based on the computation of the variance of the image gradient, and some new focus metrics are proposed and compared with existing methods. We also propose the use of multifocus fusion techniques for combining in-focus details from different focal planes into a single image, in order to obtain both a sharp valve contour and striation pattern.

The overall procedure of the automatic slide scanning system consists of three steps. First, image acquisition at low magnification is used to obtain a panoramic view of the whole slide, which allows the extraction of the position and size of all particles. Second, an intermediate resolution particle screening is carried out in order to eliminate non-diatom particles. Third, images are captured at high magnification using autofocusing and multifocus fusion. The output of the system is an annotated gallery of images that can be used for diatom identification.

This chapter is organized as follows. The next section provides a general description of the equipment and data used. Section 3 presents the basis of automatic slide scanning in brightfield microscopy. Autofocusing methods are described in Section 4, whereas multifocus fusion techniques are described in Section 5. Section 6 contains concluding remarks.

2 Material and methods

Diatom samples were analyzed using a Zeiss Axiophot photomicroscope, with a 100W halogen light source, and with 4x, 10x (low magnification), 20x (medium magnification) and 40x (high magnification) lenses. The ocular magnification was 0.6x. For image acquisition, we used the LG-3 grayscale frame grabber from Scion that includes the Scion Image software for Windows[a], which is a version of the popular Macintosh program NIH Image. Scion Image allows several image processing operations, like capture, display, analyze, enhance, measure, annotate and output images. The LG-3 is connected to a Cohu 4910 series monochrome CCD camera, which is a high quality, economical choice for grayscale scientific imaging applications. The frame grabber resolution (CCIR) is 768 × 512 pixels and the pixel depth 8 bits. Two PCs (Pentium II and III) running Windows98 were used, the one for image acquisition and the other for computations. Furthermore, for time intensive calculations a SUN Enterprise 450 with four processors was used. The microscope stage was controlled with a H101 (4" + 3") controller from Prior Scientific, with a step size of 0.1 μm for the X-Y axes and 1 μm for the Z axis.

---

[a]The Scion Imaging software can be downloaded free from http://www.scioncorp.com

Chapter 13: Automatic slide scanning

The total travel distance of the microscope stage for the X-Y-Z axes is 11 × 7 × 1 cm, respectively. Figure 1 shows a diagram of the system. The size range of the particles analyzed was between 20 and 260 μm, which can be considered part of the microplankton. Smaller diatoms (nanoplankton) cannot be observed using the available system, because they require oil-immersion lenses. The size ranges of the particles that can be observed with the lenses are as follows: 10x: range = 131–260 μm; 20x: 61–130 μm; 40x: 20–60 μm. In the future we intend to apply our methods to nanoplankton.

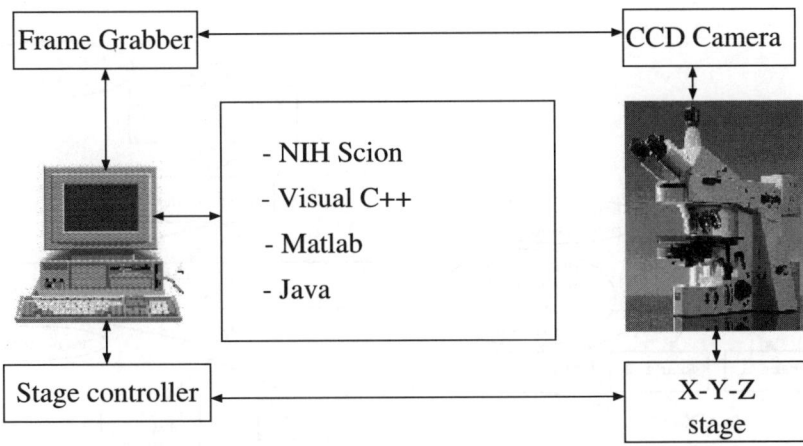

Figure 1. System configuration for automatic slide scanning. A brightfield microscope with a motorized stage attached is controlled by a PC. A CCD camera with frame grabber allows image acquisition.

3 Automatic slide scanning

3.1 Automatic localization: overall procedure

The overall procedure consists of two parts: image acquisition at low magnification and a further analysis at higher magnification. At low magnification the goal is to obtain a panoramic view of the entire slide by tiling images. The subsequent processing steps will provide the position and size of each particle, for a further analysis at high magnification. Figure 2 shows a schematic overview of the processing stages. Figure 2 (left) corresponds to the low magnification stage and Fig. 2 (right) corresponds to the higher magnification stage. Because the particles may appear at different depths, the localization method has been improved by combining different multifocus images, which allows the detection of particles that otherwise might be skipped (see Figs 9 and 10). Once the particles have been detected by applying image processing techniques (see below), we extract their positions and sizes. The size information allows us to select the correct lens for displaying the specimens at higher magnification. At high magnification (e.g. 40x) we apply the autofocusing, the particle screening and the multifocus fusion stages, which are described in the next sections. The output of the automatic system is a gallery of annotated images, some of which do not contain diatom valves, and which can be analyzed using the methods described in the other chapters of this book. We note again that at this stage of processing some of the images may contain debris. The goal of the particle screening is to remove a substantial number of particles (debris, broken valves) that are not required to be analyzed. Figure 3 shows two annotated particles from an image gallery provided by the automatic system.

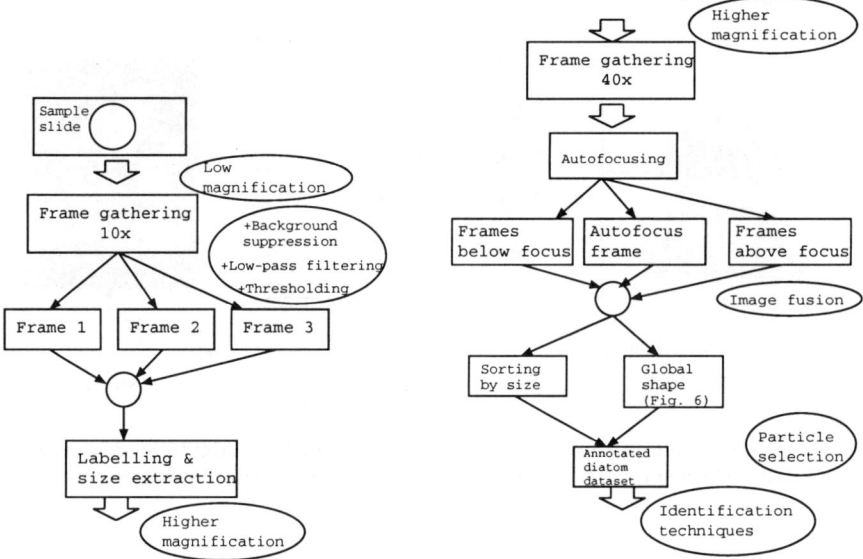

Figure 2. Schematic overview of the automatic slide scanning. Left: processing stages at low magnification. Right: processing stages at higher magnification. Note: the screening of particles using phase correlation techniques is performed at medium magnification (i.e. 20x). Screening using shape descriptors is applied at higher magnification.

3.2 Calibration

A first operation, only required once before the automatic scanning process, is the calibration of the diatom size at different magnifications. Figure 4 shows images of a microscopic ruler taken with different lenses. The ruler has a length of 1 mm and it is divided in 100 divisions of 10 μm each. Hence, it is possible to determine the number of pixels between the ruler's divisions at all magnifications. The calibration allows us to obtain real sizes by converting pixels to microns.

Chapter 13: Automatic slide scanning

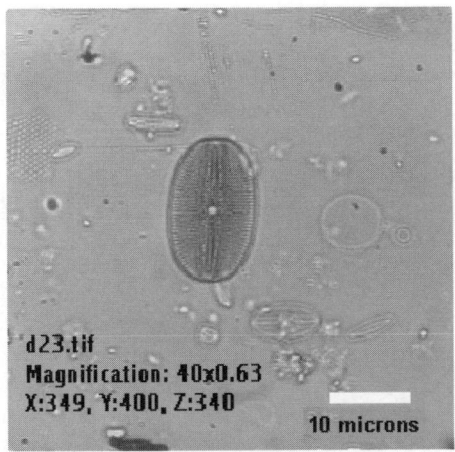 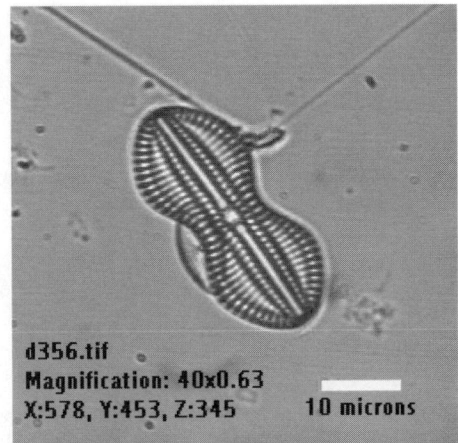

Figure 3. Annotated particles provided by the automatic system. Each particle detected at low magnification is later visualized at higher magnification and presented at the center of the image. Each particle is annotated with its spatial position, depth and magnification.

3.3 Slide scanning at low magnification

At low magnification it is necessary to arrange all the fields as a multi-frame mosaic that allows us to identify the positions of "candidate particles." With the equipment described in Section 2, an entire slide results in an array of frames, in which each frame covers an area of 500×500 μm. Figure 5 shows a subset of a mosaic at low resolution. Such a mosaic provides a panoramic view with the relative and absolute particle coordinates necessary for further analysis.

3.4 Algorithms for particle detection

The detection is based on the use of histogram thresholding, such that we obtain the centroid position and the size of all particles. We should point out that it is difficult to discriminate between diatoms and non-diatom objects at this resolution (Fig. 5). The main goal is to get an estimate of the centroid position of the particle that will be further analyzed at higher magnification. Below, we give a detailed description of the algorithms used (schematically shown in Fig. 2). The operations listed below are illustrated by a series of images shown in this section.

- Background correction: In general it is almost impossible to obtain homogeneous illumination of all frames. This effect is caused by the optics of the microscope, but can be eliminated by a background suppression method using a "rolling ball" algorithm (Sternberg, 1983; Russ, 1995). Figure 6 (top) shows one frame from a mosaic after binarization, but without background correction. We can observe that the effect of inhomogeneous illumination is more pronounced. The binarization on the basis of selecting a threshold value T located midway between the histogram extrema (see inset in the lower-left corner) is not efficient. To solve this problem, we perform two pre-processing operations on the original image. The first one is

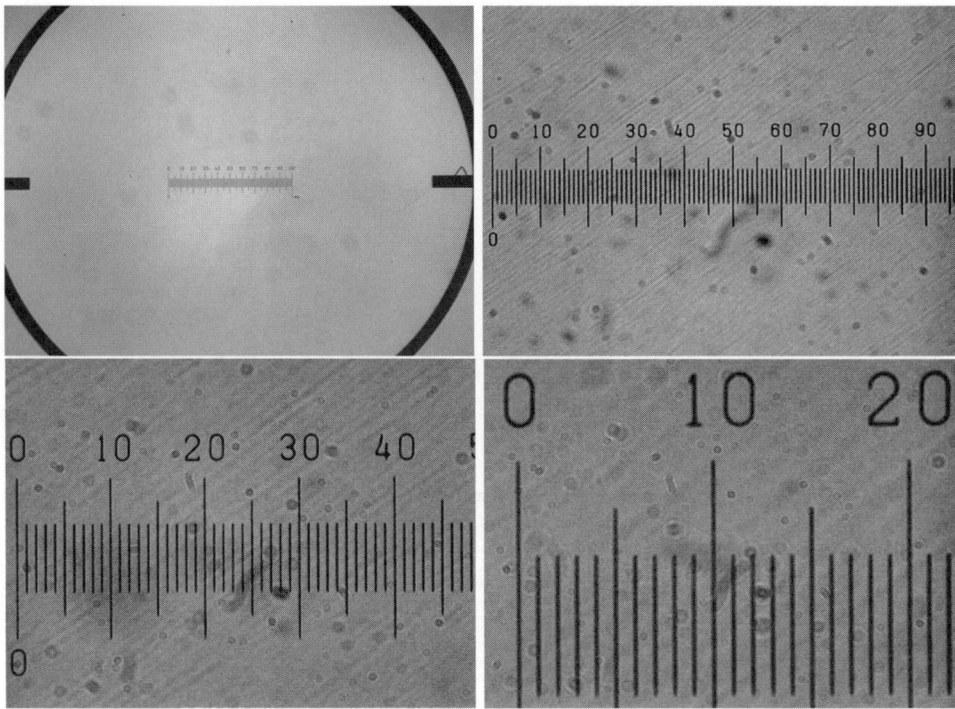

Figure 4. Images of a microscopic ruler at different magnifications. Total size of the ruler is 1 mm, with 100 divisions of 10 μm. Top-left: 2.5x lens; top-right: 10x; bottom-left: 20x and bottom-right: 40x. Ocular magnification: 0.63x.

a background correction with the "rolling ball" technique. The idea behind this algorithm is very simple. Consider the image as a 3D surface, where the height is given by the pixel values. Then imagine a ball underneath the surface, and roll the ball under the entire image. The position of the top of the ball corresponds to the background that can be subtracted from the image. An important parameter is the size of the ball. The results shown here were obtained by using a mask of 25×25 pixels containing an hemisphere. An alternative solution is to use a flat fielding technique by dividing (or substracting) the original image with a lowpass filtered version of itself, using a filter with a very large window size. See for example Seul et al. (2000).

• Lowpass filtering: Because we must discriminate between the foreground (particles) and background, a thresholding is required. However, the histogram of the image after background correction shows multiple peaks that complicate the selection of a threshold value. A simple smoothing operation, by neighborhood averaging, results in a histogram with two distinct peaks, and this facilitates the detection of the valley that segregates foreground from background. The lowpass filter used is a straightforward 3×3 convolution mask, with the nine elements equal to 1/9. We note that the resulting contours of the particles do not necessarily correspond

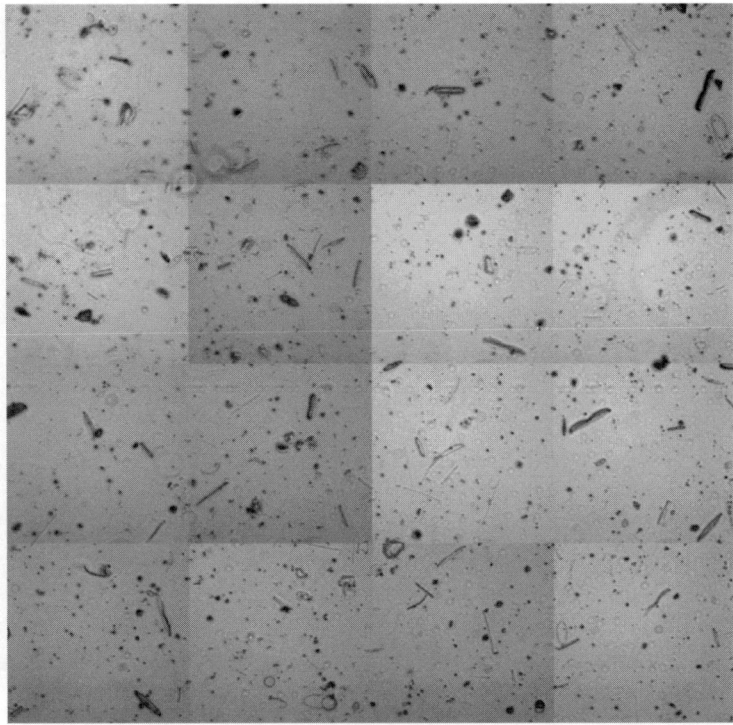

Figure 5. Example of a mosaic of images at low resolution (magnification 10x). The size of each frame is 500x500 μm; the total area is 2x2 mm. The effect of inhomogeneous illumination between frames can be observed; this can be eliminated using a background suppression algorithm.

well to the particles, which is due to the smoothing. However, this not a problem because only the particles' centroid position is required for a further analysis at higher magnification.

- Binarization: A few methods have been proposed for binarization by thresholding; for a good summary see Kindratenko (1997). The most important is to find a good threshold value, but there is no solution that suits all applications, i.e. the best method needs to be found empirically. We found that the triangle algorithm (Zack et al., 1977) provides a simple and efficient solution. The threshold value is determined from the histogram by computing the intersection of the perpendicular line to the line that joins the global maximum of the histogram with the lowest gray level (see Fig. 7). This technique is particularly effective in the case of objects that produce weak peaks in the histogram (Biemond and Merserau, 1998). Figure 8 (top) shows the result after applying the triangle algorithm.

- Combining images: In order to improve the detection of particles that are located at different focal planes, we do not only capture the best focused low magnification image, but also a small number of images above and below the best-focused one. These can be combined so that most particles in a slide can be detected in one pass. To this end we average the best-focused image with the two neighboring ones,

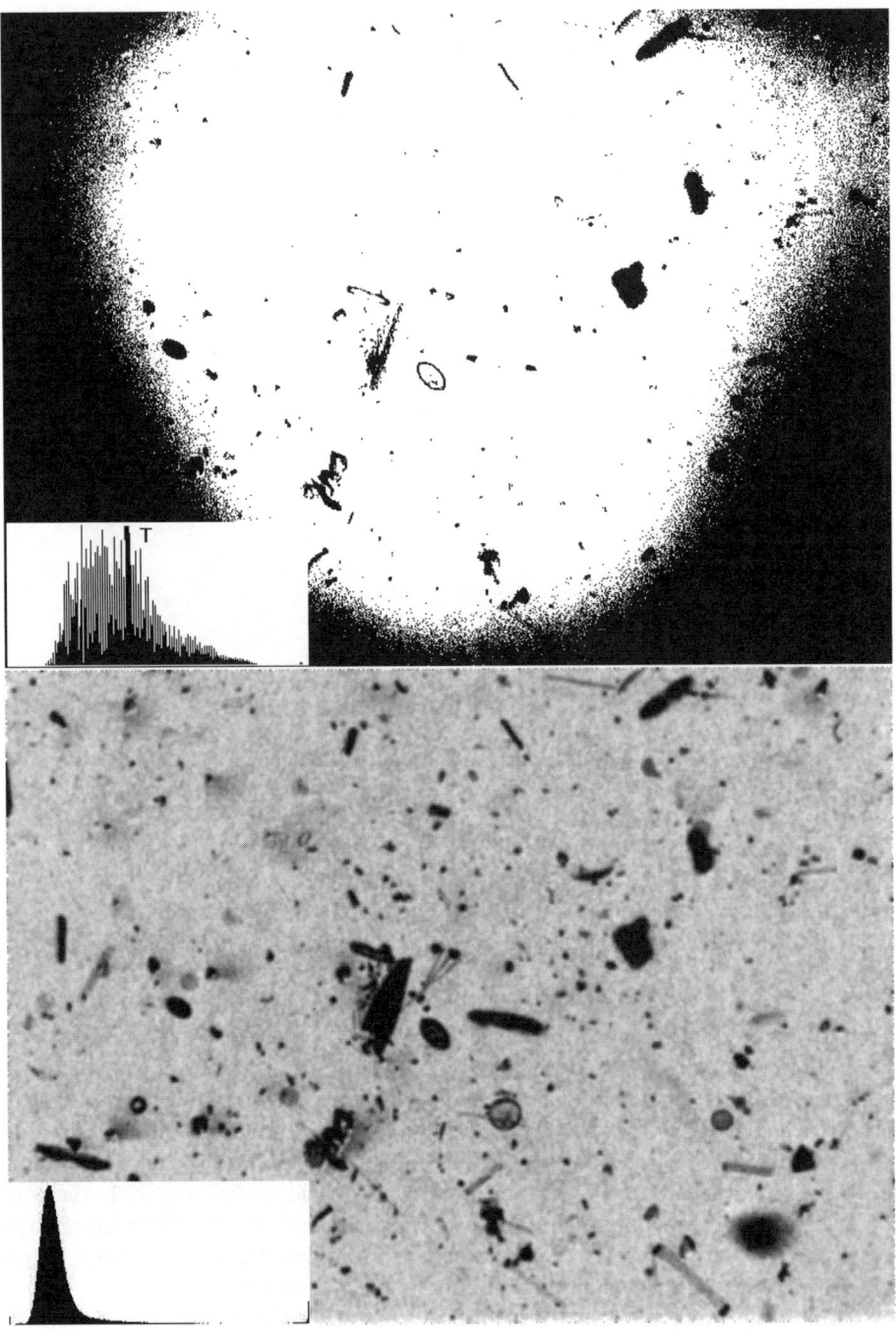

Figure 6. Top: histogram thresholding. Note the inhomogeneous illumination effect. Bottom: result of the background correction (by rolling ball) and lowpass filtering. Note the change of the histogram shown in the lower-left corner.

Chapter 13: Automatic slide scanning

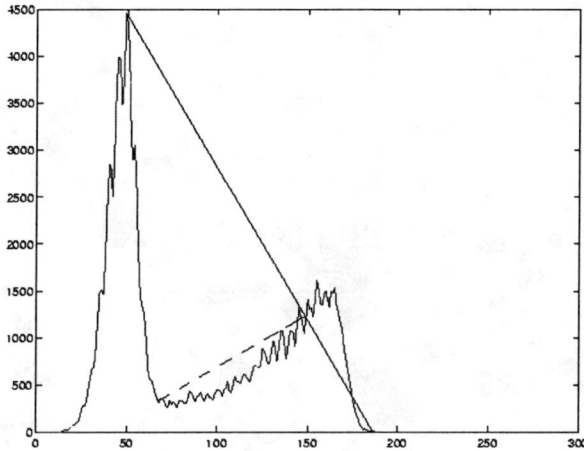

Figure 7. Triangle thresholding algorithm. A line is constructed between the maximum peak of the histogram and the highest gray level. The maximum distance from the histogram to this line (dashed) provides the threshold value, in this case 67.

separated by a distance of 10 μm. Figure 8 (bottom) shows the final result of the binarization process, after the processing described above. This procedure allows for efficient particle detection, even in the case of particles that have extreme gray levels (see the rod-shaped particle at bottom-right side of Fig. 8 (bottom), labeled with #92). The method for improving the detection by combining images from different depths is illustrated in Figs 9 and 10. Figure 9 shows the binarized low magnification images below and at the best focus plane, respectively. Figure 10 shows the images above the best focus plane as well as the fused image after averaging the three multifocus images. Hence, this very simple technique allows the detection of particles that otherwise might have been skipped.

3.5 Techniques for particle screening

Even at medium magnification it is difficult to distinguish between diatoms and debris (see Fig. 11). The smallest particles cannot be skipped, because they might correspond to tiny diatoms that cannot be resolved with the lenses described in Section 2. Even a diatom expert cannot make a decision by looking at such images. Therefore, we are considering three main strategies for removing candidate particles, and therefore debris rejection.

• Image gathering guided by size: We implemented particle localization guided by size, i.e. a selective search process for detecting particles that fall within a specific size range. This simple method allows us to create different image catalogues indexed by size, but it does not allow us to eliminate debris. It should be combined with other methods based on shape or symmetry, for example, in order to improve the discrimination. Figure 12 shows an example of indexing by size.

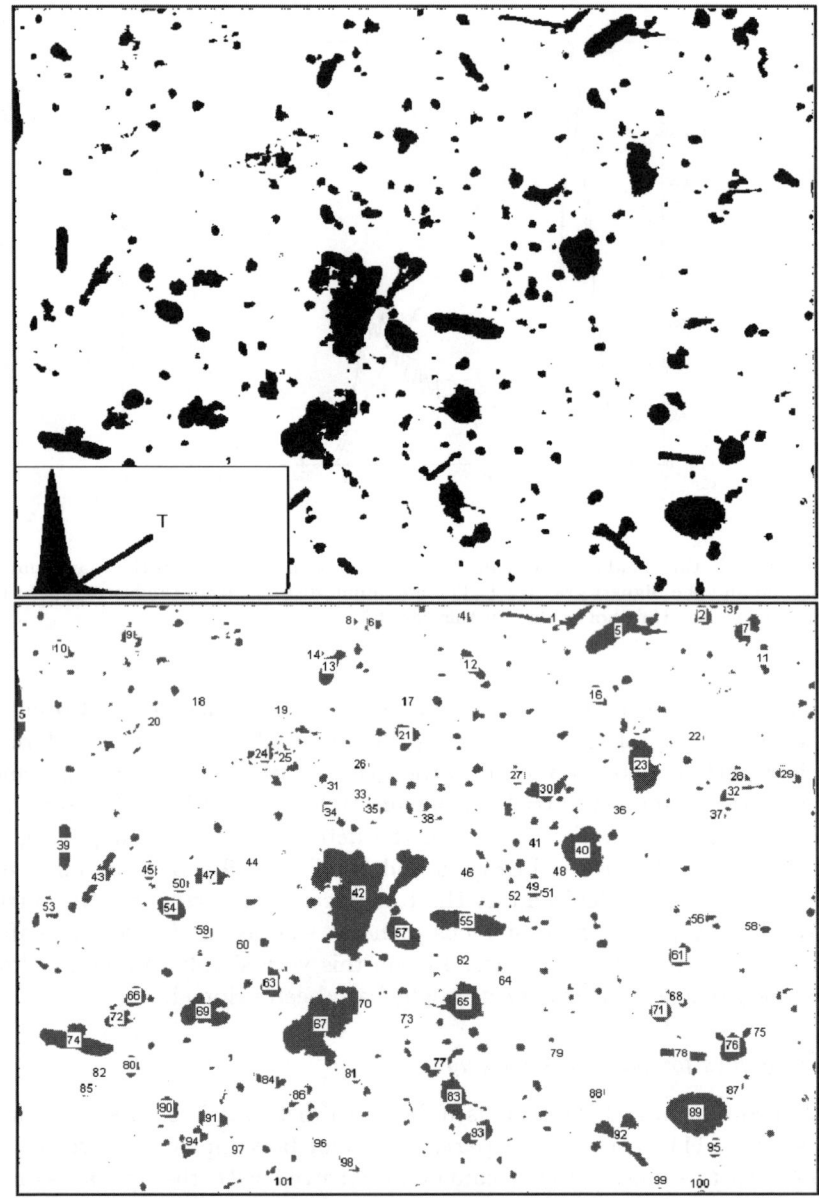

Figure 8. Top: binarized image using the "triangle" thresholding. Note the adaptive threshold value marked by an arrow line (lower-left corner). Bottom: labeled particles.

- Image gathering using shape descriptors: The main difficulty for automating slide scanning is that at low (or even medium) magnification it is difficult to distinguish candidate diatoms from debris. One possible solution is to apply an analysis at high magnification, e.g. using a combination of shape and symmetry descriptors,

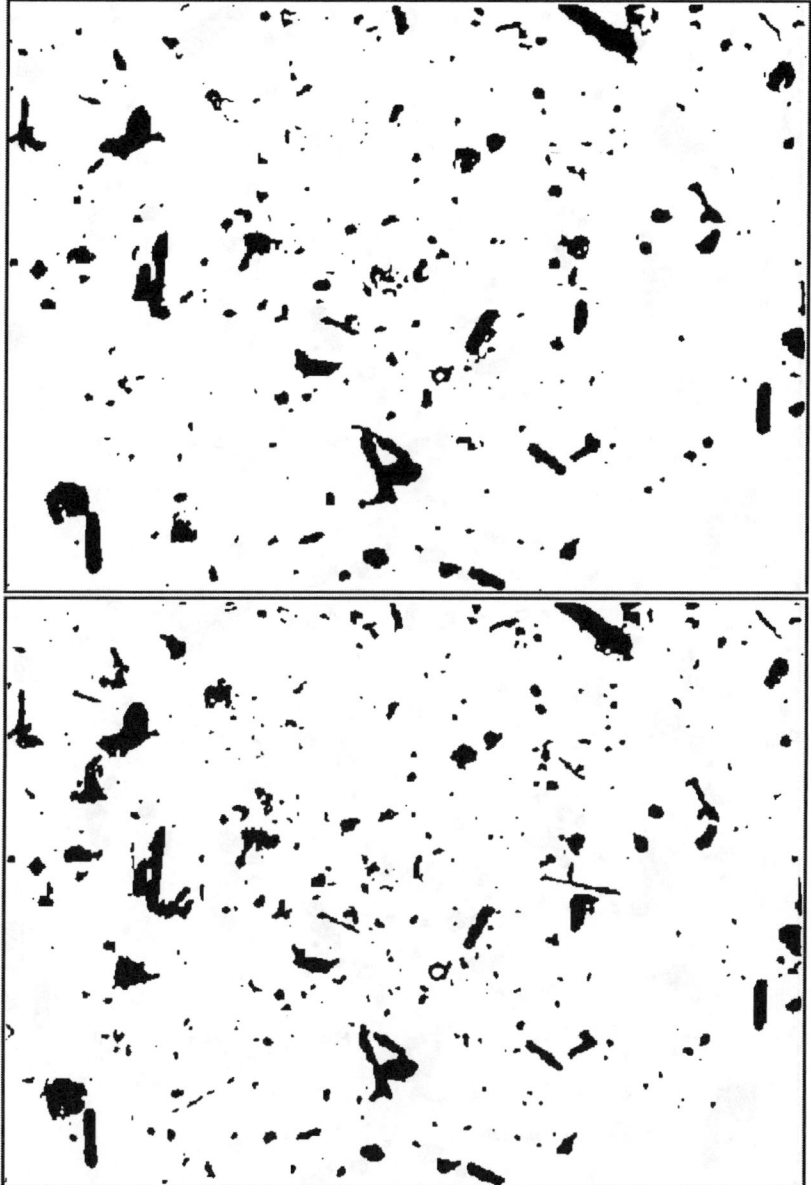

Figure 9. Top: binarized, low magnification image from below the best focused plane; bottom: binarized autofocus plane.

and only acquiring the particles that pass the tests. This approach provides a more effective and flexible screening technique, because we can establish a multicriteria decision-rule system based on both local and global shape descriptors. There are some methods based on symmetry detection (e.g. based on the gradient of the

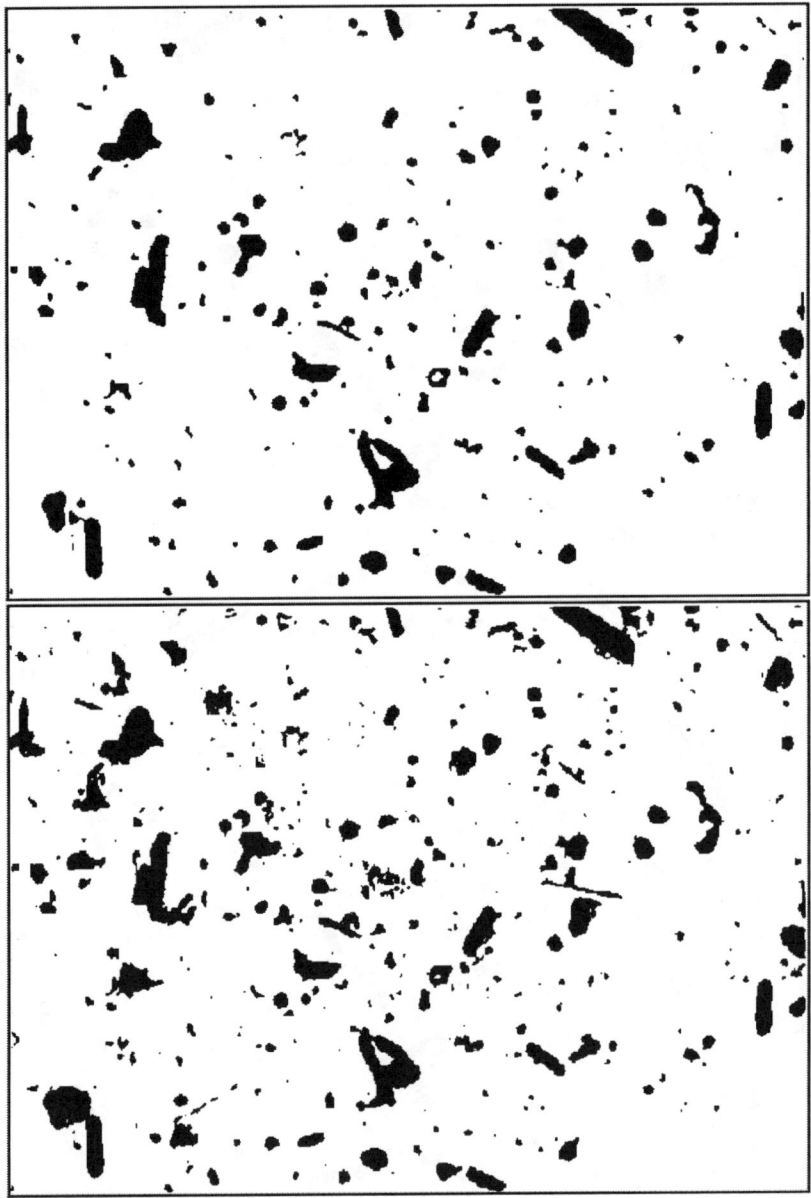

Figure 10. Top: binarized, low-magnification image from above the best focus plane; bottom: the fused image obtained by averaging the three images shown in Fig. 9 and here (top).

histogram, see Sun, 1995), geometric descriptors (circularity, elongation, etc., see Jain and Hong, 1996), and global shape descriptors (e.g. Hough transform). We have tested an algorithm for the detection of elliptic shapes using the Randomized Hough Transform as described by McLaughlin (2000). The main idea here is to

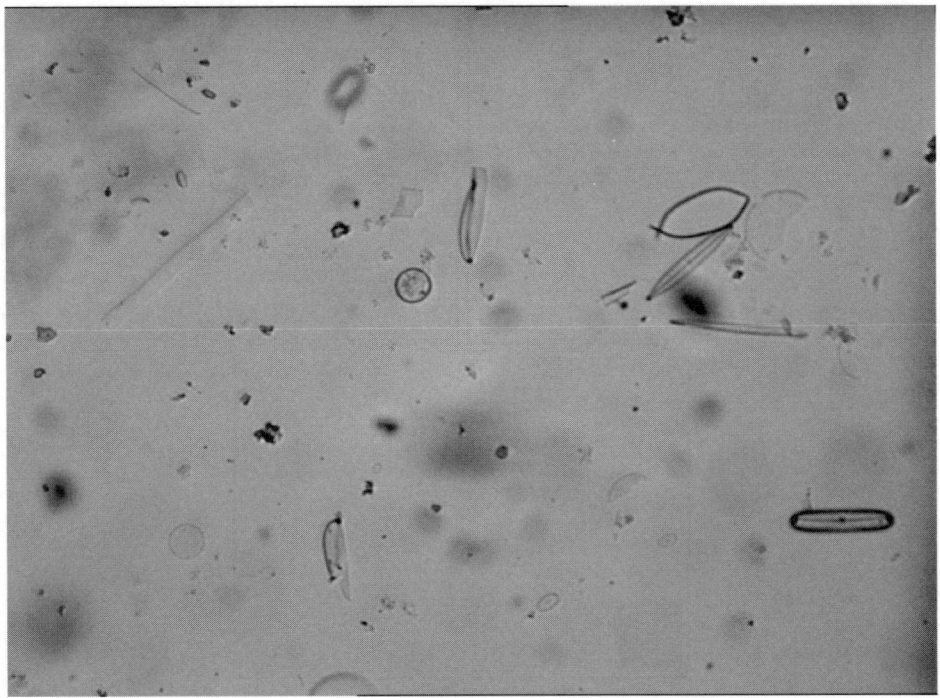

Figure 11. Example of a microscopic image at medium magnification (20x). Note the difficulty in discriminating between diatoms and debris. Image size: 768 × 576 pixels.

apply an edge detector, such as the Sobel gradient, which will be described in Section 4, and to detect (pseudo-)elliptical shapes via the Hough transform. Figure 13 shows examples. Purely elliptical pennate diatoms can be detected and circular (centric) ones as well. One problem of the Hough transform is that it costs a lot of CPU time. The percentage of usable valves that can be detected with this method remains to be estimated. There are many diatoms with shapes that do not resemble simple ellipses. In such cases the screening process may require some of the analysis methods described in the identification chapters in this book. A very recent book about shape analysis has been written by da Fontoura and Marcondes (2001).

• Image screening using phase-only correlation filtering: this is the only technique that can be directly applied to medium magnification images (i.e. 20x), see Fig. 11. Each particle detected at this magnification is padded with a uniform gray level to obtain an image of 256 × 256 pixels (Fig. 14). The screening is accomplished by matching a prototype image and the test image with the particle. Image matching is defined as the problem of evaluating the similarity of objects in different images (Chen and Defrise, 1994). This technique has been used in many applications. Here, it will allow us to verify the presence of a diatom in a captured image. Let $T(x,y)$ be the template image and $S(x,y)$ a sample image to be analyzed. The technique that we are using here is referred to as symmetric phase-only matched filtering (SPOMF) in the literature. It is based on computing the

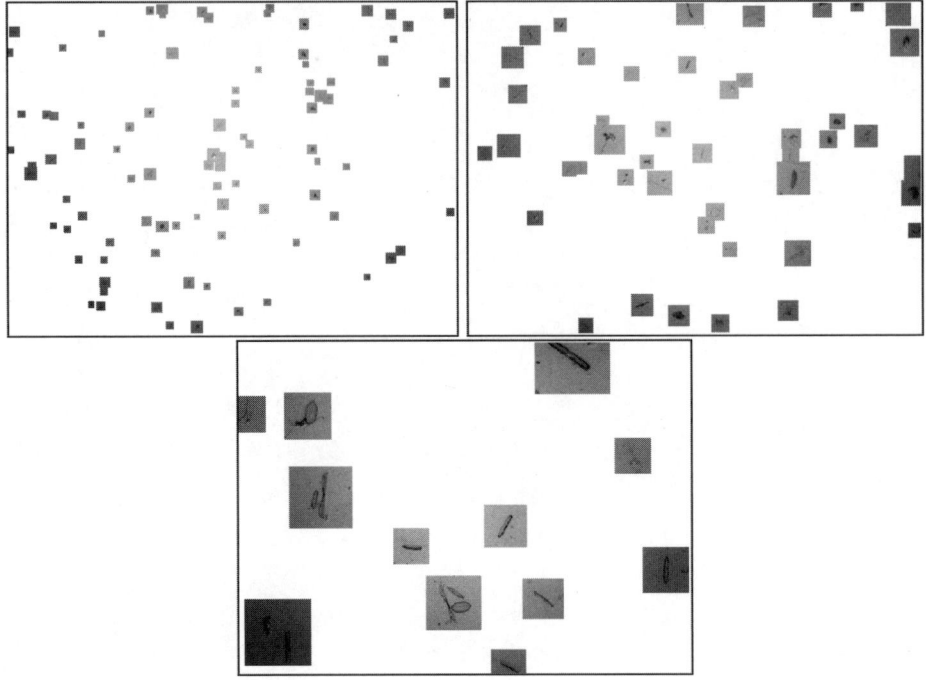

Figure 12. Image indexing by size. Size ranges between 10–20 μm (top-left), 20–50 μm (top-right) and 50–120 μm (bottom).

phase spectra of the two images and then the phase-only matched filtering (e.g. de Castro and Morandi, 1987). However, the spectral phase is not scale and rotation invariant. Matching of scaled and rotated images can be done by substituting the Fourier transform with the Fourier-Mellin transform, as has been shown by Chen and Defrise (1994), or with the scale transform as proposed by Pech-Pacheco et al. (2000). The phase-only matched filtering algorithm using the Fourier transform is as follows:

Algorithm 1: Phase-only correlation filtering

1. Compute the Fourier transform $\mathcal{T}(u,v)$

2. Extract the spectral phase $\mathcal{P}(\mathcal{T}(u,v)) = \exp[j\phi_T(u,v)]$

3. Compute the Fourier transform $\mathcal{S}(u,v)$

4. Extract the spectral phase $\mathcal{P}(\mathcal{S}^*(u,v)) = \exp[-j\phi_S(u,v)]$

5. Compute the inverse Fourier transform of the phase product:

$$C(x,y) = \mathcal{F}^{-1}[\frac{\mathcal{T}(u,v)}{|\mathcal{T}(u,v)|}\frac{S^*(u,v)}{|S^*(u,v)|}] = \mathcal{F}^{-1}\left[\exp\{j[\phi_T(u,v) - \phi_S^*(u,v)]\}\right]$$

Chapter 13: Automatic slide scanning

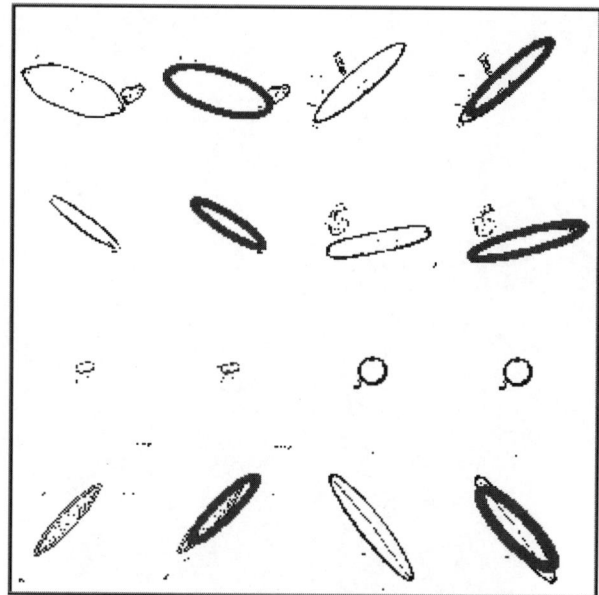

Figure 13. Ellipse detection through the Hough transform. The input for the Hough transform (first and third columns) are edges detected by the Sobel masks (see Section 4). The second and fourth column show the results of the ellipse detection. The third particles in the first and third column are debris and the method does not detect any feature.

6. Threshold the magnitude of $C(x,y)$ to detect the maximum peaks (eliminating false alarms).

Here ϕ_S^* represents the complex conjugate of the phase ϕ_S.

As a result, the SPOMF method produces a sharp correlation peak at the position where the images match. Figure 14 shows some of the particles detected by the method of indexing by size, as described in Section 3.5. Figure 15 shows the results of the SPOMF correlation method using as a template the top-left image in Fig. 14. A significantly higher correlation peak can be observed in the case of a diatom (Fig. 15, middle) if compared with the case of debris (Fig. 15, bottom). A first estimate is that we can detect about 80% of usable valves. This is a very significant rate in view of the magnification of 40x. The same process can be applied at higher magnifications, but would cost more CPU time.

4 Autofocusing

Several autofocusing methods have been proposed in the literature (Groen et al., 1985; Krotkov, 1987; Firestone et al., 1991; Subbarao et al., 1993; Yeo et al., 1993; Bocker et al., 1996). These methods use different focus functions, which measure the relative sharpness of images taken at different depths. The depth at which the function has the largest value will give the best-focused image. In the next sections we will review different focus functions. Let $I(n,m)$ be an image of size $N \times M$

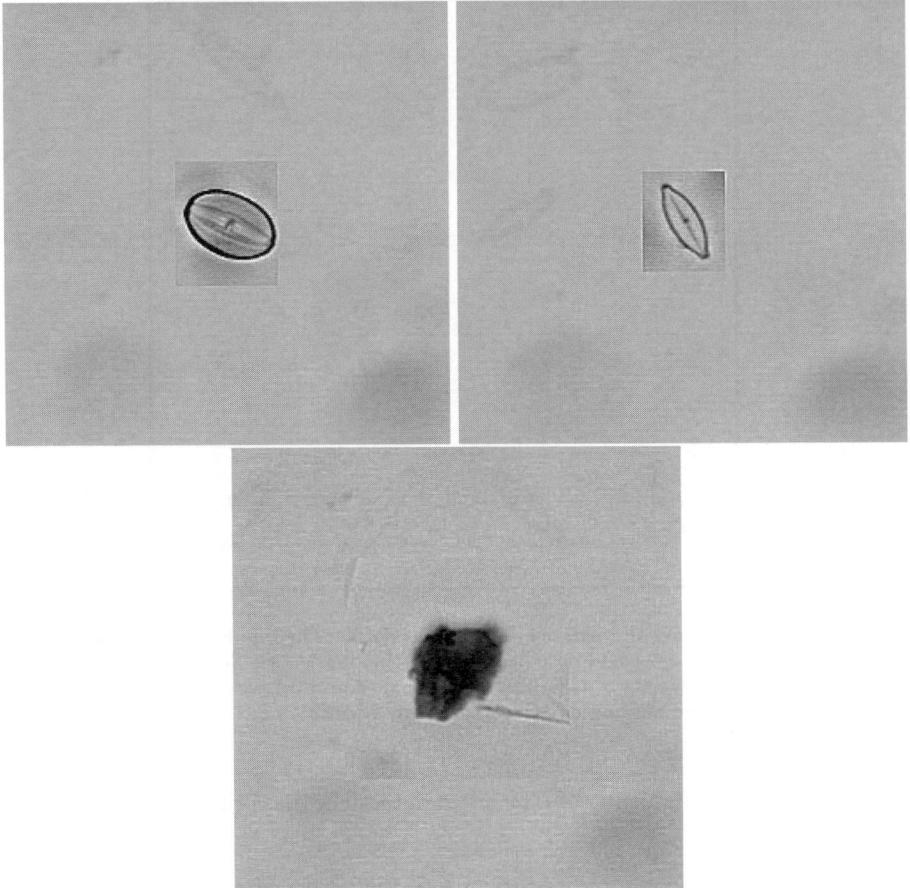

Figure 14. Top-left: diatom image used as template for the phase-correlation process. Top-right and bottom: two particles detected by the method of indexing by size as described in Section 3.5.

and let us assume a stack of k images taken by changing the focal depth in steps of 1 μm. Figure 16 shows images from a stack above and below the best-focused image shown at the center.

4.1 Focus function based on image contrast

A well-focused image can be expected to show strong variation in gray levels. The local variance at point (n, m), with $n = 1 \ldots N$ and $m = 1 \ldots M$, is given by

$$F_{lv}(n, m) = \frac{1}{w_x w_y} \sum_{i}^{w_x} \sum_{j}^{w_y} \left[I(n+i, m+j) - \overline{I} \right]^2, \qquad (1)$$

Chapter 13: Automatic slide scanning 275

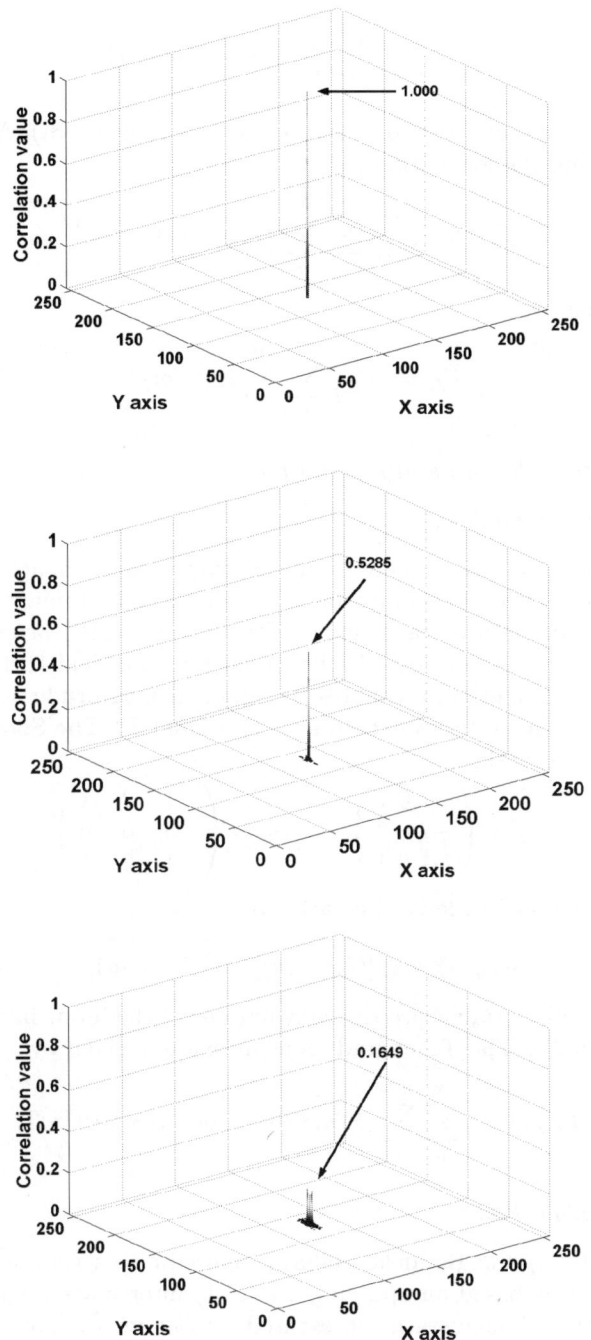

Figure 15. Top: the phase-only autocorrelation peak of the diatom template (Fig. 14, top-left) has a value of 1.0. Middle: correlation peak of the template (Fig. 14, top-left) with the diatom shown in Fig. 14 (top-right). Bottom: correlation peak corresponding to the particle shown in Fig. 14 (bottom).

where $\bar{I}$ is the mean gray level value

$$\bar{I} = \frac{1}{w_x w_y} \sum_{i}^{w_x} \sum_{j}^{w_y} I(n+i, m+j) \qquad (2)$$

and $w_x \times w_y$ is the size of a window centered on the point (n, m). A *global* variance can be defined from the local variance by

$$F_{var}(I) = \frac{1}{NM} \sum_{n}^{N} \sum_{m}^{M} \left[F_{lv}(n,m) - \overline{F}_{lv} \right]^2, \qquad (3)$$

where $\overline{F}_{lv}$ is given by

$$\overline{F}_{lv} = \frac{1}{NM} \sum_{n}^{N} \sum_{m}^{M} lv(n,m). \qquad (4)$$

4.2 Focus functions based on differentiation

4.2.1 First derivative methods

A well-focused image is also expected to have sharper edges, hence image gradients are applied for computing focus measures. Given a gradient image, the focus measure has to pool the data at each point. Tenenbaum (1970) proposed a method, called Tenengrad, that is considered a benchmark in this field (Schlag et al., 1983; Krotkov, 1987). This method estimates the gradient magnitude at each image position and sums all magnitudes exceeding a threshold T. The Sobel operator uses the two convolution masks

$$S_x = \begin{pmatrix} -1 & 0 & 1 \\ -2 & 0 & 2 \\ -1 & 0 & 1 \end{pmatrix} \quad S_y = \begin{pmatrix} 1 & 2 & 1 \\ 0 & 0 & 0 \\ -1 & -2 & -1 \end{pmatrix} \qquad (5)$$

and the gradient's magnitude is calculated as

$$S(n,m) = \sqrt{[G_x(n,m)]^2 + [G_y(n,m)]^2}, \qquad (6)$$

where $G_x(n,m)$ and $G_y(n,m)$ are the convolutions of the input image $I(n,m)$ with the masks S_x and S_y. The Tenengrad focus measure is given by

$$F_{ten}(I) = \sum_{n}^{N} \sum_{m}^{M} [S(n,m)]^2 \quad \text{for} \quad S(n,m) > T. \qquad (7)$$

4.2.2 Boddeke's algorithm

Boddeke et al. (1994) and Boddeke (1998) proposed also a very simple focus measure. This method is based on applying a (-1,0,1) filter mask *only* along the horizontal (n) axis of an image stack represented by $I(n,m,z)$. The focus function is defined by squaring and adding all the filtered pixel values, i.e.

$$F(z) = \sum_{n,m} [I(n+1,m,z) - I(n-1,m,z)]^2. \qquad (8)$$

Chapter 13: Automatic slide scanning

The method is extremely simple and provides a single, sharp peak (Ellenberger and Young, 1998).

4.2.3 Second derivative methods

The use of a second derivative operator is another technique for analyzing the high spatial frequencies associated with sharp edges. As a second derivative operator we use the Laplacian, which can be approximated using the filter mask

$$L = \frac{1}{6} \begin{pmatrix} 0 & -1 & 0 \\ -1 & 4 & -1 \\ 0 & -1 & 0 \end{pmatrix}. \tag{9}$$

We use two methods for pooling the data at all positions. The first one is the sum of all the absolute values, giving the following focus measure:

$$F_{lap}(I) = \sum_{n}^{N} \sum_{m}^{M} |L(n,m)|, \tag{10}$$

where $L(n,m)$ is the convolution of the input image $I(n,m)$ with the mask L. The second method consists of computing the variance, see below.

4.3 New focus functions based on differentiation

4.3.1 Variance of gradient

The alternative for pooling the gradient information is to calculate the variance of gradient magnitudes, i.e. the variance of the magnitude of the Sobel gradient. The reason for this approach is to define a more discriminative measure, similar to a second derivative (Laplacian), but increasing the robustness to noise (Pech-Pacheco et al., 2000; Pech-Pacheco and Cristobal, 2001). Figure 16 shows a comparison of focus measurements without and with noise. The new focus measure is given by

$$F_{sob\_var}(I) = \sum_{n}^{N} \sum_{M}^{m} \left[S(n,m) - \overline{S}\right]^2 \quad \text{for} \quad S(n,m) > T, \tag{11}$$

where $\overline{S}$ is the mean of magnitudes given by

$$\overline{S} = \frac{1}{NM} \sum_{n}^{N} \sum_{m}^{M} S(n,m). \tag{12}$$

4.3.2 Variance of Laplacian

In the second new method we calculate the variance of the absolute value of the Laplacian, i.e.

$$F_{lap\_var}(I) = \sum_{n}^{N} \sum_{m}^{M} \left(|L(n,m)| - \overline{L}\right)^2, \tag{13}$$

where $\overline{L}$ is the mean of absolute values given by

$$\overline{L} = \frac{1}{NM} \sum_{n}^{N} \sum_{m}^{M} |L(n,m)|. \tag{14}$$

4.3.3 Window size selection

An important parameter is the window size for computing the variance and gradient. We have experimented with 23 different stacks of multifocus images taken at high magnification (40x). Each stack is composed of 100–150 images. Figure 17 (top) shows the focus estimate of all the stacks tested as a function of the window size. Level zero in the plot indicates the optimum focus plane. The variance-only method provides a better focus estimate for small windows, but it costs more CPU time for large windows. The vertical bars are shorter for small windows, meaning a better focus estimate. However, the variance-only method lacks robustness, because, as can be seen from the dash-dotted plot in Fig. 16, there are situations in which it yields a wrong focus estimate. Figure 17 (bottom) shows that the variance of Sobel provides better results for large windows, which is an advantage because of its smaller CPU time. The vertical bars are shorter for large window sizes, hence it provides a better focus estimate. Therefore, we decided to use the variance of Sobel for the full image size (global variance), because it has the lowest computational cost.

4.4 Assessment of focus measures

Here we will assess the shape of the focus measures shown in Fig. 16. Due to the lack of a quantitative, standard metric for assessing the shape of the focus curves, we propose two new measures based on sharpness and smoothness.

• Sharpness: The sharpness of the peak is not the only criterion for selecting a good focus measure. In fact, as Subbarao et al. (1993) have pointed out, any focus measure can be artificially sharpened, simply by squaring the focus measure. Kurtosis is a candidate for "peakedness." However, it suffers from a high noise sensitivity. We are more interested in having a sharp autofocus curve, in order to reduce the number of images to fuse by means of multifocus fusion techniques described in the next section. We define a sharpness metric as the average value of the width d of the focus curves at three equidistant focus levels from the maximum value, i.e. the width of the focus curves at values of 0.25, 0.5 and 0.75 (see Fig. 16):

$$F_1 = k_1 \sum_{i=1}^{3} d_i, \tag{15}$$

where d_i is the width at the three levels mentioned above and k_1 is a normalizing constant. Table 1 shows sharpness measures (normalized with respect to the Sobel+variance value). It can be seen that the Sobel+variance and Tenengrad methods provide the smallest F_1 values.

• Smoothness and noise sensitivity: We define a relative measure based on the accumulative sum of the absolute gradient value of the focus measure (Fig. 16) with

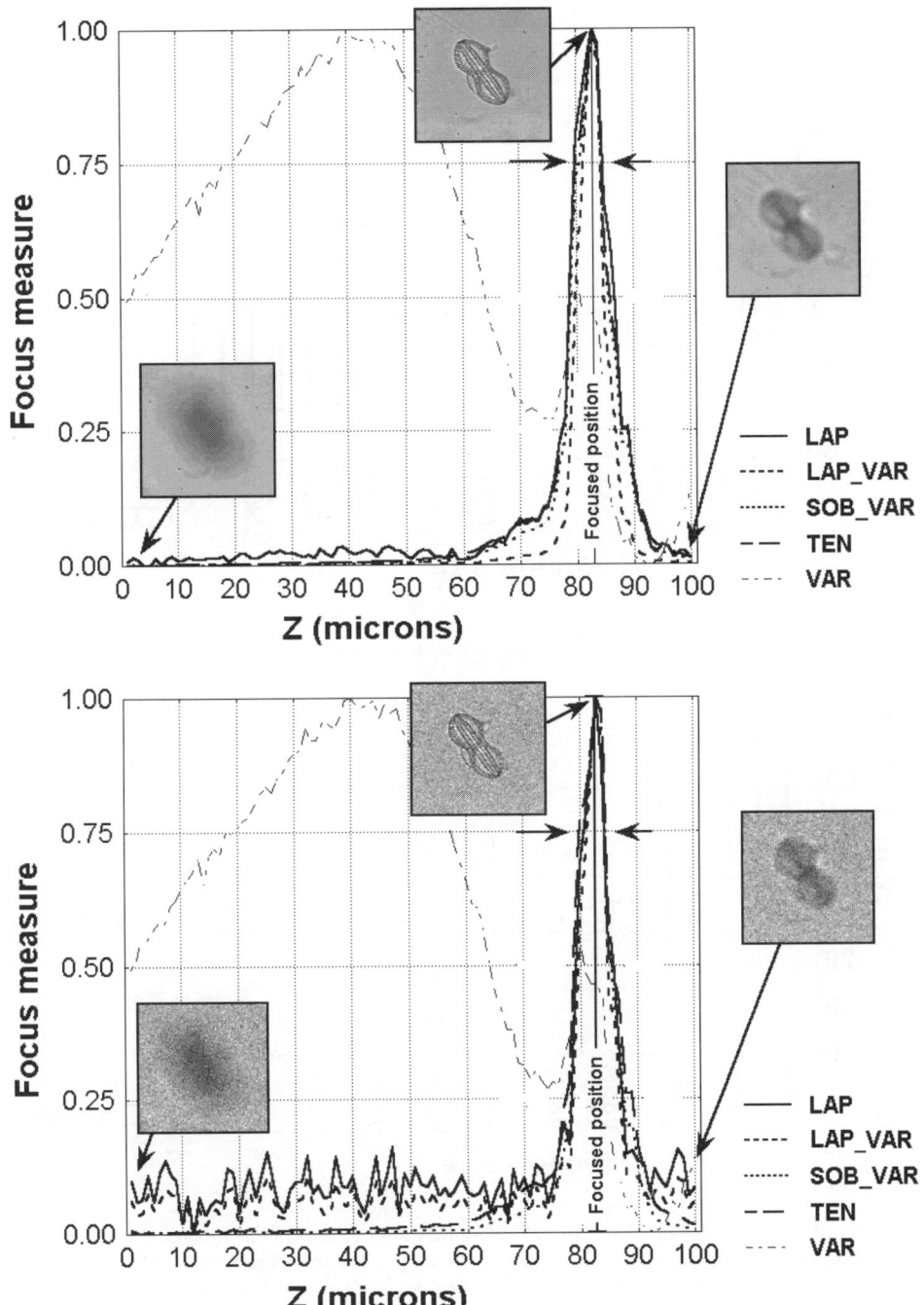

Figure 16. Top: comparison of focus measurements; no noise. Bottom: the same for noisy data. LAP: Laplacian; LAP_VAR: variance of Laplacian; SOB_VAR: variance of Sobel; TEN: Tenengrad; VAR: global variance. Note that the global variance method (VAR) provides a wrong estimate of the best-focused image for this image stack. The images located above the horizontal arrows are the ones to be selected for multifocus fusion; see Fig. 18.

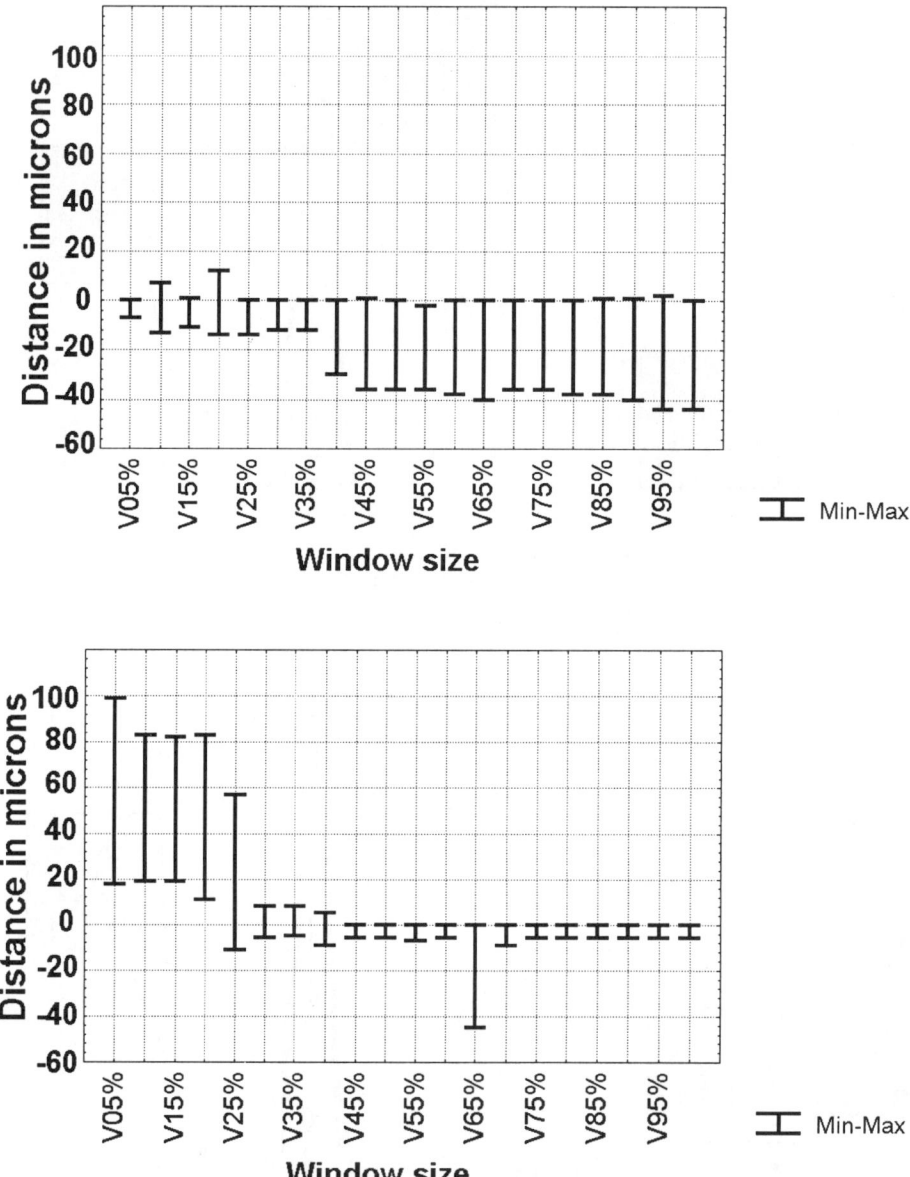

Figure 17. Window size evaluation from 23 different stacks of multifocus images. Each stack consists of 100–150 multifocus images. The horizontal axes of the plots represent the size percentage of the window used for the computation in relation to total image size. Level 0 at the vertical axes means the best focusing plane. Top: variance measure; bottom: variance of Sobel measure.

Chapter 13: Automatic slide scanning

Table 1. Sharpness and smoothness results.

function	Sobel+var	Tenengrad	Laplacian	Lapl+var
F_1	1	1.21	3.79	3.28
F_2	1	1.02	2.41	2.02

respect to fitting a Gaussian with $\sigma = 1$:

$$F_2 = \frac{\sum_k |AF'_k|}{\sum_k |AF'_k|_G}, \qquad (16)$$

where $|AF'_k|$ represents the absolute value of the first derivative of the focus measure and $|AF'_k|_G$ the corresponding value for the fitting Gaussian. Table 1 shows that, again, the Sobel+variance and Tenengrad methods provide the smaller values of F_2, which correspond also to a better noise tolerance.

In conclusion, the sharpness measure based on Sobel+variance is clearly better than that based on Tenengrad, whereas the smoothness measures are comparable.

5 Multi-focus visualization techniques

In this section we present two methods for visualizing the 3D structure of diatoms. The first method is based on image fusion techniques, by combining the most important structures of a specimen from a multifocus image stack into a single fused image. The second approach is based on extracting the 3D information by deconvolving the multifocus stack.

5.1 Multifocus fusion

Image fusion produces a single image from a set of images. In this case, the images to be fused have the same origin, i.e. they are from a multifocus image stack. The main advantages of the fusion process are: (1) the capability of gathering complementary information, and (2) improved reliability by combining redundant information. Since most diatom valves are not flat, a sharp contour *and* a sharp ornamentation cannot appear together in a single image captured, unless different images taken at different depths are combined. A large number of image fusion techniques has been proposed (e.g. Rockinger, 1999; Wang and Lohmann, 2000). We are only interested in combining multifocus images, but there are also other applications. For example, Seales and Dutta (1996) applied fusion techniques to images taken with inexpensive cameras, with the aim to create focused images in different situations.

Here, we use fusion methods based on simple operations like averaging. The main considerations in the design of the fusion algorithm are its implementation efficiency and portability to other applications, see also Kundur et al. (2000). The most difficult part of the fusion process is the selection of the range of images to be fused. In the case of diatoms we found empirically that it is sufficient to use images in the upper fourth part of the focusing curves, i.e. with values between 0.75

and 1. These are the images located above the horizontal arrows in Fig. 16, shown in more detail in Fig. 18. After normalizing these images, we apply a binarization using the triangle algorithm, in order to extract the regions that are better focused. This provides a set of binary images that are used to extract the regions that are focused in all images. Each binary image is multiplied with its associated gray level image; hence we can extract the most salient features from each image of the stack (Fig. 20). Figure 18 shows an example of the use of the multifocus fusion technique described here. This figure represents a detailed view of the peak of the focusing curves of Fig. 16, with the eight images to be fused. Figure 19 shows the fused result. This image is computed by averaging the salient features extracted from the image stack, showing a sharp contour and striation. The multifocus fusion algorithm is as follows:

Algorithm 2: Multifocus fusion algorithm

1. Capture a multifocus image stack
2. Select the range of images to be fused (upper quarter of the focus curve)
3. Background suppression of the selected images, using "rolling ball" method
4. Clip each image between gray levels 0 and 255
5. Threshold each image using the triangle algorithm (Fig. 20)
6. Multiply the binary images of step 5 with the images after step 3
7. Average and normalize the results of step 6.

5.2 Optical sectioning techniques

Defocusing in optical microscopy can be modeled as a linear, 3D convolution of the optical density of a specimen with the point spread function of the optics (the microscope lens). There are a number of methods in the literature for inverse filtering or deconvolution, using either measured or estimated optical point spread functions (Preza et al., 1992). It is also possible to estimate the in-focus information—sharp detail and its depth—from the degree of blurring. As for multifocus fusion, a stack of images is taken at equidistant depths, normally 10 to 15 images. Some authors have determined the in-focus depth after applying a highpass filter to each image in the stack, maximizing the result. In this way, a set of focus measures is obtained at each position (x,y). Then, the depths of surface points at each position are obtained by detecting the peaks (Noguchi and Nayar, 1994). Glasbey (1996) has applied this technique to recover the 3D valve surfaces of diatoms. Figure 21 shows four images of a diatom taken at different focal planes, and Fig. 22 shows the recovered 3D surface of the valve, obtained from a total of fourteen images. Such a 3D reconstruction can be used to create one single image with a sharp contour as well as striation pattern. It can also be used to compute new surface features concerning the 3D shape of a valve. These features can complement the ones described

Chapter 13: Automatic slide scanning

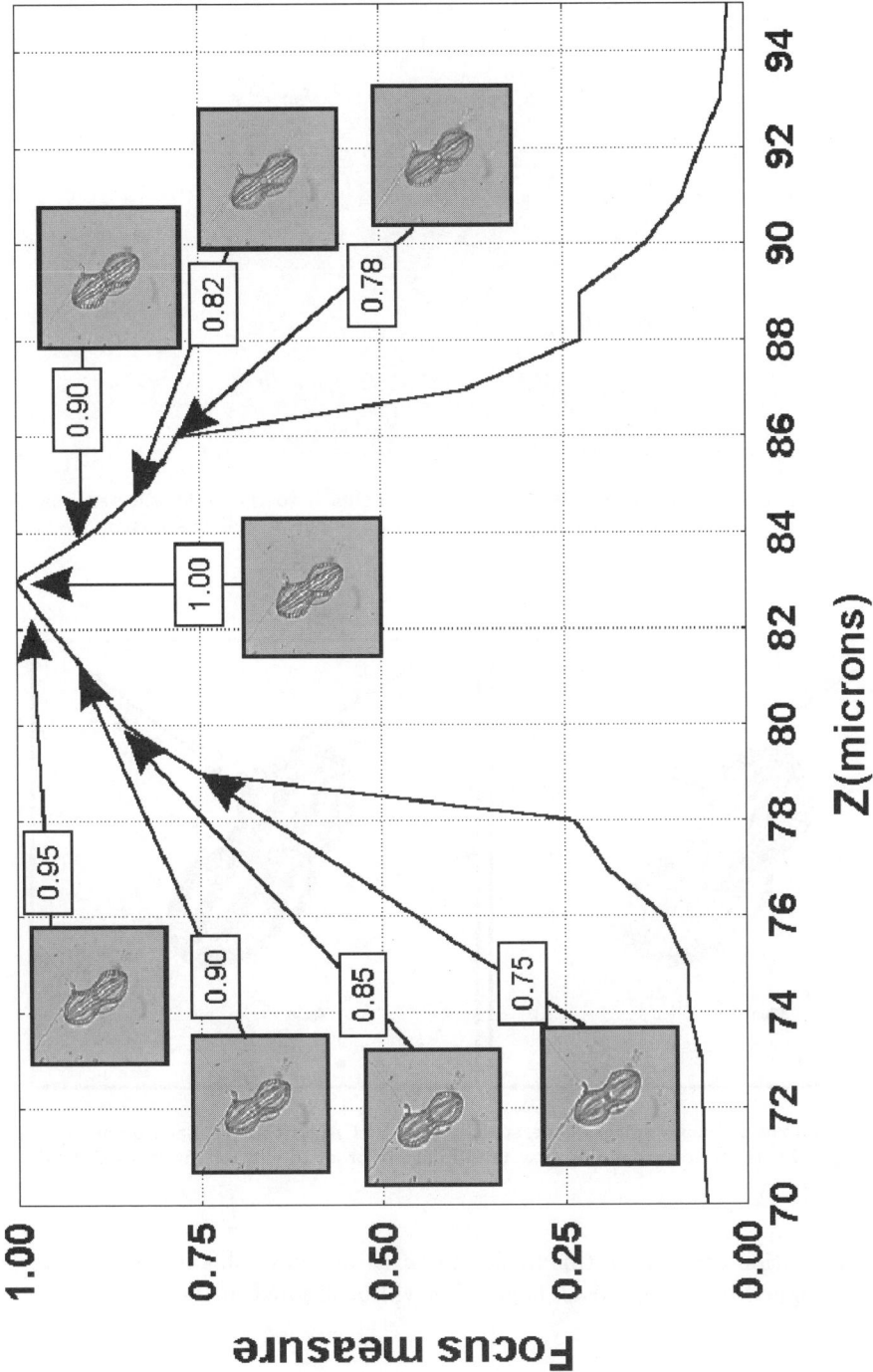

Figure 18. Schematic representation of the multifocus fusion technique. Only the images located at the upper quarter of the autofocusing curve (Fig. 16) will contribute to the fusion process. See Fig. 19 for the fused image

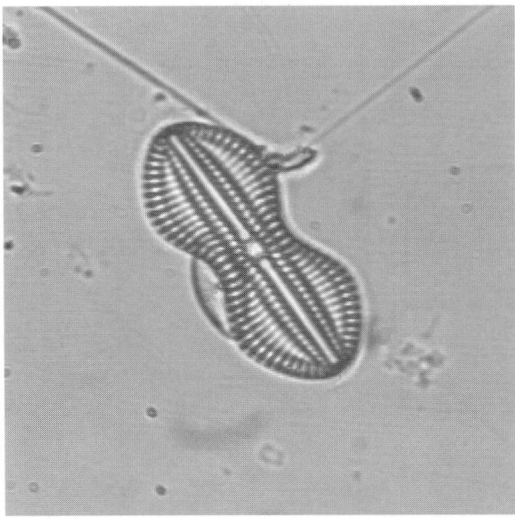

Figure 19. Fused image that results from applying Algorithm 2 to the eight selected images (Fig. 18). The goal is to obtain a single image with a sharp contour as well as striation.

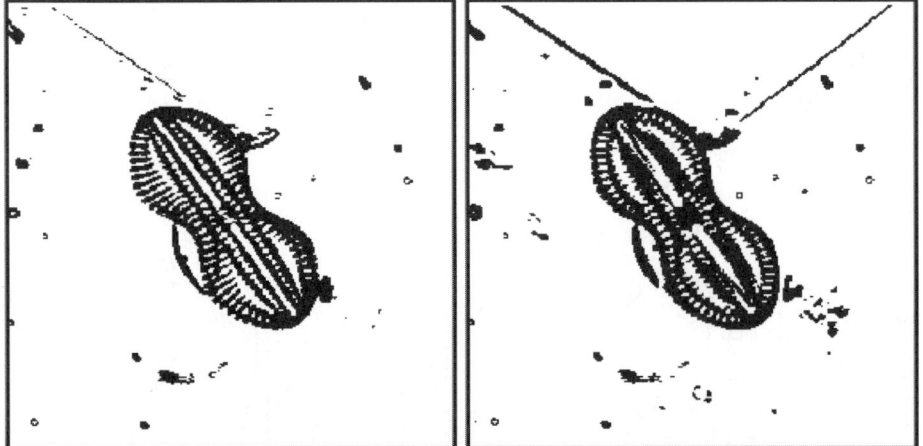

Figure 20. Examples of binary images generated by step 5 of Algorithm 2. Each binary image will be multiplied with its corresponding gray level image in order to extract the focused detail.

in the identification chapters in this book. As far as we know, this information has never been applied by taxonomists, hence this will be studied in the future.

Chapter 13: Automatic slide scanning

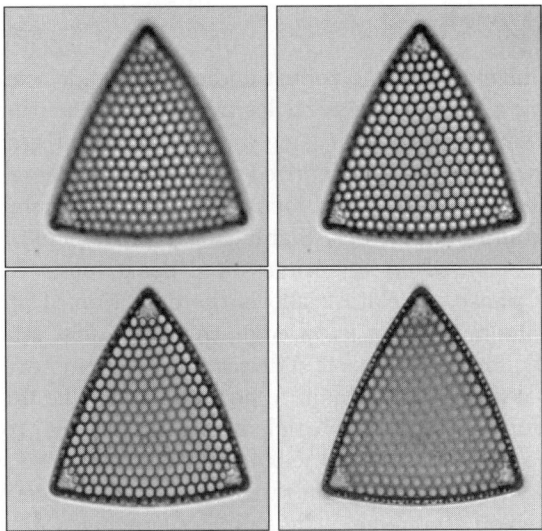

Figure 21. Diatom images taken at different depths. Reprinted with kind permission of Glasbey (1996).

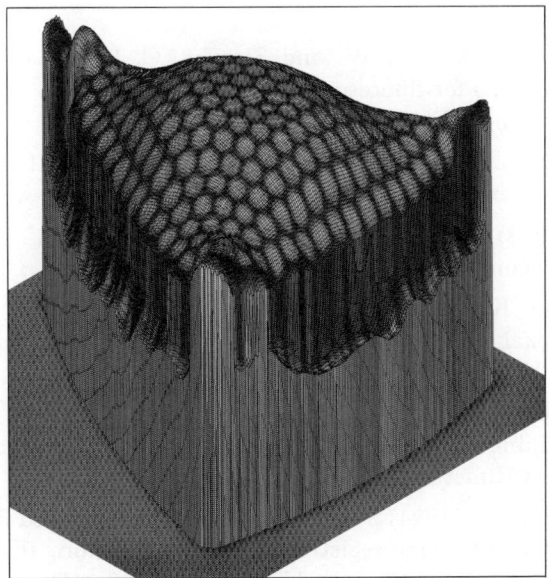

Figure 22. 3D diatom reconstruction from a series of fourteen focal planes. Reprinted with kind permission of Glasbey (1996).

6 Conclusions

In this chapter we have described a robust and efficient slide scanning technique for accurately detecting all candidate particles at low magnification. We have shown two new methods for autofocusing, using a combination of gradient and variance operators. Extensive work has been carried out on autofocusing for a large number of different specimens. The proposed focus measures were compared with a number of techniques from the literature. Significant improvements have been obtained with focusing metrics based on smoothness and peakedness. The particle screening on the basis of phase correlation allows the detection of about 80% of usable diatom valves. To the best of our knowledge, this is the first study in which all the components required for an automatic slide scanning system have been investigated.

There are several issues that remain to be addressed. The first is to improve the discrimination performance of the particle screening process, together with a systematic statistical analysis, assisted by expert diatomists. The second is to extend the present study to the nanoplankton range. The third concerns an improvement of the multifocus fusion technique. The simple averaging applied can be replaced by a weighted averaging, using principal component analysis of all images to be fused, as suggested by Wang and Lohmann (2000).

References

Bocker, W., Rolf, W., Muller, W. and Streffer, C. (1996) Investigations about autofocus-algorithms for fluorescent-microscopy. In: SPIE Applications of digital image processing XIX, Vol. 2847, pp. 445-456.

Boddeke, F.R., van Vliet, L.J., Netten, H. and Young, I.T. (1994) Autofocusing in microscopy based on the OTF and sampling. Bioimaging, Vol. 2, pp. 193-203.

Boddeke, F.R. (1998) Quantitative fluorescence microscopy. PhD Thesis, Delft University of Technology, The Netherlands.

Cairns, J., Dickson, K., Lanza, G., Almeida, S. and del Balzo, D. (1972) Coherent optical spatial filtering of diatoms in water pollution monitoring. Archiv Mikrobiologie, Vol. 83, pp. 141-146.

Cairns, J., Dickson, K., Pryfogle, S.P., Almeida, S., Fournier, J.M. and Fuji, H. (1979) Determining the accuracy of coherent optical identification of diatoms. Water Resources Bulletin, Vol. 15, pp. 1770-1775.

Chen, Q. and Defrise, M. (1994) Symmetric phase-only matched filtering of Fourier-Mellin transforms for image registration and recognition. IEEE Trans. Pattern Analysis and Machine Intelligence, Vol. 16, pp. 1156-1168.

Culverhouse, P.F., Simpson, R.G., Ellis, R., Lindley, J.A., Williams, T., Parisini, T., Reguera, B., Bravo, I., Zoppoli, R., Earnshaw, G., McCall, H. and Smith, G. (1996) Automatic classification of field-collected dinoflagellates by artificial neural network. Marine Ecology Progress Series, Vol. 139, pp. 281-287.

da Fontoura, L. and Marcondes, R. (2001) Shape analysis and classification. CRC Press, Boca Raton (FL), USA.

de Castro, E. and Morandi, C. (1987) Registration of translated and rotated images using finite Fourier transforms. IEEE Trans. Pattern Analysis and Machine Intelligence, Vol. 9, pp. 700-703.

Ellenberger, S.L. and Young, I.T. (1998) Microscope image acquisition. In: Image processing and analysis, R. Baldock and J. Graham (eds), Oxford University Press, Oxford, UK.

Firestone, L.K., Cook, K., Culp, K., Talsania, N. and Preston, K. (1991) Comparison of autofocus methods for automated microscopy. Cytometry, Vol. 12, pp. 195-206.

Glasbey, C.A. (1996) Problems in digital microscopy. Proc. XVIII Int. Biometric Conference, Amsterdam, pp. 183-200.

Groen, F., Young, I.T. and Ligthard, G. (1985) A comparison of different focus functions for use in autofocus algorithms. Cytometry, Vol. 6, pp. 81-91.

Jain, A.K. and Hong, L. (1996) Automatic classification of bacteria culture images. Technical Report, Michigan State University.

Kindratenko, V. (1997) Development and application of image analysis techniques for identification and classification of microscopic particles. PhD Thesis, Universiteit Antwerpen, Belgium.

Krotkov, E. (1987) Focusing. Int. J. Computer Vision, Vol. 1, pp. 223-237.

Kundur, D., Hatzinakos, D. and Leung, H. (2000) Robust classification of blurred imagery. IEEE Trans. Image Processing, Vol. 9, pp. 243-255.

McLaughlin, R.A. (2000) Intelligent algorithms for finding curves and surfaces in real world data. PhD Thesis, Univ. of Western Australia, Dept. Computer and Electronic Engineering.

Nayar, S.K. and Nakagawa, Y. (1994) Shape from focus. IEEE Trans. Pattern Analysis and Machine Intelligence, Vol. 16, pp. 824-831.

Neveu, C. (1999) Formal models of blur detection. Perception, Supplement, Vol. 28, p. 147.

Noguchi, M. and Nayar, S.K. (1994) Microscopic shape from focus using active illumination. Proc. Int. Conf. Pattern Recognition, IEEE Computer Society, Los Alamitos (CA), USA, pp. 147-152.

Pech-Pacheco, J.L. and Alvarez-Borrego, J. (1998) Optical-digital system applied to the identification of five phytoplankton species. Marine Biology, Vol. 132, pp. 357-365.

Pech-Pacheco, J.L., Alvarez-Borrego, J., Orellana-Cepeda, E. and Cortés-Altamirano, R. (1999) Diffraction patterns applicability in identification of *Ceratium* species. J. Plankton Research, accepted for publication.

Pech-Pacheco, J.L. and Cristóbal, G. (2000) Diatom autofocusing in brightfield microscopy: a comparative study. Proc. Int. Conf. on Pattern Recognition, Barcelona, IEEE Computer Soc., Los Alamitos (CA), USA, Vol. 3, pp. 318-321.

Pech-Pacheco, J.L., Cristóbal, G., Alvarez-Borrego, J. and Cohen, L. (2000) Power cepstral image analysis through the scale transform. Proc. SPIE, Vol. 4113, pp. 68-79.

Pech-Pacheco, J.L. and Cristóbal, G. (2001) Automatic system for phytoplanktonic algae identification. Limnetica, submitted.

Preza, C., Miller, M.I., Lewis, J.T. and McNally, J.G. (1992) Regularized linear method for reconstruction of three-dimensional microscopic objects from optical sections. J. Optical Society of America A, Vol. 9, pp. 219-228.

Rockinger, O. (1999) Multiresolution-Verfahren zur Fusion dynamischer Bildfolgen. PhD Thesis, Technische Universität Berlin, Germany.

Russ, J.C. (1995) The image processing handbook. CRC press, Boca Raton (FA), USA.

Schlag, J.F., Sanderson, A.C., Neuman, C.P., Wimberly, F.C. (1983) Implementation of automatic focusing algorithms for a computer vision system with camera control. Technical Report CMU-RI-TR-83-14, Robotics Institute, Carnegie Mellon University.

Seales, W.B. and Dutta, S. (1996) Everywhere in focus image fusion using controllable cameras. Proc. SPIE, Vol. 2905, pp. 227-234.

Seul, M., O'Gorman, L. and Sammon, M.J. (2000) Practical algorithms for image analysis. Description, examples and code. Cambridge University Press, Cambridge, UK.

Sternberg, S.R. (1983) Biomedical image processing. Computer Magazine, Vol. 16, pp. 22-34.

Subbarao, M., Choi, T. and Nikzad, A. (1993) Focusing techniques. Optical Engineering, Vol. 32, pp. 2824-2836.

Sun, C. (1995) Symmetry detection using gradient information. Pattern Recognition Letters, Vol. 16, pp. 987-996.

Tenenbaum, J.M. (1970) Accomodation in computer vision, PhD Dissertation, Stanford University, USA.

Vollath, D. (1987) Automatic focusing by autocorrelative methods. J. Microscopy, Vol. 147, pp. 279-288.

Wang, Y. and Lohmann, B. (2000) Multisensor image fusion: concept, method and applications. Technical Report, Inst. of Automation Technology, Univ. Bremen, Germany.

Yeo, T., Ong, S., Jayasooriah and Sinniah, R. (1993) Autofocusing for tissue microscopy. Image and Vision Computing, Vol. 11, pp. 629-639.

Young, I.T., Gerbrands, J.J. and van Vliet, L.J. (1998) Image processing fundamentals. In: The digital signal processing handbook, V.K. Madisetti and D.B. Williams (eds), CRC Press, Boca Raton (FA), USA.

Zack, G.W., Rogers, W.E. and Latt, S.A. (1977) Automatic measurement of sister chromatid exchange frequency. J. Histochemistry and Cytochemistry, Vol. 25, pp. 741-753.

CHAPTER 14

ADIAC ACHIEVEMENTS AND FUTURE WORK

HANS DU BUF AND MICHA M. BAYER

As already mentioned in Chapter 1, it is unrealistic to expect a 3-year pilot project to resolve every single problem associated with automatic identifcation, and that any software developed will be ready to be converted into products for real-world applications. Some readers may be under the impression that ADIAC software is ready to be marketed commercially or otherwise, because identification rates for most methods are excellent and even exceed those achieved by most human experts. However, one must bear in mind that feature extraction and classification methods need adequate amounts of training data before they can be put to work in a real-life situation: any application, such as climatology or the monitoring of water quality, requires that all the taxa likely to occur in the respective samples are represented adequately, i.e. to an extent that will reflect their morphological variability across different populations and environments. The two large data sets used here, i.e. the *Sellaphora pupula* and mixed genera sets, both represented each taxon with at least 20 specimens, which is probably near the minimum required, and they represent only two relatively specific test scenarios. There are many more taxa in the ADIAC databases, but most are represented by far fewer specimens. Future work might either focus on one particular application, and aim to build an adequate training set of taxa specific to this application (usually several tens to a few hundreds of taxa, rather than thousands), or provide users with the opportunity to build their own training sets for their specific application. The latter scenario could—in theory—open up automatic identification to any group with the relevant equipment and sufficient manpower to acquire the necessary training images.

The results achieved to date might raise the expectation that one can select the best features from a number of feature sets, and that this selection will then perform equally well on other sets of taxa in the real-world applications mentioned above. Many experiments in which different feature sets and classifiers were combined have shown that the results are often unpredictable. This implies that further research, involving improvements of methods as well as extensions to the databases, will be necessary in order to consolidate the results. This research should preferably be done by teams composed of pattern recognition experts, taxonomists and researchers representing applications. The latter must collect representative specimens, the taxonomists must check these, and only when sample sizes are sufficient should the methods be tested and, if necessary, improved. Below, we highlight ADIAC's main achievements, but also discuss the problems encountered, which offer numerous possibilities for future research.

1 Databases

Chapter 4 is a review of the protocols necessary in order to obtain images suitable for automatic identification. Our experience with ADIAC has shown that, instead of two or more groups working separately with standard equipment, it may be a better idea to share high quality equipment, especially if automatic slide scanning will be available in the near future. The main concern here is the quality of the microscope optics and the CCD camera, including the frame-grabber. Sometimes images can be improved by adaptive filtering or sharpening, but if images of inferior quality are acquired, post-processing can lead to an amplification of noise. With or without post-processing, contour and feature extraction methods can be hampered by poor image quality, and this can have a negative impact on the identification rates. Image quality is really critical in creating databases.

At the end of the ADIAC project, the PANDORA database (represented by the WWW diatom browser at RBGE) included approximately 500 diatom taxa, represented by approximately 4700 images. About 2300 of these are available to the general public on the FTP server at RBGE. Two specific test sets with at least 20 specimens per taxon have been created to test algorithms: the *Sellaphora pupula* set, with 6 taxa and 120 images, and a mixed genera set with 48 taxa and 1009 images. These sets were intended for different purposes, with the former representing fine-grain morphological differences between taxa, and the latter reflecting major morphological differences. However, an additional challenge was that different taxa in the mixed genera set can have quite similar shapes, while different specimens from the same taxon can differ drastically in shape, largely as a consequence of the diatom life cycle (Chapter 2).

It is a considerable achievement that relatively few staff were able to create these databases in a very limited amount of time. It demonstrates that, in future projects and collaborations, it must be possible to create a realistic database that covers a real-world application with sufficient specimens to train and test classifiers.

2 Summary of identification results

In ADIAC's last annual workshops, we tried to establish rules in order to be able to compare results obtained with different feature sets. This is not a trivial task: a standard technique like splitting a data set into equally-sized training and test sets, training a classifier on the training set and then applying the trained classifier to the unseen test set, may not produce the best identification rates, especially when the image sets have comparatively low levels of replication (mostly approx. 20 images per taxon). It is tempting to use statistical tricks in order to make identification rates appear better than they actually are.[a] However, we tried to use different methods in parallel wherever possible, in order to compare outcomes, and in some chapters (8, 9 and 10), decision trees (C4.5 and/or C5.0) with leave-one-out or other strategies, and with or without windowing or boosting, are compared with other classifiers using the 50/50 training/testing approach. In the other chapters only

---

[a] Or, as British politician and author Benjamin Disraeli (1804-1881), Earl of Beaconsfield, put it: "There are lies, damned lies, and statistics."

Table 1. Summary of identification rates (%) for the two test sets, using decision trees C4.5 or C5.0 applied to entire feature sets. Where two numbers are given in the second or third column, they apply to the different data sets. "Bag" implies bagging, whereas "boost" means boosting. Notes: (1) contour features only; (2) using the full mixed genera set of 48 taxa rather than only 37; (3) average of several repeats which varied strongly in the case of the *Sellaphora pupula* set; (4) using preliminary feature sets, excluding convex/concave curvature; (5) results using the mixed genera set are shown here but should not be directly compared with the computer's results because testers had no prior knowledge of which taxa to expect, unlike the identification software. See text for further details.

features (chapter)	number features	classifier used	*Sellaphora pupula*	mixed genera	notes
classical + new (7)	149	C4.5 bag	92.5	94.9	
conv./conc. curv. (8)	40/47	C5.0/C4.5	88.3	77.6	
charact. profile (9)	40	C4.5	89.2	60.2	1,2
Legendre polyn. (9)	18	C4.5	80.8	71.5	1,2
Gabor + Legendre (10)	84/117	C5.0 boost	85.0	88.0	
math. morphology (11)	46	C4.5 bag	83.7	89.6	3
all sets (12)	329	C4.5 bag	—	96.9	4
human experts (5)	N/A	N/A	82.0	63.3	5

the C4.5 classification algorithm is used (the C5.0 demo version has a limit on the number of features). The main statistical *hocus-pocus* concerns the use of bagging, i.e. multiple decision trees in Chapters 7, 11 and 12, and leave-N-out: in Chapter 7 leave-one-out is applied, which means that almost all available data are used to train classifiers, whereas in Chapters 11 and 12 only 75% of available data are used for training. In the other Chapters 8, 9 and 10, tests with leave-half-out and the 50/50 training/testing approach are also reported. This implies that, apart from differences due to bagging, results presented in Chapter 7 are most biased towards better ID rates, whereas those presented in Chapters 11 and 12 are probably more realistic. But we need to keep in mind that e.g. the leave-one-out method is used to obtain an estimate of the ID rate for *new* data while using all *existing* data. In addition, the use of multiple classifiers, in this case decision trees or forests, is becoming standard practice in pattern recognition because they can achieve higher confidence rates. Either way, the results obtained here are preliminary, rather than final results, and work will continue as part of ongoing PhD projects and, hopefully, new projects in the future.

The purpose of having multiple approaches to creating auto-ID software was essentially to stimulate competition between groups. We also aimed to implement and to test established as well as new methods and, using the same image data sets, to compare the results and to draw some general conclusions. Table 1 summarizes the results obtained with C4.5 and C5.0 decision trees in combination with entire feature sets. These results are based on the classifier listing the correct taxon as the best (rather than second or third best) match. In most tests, contour and striation features were combined.

In the case of the *Sellaphora pupula* set, the results were stable and all exceeded 80%, in one case even 90%. Most methods outperformed the human experts' av-

Table 2. Summary of identification results (%) using other classifiers or feature set selections. Here 50/50 refers to equally-sized training and test sets, BMD is Bayes minimum distance or nearest mean, and NN stands for neural network. Notes: (1) reduced or selected feature set; (2) only 17 unambiguous taxa from the full size mixed genera set; (3) 75% for 50/50 training/testing, 99.17% for training using all 120 images, (4) only contour features, (5) using preliminary feature sets, (6) see note 5 in Table 1.

features (chapter)	number features	classifier used	*Sellaphora pupula*	mixed genera	notes
classical + new (7)	10	C4.5 bagging	90	93.7	1
conv./conc. curv. (8)	4/10	50/50 BMD	100	86.2	1
charact. profile (9)	≤ 62	syntactical	75/99	—	3
Legendre polyn. (9)	18	50/50 NN	85	91.5	2,4
robust only (12)	17	C4.5 bagging	—	95.5	1,5
human experts (5)	N/A	N/A	82	63.3	6

erage of 82%. Individual experts' results ranged from 60 to 98.3% (Chapter 5), with two experts outperforming the best computer algorithm. In general, the results obtained with the mixed genera data set were worse, although two methods produced identification rates close to or above 90%. The excellent result of 96.9% was obtained by combining all feature sets, but was based on preliminary data sets from earlier in the project, i.e. the same experiment could improve the identification rate when using the data sets available at the end of the project. The result from the human experts' test on the mixed genera set (Chapter 5) cannot directly be compared to the result obtained with the computer, because—unlike the computer—testers had no prior knowledge of which species were represented in the test. This particular experimental design was chosen because the data were intended as a long-term benchmark to compare computer ID rates against, rather than restricting the comparability of results to this particular data set only. Nevertheless, individual experts' ID rates ranged from 43 to 86.5% (Chapter 5), which is an indication that well-trained experts can still compete with some of the best ADIAC methods.

If the results for the two test sets had shown the same trends, their interpretation would have been easy, but unfortunately the results are inconclusive in that different methods scored differently with the two data sets. In addition, experiments involving different combinations of features and classifiers showed that particular combinations can increase identification rates, but they also showed that results from tests such as these need to be interpreted with caution because they cannot necessarily be generalized. For example, using all Gabor contour features and C4.5 (with the leave-one-out method) results in an ID rate of 41.7% for the *Sellaphora pupula* set, but a linear feature search boosts this to 78.4%.

Table 2 shows some results obtained with different combinations of feature sets, subsets and classifiers. A comparison with Table 1 shows that some methods (classical and new, robust features of all) produced almost the same results even when far fewer features were used. One method (convex/concave curvature) even resulted in an identification rate of 100% for the *Sellaphora pupula* set when only 4 features were used with the minimum distance classifier. The hand-optimized syntactical classifier correctly identified 99% of the *Sellaphora pupula* specimens if all 120 images are used for training, and is expected to perform very well on new, unseen samples. Legendre polynomials, which use only the diatom contour, in conjunction with a neural network classifier, were able to correctly identify 91.5% of the full-size mixed genera set (Chapter 4), but only if taxa that share similar shapes were excluded (the identification rate was reduced to 82.5% when all 48 taxa were included). One possible conclusion from this is that more effort should be spent in selecting the most robust features.

One aspect not shown here in the tables is the ranking of the outputs of classifiers. Results described in Chapters 7, 11 and 12 show that if the first three or five possible taxa are listed, the ID rates improve significantly. Although the user then has to check the three or five offered possibilities, he or she can assume that the correct taxon is listed with a certainty of almost 100%. This is an example of semi-automatic identification in which the computer assists the user.

The results obtained in ADIAC were surprisingly good, given the complexities of the two data sets. Some algorithms were able to compete with the best human experts, and outperformed the majority of testers (Chapter 5). It is fair to say that ADIAC, although officially only a pilot project, has achieved identification results beyond anyone's expectations, and future research will undoubtedly further improve the results. But, as mentioned above, care must be taken in generalizing from the present results, because the same feature sets and classifiers might not perform equally well when applied to other data sets.

3 Slide scanning and autofocusing

The slide scanning method developed here (Chapter 13) initially selects all potential target particles at low magnification. This process is reasonably fast: scanning a slide takes approximately 1.5 minutes. However, the level of detail at this magnification is insufficient for discriminating between usable diatom valves and debris (the latter includes broken or slanting valves), and target particles have to be recaptured at higher magnification for this task.

Image capture at high magnification requires autofocusing, for which an improved focus criterion has been developed. Using a multi-focus fusion technique, in which a set of images taken at a range of depths around the optimum focus plane are combined automatically, it is possible to create a single, all in-focus image. Finally, several methods for the discrimination between diatoms and debris at medium magnification have been implemented, and while the present results are promising—about 80% of usable valves can be selected—further work still needs to be done before this method is fully usable.

It is relatively easy to sort particles by size, and to reject those smaller or larger than predetermined sizes. Ellipse detection by the Hough transform works well but many diatoms are not elliptic. The best method uses cross-correlation of phase spectra, and allows detection of approximately 80% of usable valves. Hence there is still room for improvements, but we should stress that even a system that can detect 80% of usable valves will probably provide sufficient data for most diatom applications, and could still lead to very significant time savings.

A complete integration of all system components could not be tested, because the microscope stage controller did not have the necessary repetition accuracy to locate the particles at high magnification that had been detected at low or medium magnification. However, this work is still significant and novel in that all the individual components required for automatic slide scanning have now been developed. To the best of our knowledge, there still are no commercial scanning packages available.

4 Problems and future research

Most of the problems encountered during the ADIAC project appear to be resolvable, although some are approaching the limits of what is possible with existing image processing techniques. We will discuss these briefly, in the hope that we—or others—can address at least some of them in the future. Some may even be suitable for a PhD project.

(1) Automatic contour extraction: If debris or other diatoms touch or overlap a contour, or if there are gaps due to inconsistent contour grey levels, automatic contour extraction is a serious problem. Gaps could be closed by postprocessing if they occur in a convex part of the contour, but for concave parts this is a problem (Chapter 6). Occlusions frequently result in contour deformations; in these cases, contour information is present, but usually with reduced contrast. Two possible solutions to this are (a) to apply a deformable model (snake and/or splines) with shape restrictions, although this is difficult to define due to the large diversity of diatom shapes, or (b) to first detect the striae, for example with the grating cell method (Chapter 10), which can then be used to guide the contour extraction (e.g. snake). The same solutions could be applied to the forensics images, which are heavily contaminated with organic debris (Chapter 4). Even the detection of striation can be difficult here, but perhaps an iterative procedure, in which striations and contours are detected separately, could lead to a satisfactory solution. This type of image remains a real challenge.

(2) Classifiers: Graph matching has not been covered during ADIAC, although it was part of the original work plan. At the beginning of the project, our intention was to implement a system that mimicks the way diatomists work, using sets of rules and "traditional" diatom characters. This has not been implemented as such, because some of the "traditional" characters are virtually impossible to compute. Instead, a sort of *brute-force* approach has been followed, by implementing different feature sets, most of them consisting of features that describe a contour but which cannot be easily interpreted by humans (with some exceptions). Graph matching could be a good alternative here, because it allows the implementation of relationships between features, as opposed to decision trees which select features and criteria on the basis of entropy. Generally, it may pay to explore different classifiers. Some of the possibilities are decision trees, decision forests, multiple (different) classifiers with a vote system, even including long-established methods such as nearest-mean (Chapter 8) or maximum-likelihood (Chapter 10), which could compete with decision trees. There was insufficient time to explore, for example, IsoData. If sufficient human resources are available, it may be a good idea to hand-optimize a syntactical classifier (Chapter 9) as a basis for graph matching.

(3) **Feature sets:** It has been shown that a careful selection of the most robust features can yield extremely good results. However, the main goal still is to develop a small feature set that produces optimum results for most applications, i.e. both for morphologically very similar taxa (e.g. *Sellaphora pupula* demes) *and* taxa with a wide variety of characteristics (e.g. the mixed genera data set). There are several good candidate features for this, but further work is required first. Tests using curvature features of convex and concave contour segments resulted in a 100% ID rate for the *Sellaphora pupula* set with just 4 selected features, but this selection process is extremely time consuming (CPU time) when dealing with many taxa and images. Gabor features can complement Legendre polynomials, but if they are computed on the basis of the characteristic profile after ellipse fitting, and a diatom does not have an elliptical central part, they tend to fail (Chapters 9 and 10). Instead, these should be computed directly from a contour, as it is done by scale-space curvature (Chapter 11), which gave very reasonable results.

Hence, further improvements to the existing methods are possible, and even new methods may be developed, but all of them must be able to stand up to rigorous testing, for example with our existing data sets, or new sets that provide even tougher challenges. The latter clearly should include several hundreds of taxa, all represented by a large number of specimens. The set of selected features should be suitable for high-throughput work: if automatic slide scanning produces, say, 1000 images per day, all feature extractions and identifications should also be done within a day. One could imagine a relatively small feature set composed of some morphometric characters, some Legendre polynomials, and curvature and striation features, all fast to compute, and this could be sufficient for achieving extremely good results. This scenario appears to be supported by experiments described in this book, in which results obtained with complete feature sets are compared to others carried out with reduced or selected sets (Chapters 7, 8, 10 and 12).

(4) **Data sets:** It took three years to create the image databases for ADIAC, and however useful they may be for diatomists, the number of specimens is still relatively small for large scale testing in pattern recognition. More data sets with at least 20 specimens per taxon are badly needed in order to validate results. Most of the previous work on automatic identification in biology (Chapter 1) has been very limited. One notable exception to this is the DiCANN (MAS2) project, which deals with dinoflagellates (http://www.dist.unige.it/dicann/; see also Culverhouse et al., 1996). The DiCANN data are publicly available to all interested researchers. Once big data sets such as this are available, they can be used by other research groups in pattern recognition for the testing of existing and new algorithms. One can only hope that new projects in biology attract the funding to produce other large scale image databases; not only diatoms and dinoflagellates, but also organisms such as coccolithophorids, foraminiferans and radiolarians, to name just a few groups.

(5) **Slide scanning:** The two main goals in this area, the scanning of slides at low magnification and autofocusing at high magnification, have both been accomplished. For the latter, it will be difficult to make further improvements, given

that only very few diatom valves are completely flat, and that the majority therefore cannot be adequately represented in a single focal plane. This problem can be addressed by creating a single output image through multi-focus fusion techniques, such that both contour and striation are well defined for feature extraction. Further improvements are needed on the methods for discrimination between usable valves and rejects or debris, which currently detect only 80% of usable valves (Chapter 13). The best current method for this is based on cross-correlations of phase spectra, but requires the development of a template, i.e. some sort of *eigen-diatom*, like the eigen-face used in face recognition. All faces have fixed characters in common, e.g. two eyes and one nose, but diatoms have varying ornamentation patterns, and the contour does not contribute much in correlating phase spectra. It might be possible to construct a template that consists of a circular area filled with a regular pattern of points such that virtual lines with different orientations and frequencies are formed. Scaling and rotation have not been considered in Chapter 13, but the use of the Fourier-Mellin transform could provide a good solution.

Finally, there are issues relating to the hardware used for slide scanning, including microscopes. The motorized stage controller must be of sufficient quality to allow positions logged at low magnification to be relocated at high magnification. Microscopes with automatic lens changers are available on the market but these are extremely expensive, which means that probably only the larger research institutes could afford a completely automated, unsupervised system. Smaller laboratories might have to restrict themselves to a system that requires users to change lenses manually between the low and high magnification steps in the scanning process. However, if the scanning software is to be distributed to users other than the project partners, a commercially available stage controller must be integrated into the system which meets the precision requirements and is affordable for all users. This requires further testing of a variety of products. The same holds for CCD cameras and frame grabbers, if end users wish to purchase a complete package with embedded image acquisition software, a so-called "plug-and-play" solution.

(6) *Der Teil und das Ganze* **(Werner Heisenberg):** When concentrating on the various project components it is easy to lose sight of the real-world applications that are the driving force behind ADIAC. In Chapter 1, we referred to the fact that many researchers spend significant amounts of their time on routine microscopy tasks, and that there are biologists who have to do diatom identifications only occasionally, and who probably do not regard themselves as diatomists. Given that ADIAC's computer algorithms can already compete with good diatom experts, it is probably timely to start applying our technology to real-life applications. Even just a simple tool for slide scanning alone could save a tremendous amount of time. In the section following below, we explore some of the scenarios for possible products resulting from ADIAC.

Also, there is the issue of the transferability of methods to other taxonomic groups. If classifiers and feature sets performed well irrespective of the taxonomic group they were applied to, a modular package could be developed that could serve many research areas, provided that training data are available. In such a scenario it is conceivable that core modules perform common tasks (image acquisition, databasing, classifiers) and optional modules perform the feature extraction,

Chapter 14: ADIAC achievements and future work   297

optimized for diatoms or coccoliths etc. This scenario requires a much larger-scale approach, which is impossible if funding is routinely restricted to small research consortia and short periods.

5 Potential tools for routine diatom work

To conclude, we will present some of the tools that could potentially be developed using the software generated in ADIAC. There are a number of these, and they assume a variety of different user requirements. Undoubtedly, it will take some time before diatomists are convinced that this technology can be trusted, as is always the case with new technologies, and this should be taken into account by offering versions that allow more user control than others. However, all tools need to allow at least facultative user control, to ensure that data output conforms to the standards expected, rather than taking a "black-box" approach. Also, some of the tools are clearly futuristic scenarios at this stage, since the software available at present may require further improvements before it can function reliably.

1. Image databases with taxonomic and ecological information. These allow diatomists to check identifications obtained by classical means (e.g. identification keys), if additional assurance is needed and more than one source of taxonomic information is required. This has already been implemented with the release of the online ADIAC image database (see http://www.rbge.org.uk/ADIAC/db/adiacdb.htm), although the amount of taxonomic information available at present is limited. However, in the future this could easily be extended to include, for example, morphological data.

2. A character extraction tool which automates character measurements on images supplied by the user, and produces a list of character values measured, e.g. length, width and striation density. The output can then be compared by the user against values listed in traditional identification keys.

3. A slide scanning system that is restricted to finding specimens on a slide. This would be especially suitable for slides prepared from dilute samples which contain few diatoms, and which would otherwise require a visual inspection of many empty fields of view. Users who prefer to carry out their own identifications can then quickly find pre-located diatoms using the stored coordinates.

4. A semi-automatic system that combines the functionality of the image databases with automated identification. In this case, the user submits an image of an unknown diatom which has been acquired with a CCD camera mounted on a microscope, and the software—which runs locally on the user's computer—then outputs a sorted list of matching taxa. A useful option for users would be the opportunity to create their own training image databases of taxa that are specific to their application (e.g. water quality monitoring).

5. Users who only occasionally need to confirm identifications could query a centrally located, semi-automatic identification system via the Internet, by submitting a digital image of the diatom to be identified. This system would

most likely take the shape of an Internet server connected to a powerful computer or even a supercomputer. The server would act as the frontend to the (super)computer, which would allow multiple users to submit image queries simultaneously. Within this scenario, two options exist: (a) sending an email message with keywords (e.g. "identify diatom") with an image attached, which the server will respond to by email, or (b) an interactive web-page with predefined queries and fields for entering local file names. Both options can be realized such that no operator intervention is necessary on the side of the identification system. The ADIAC online web-demo provides an impression of how this might look: http://www.cs.rug.nl/hpci/demos/adiac/

6. A fully automated system which would combine slide scanning and automatic identification of diatoms, and which would not require any user input, apart from problem cases such as unknown diatoms and debris that passed the diatom selection. In this scenario, slides are scanned automatically, diatoms are located and their images captured, and these are identified to the lowest taxonomic level possible.

As mentioned above, some of these ideas may seem very futuristic, but most of the technology is already available and only needs further finetuning. One bigger goal might be to develop databases for various diatom applications, such that researchers can specifically select the taxa that they are interested in. Instead of building their own databases of training images, which costs a lot of time, users could download ready-prepared training sets for individual taxa and start using their customized system immediately.

This leads on to yet another futuristic scenario: groups of researchers around the globe could upload diatom images they have gathered as part of their work to a central server via the Internet, and so create a single, vast image database, which would be absolutely invaluable in terms of taxonomic knowledge, and could serve as a data repository for training sets that could in turn be downloaded by others. So, diatomists, unite!

APPENDIX

THE MIXED GENERA DATA SET

Example images of the taxa used in the mixed genera data set (full version, see Chapter 4). Images are not to the same scale, and scale bars of 10 μm length have been included for each image.

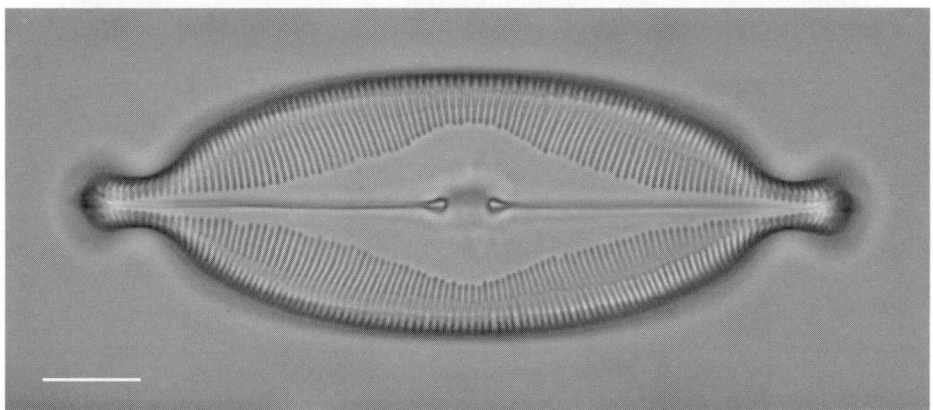

Figure 1. *Caloneis amphisbaena* (Bory) Cleve

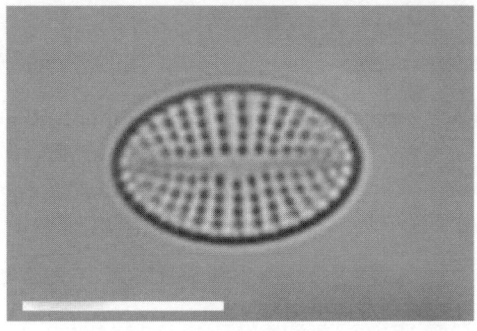

Figure 2. *Cocconeis neothumensis* Krammer

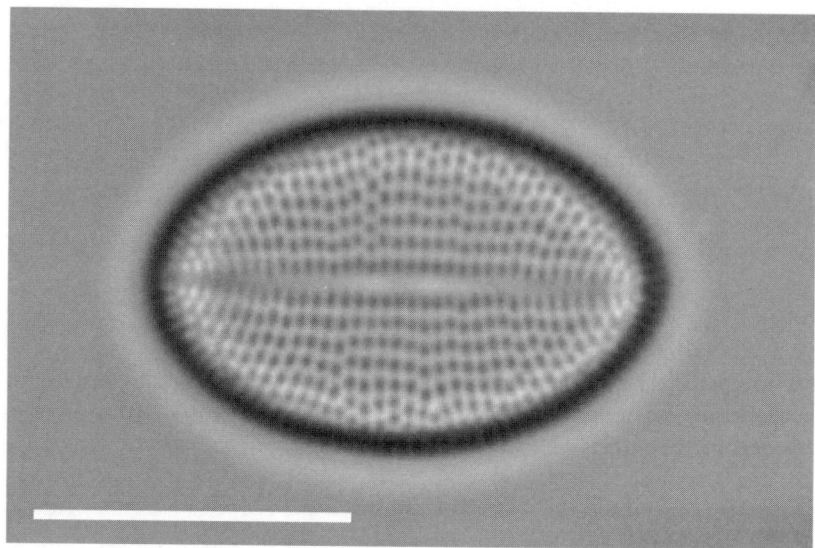

Figure 3. *Cocconeis placentula* var. *placentula* Ehrenberg

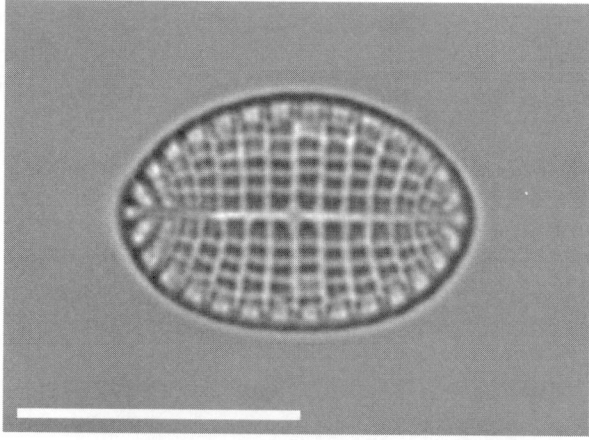

Figure 4. *Cocconeis stauroneiformis* (W. Smith) Okuno

Appendix: The mixed genera data set

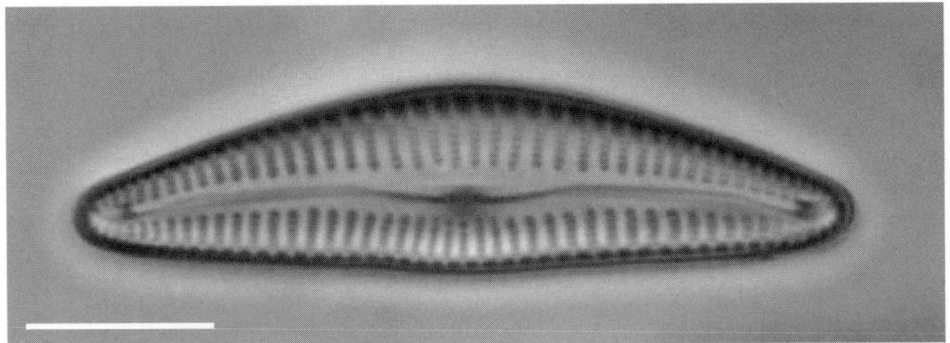

Figure 5. *Cymbella helvetica* Kützing

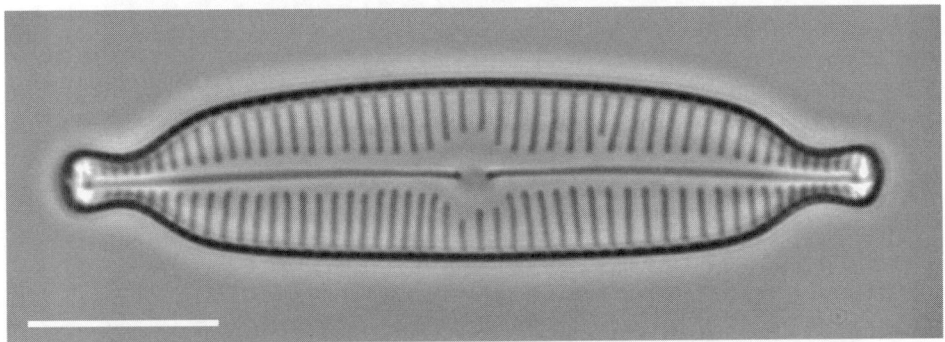

Figure 6. *Cymbella hybrida* var. *hybrida* Grunow in Cleve & Möller

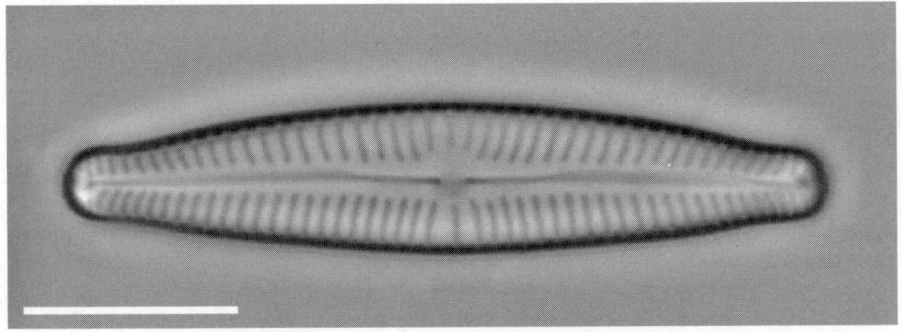

Figure 7. *Cymbella subequalis* Grunow in Van Heurck

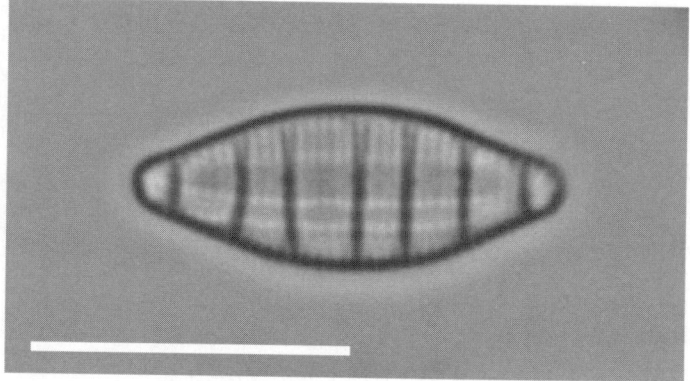

Figure 8. *Denticula tenuis* Kützing

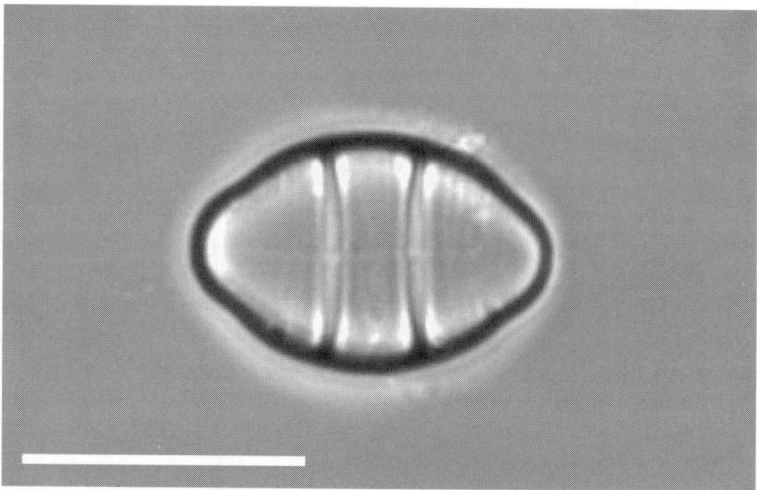

Figure 9. *Diatoma mesodon* (Ehrenberg) Kützing

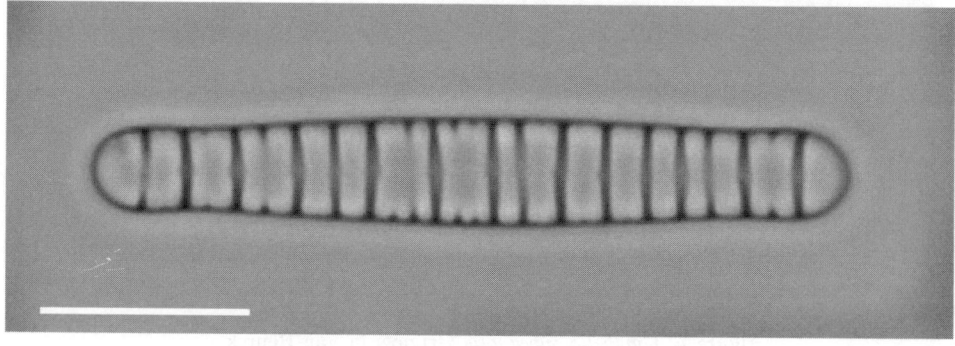

Figure 10. *Diatoma moniliformis* Kützing

Appendix: *The mixed genera data set*   303

Figure 11. *Encyonema silesiacum* (Bleisch in Rabenhorst) D.G. Mann

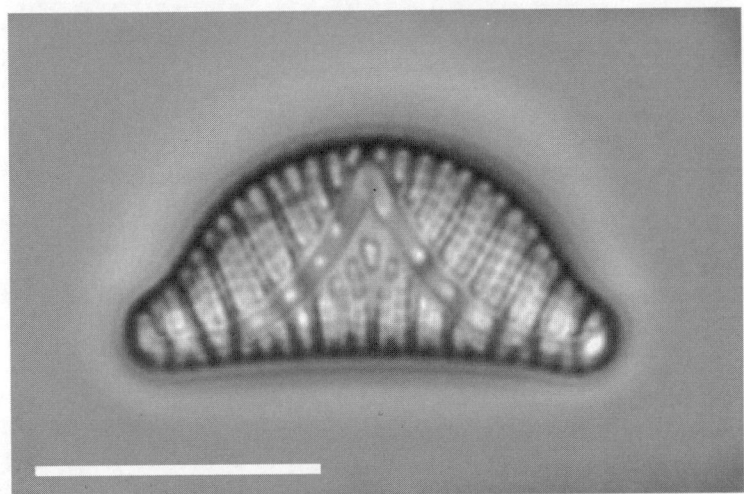

Figure 12. *Epithemia sorex* var. *sorex* Kützing

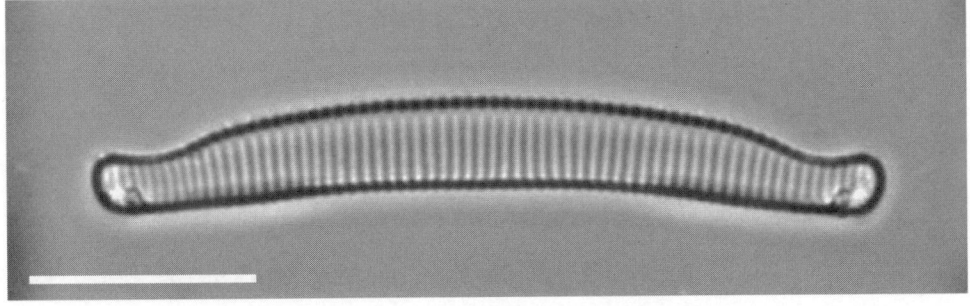

Figure 13. *Eunotia denticulata* (Brébisson) Rabenhorst

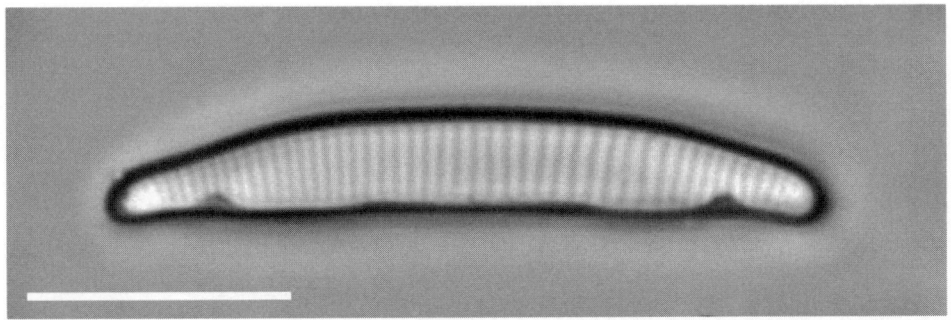

Figure 14. *Eunotia incisa* Gregory

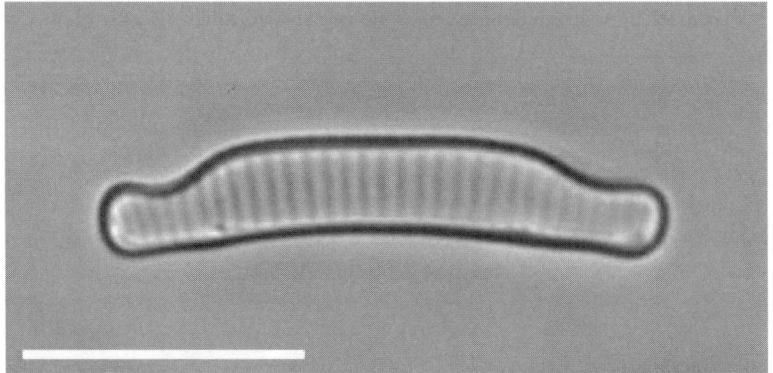

Figure 15. *Eunotia tenella* (Grunow) Hustedt in Schmidt

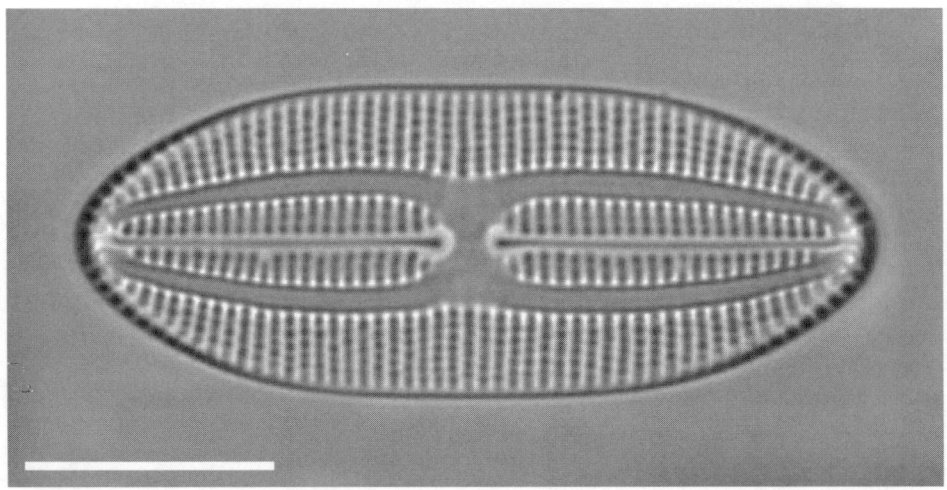

Figure 16. *Fallacia forcipata* (Greville) Stickle & Mann

Appendix: The mixed genera data set

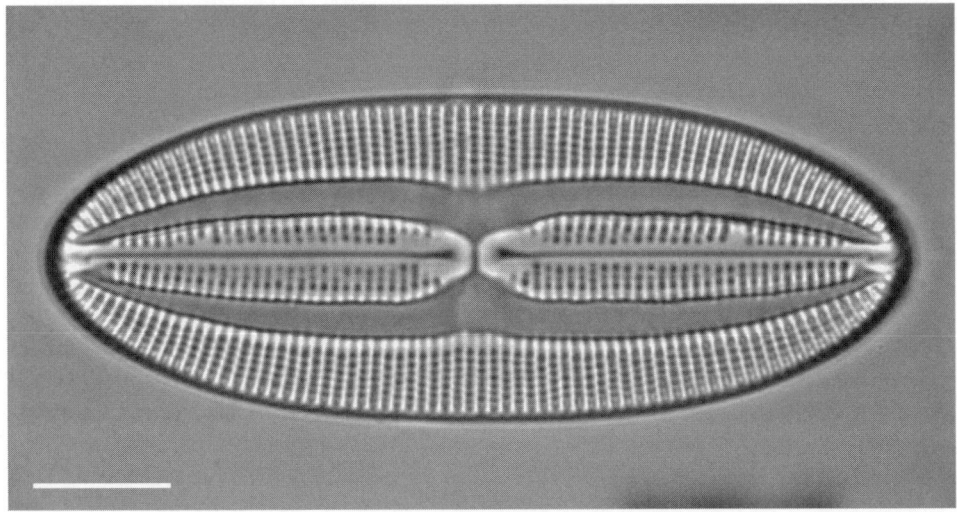

Figure 17. *Fallacia* sp. 5

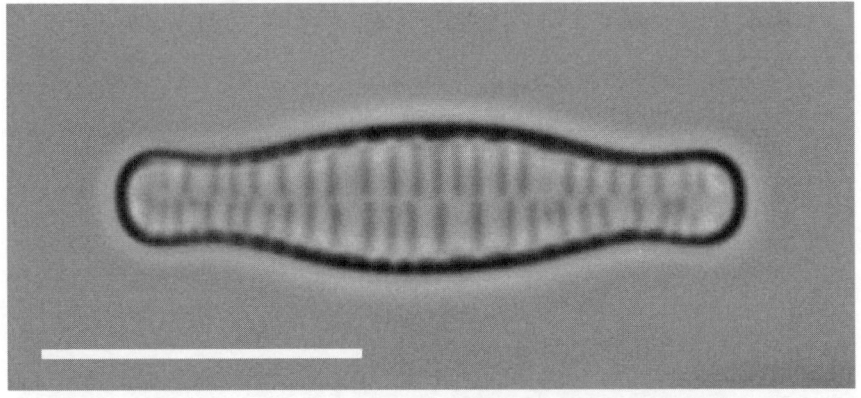

Figure 18. *Fragilariforma bicapitata* Williams & Round

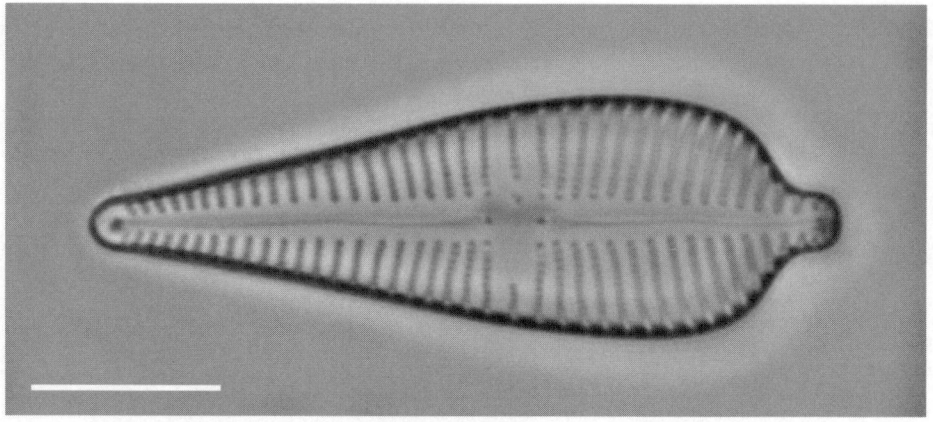

Figure 19. *Gomphonema augur* var. *augur* Ehrenberg

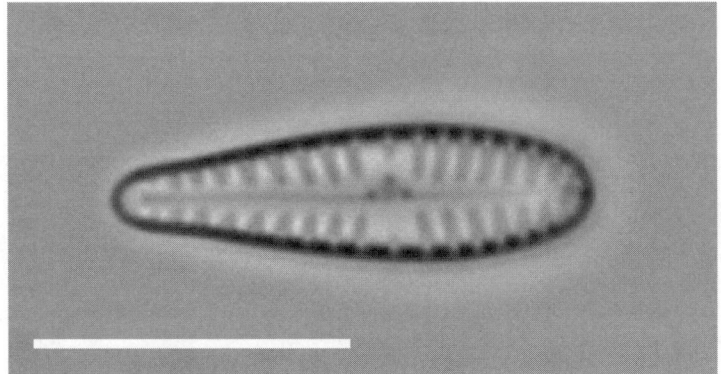

Figure 20. *Gomphonema minutum* (Agardh) Agardh

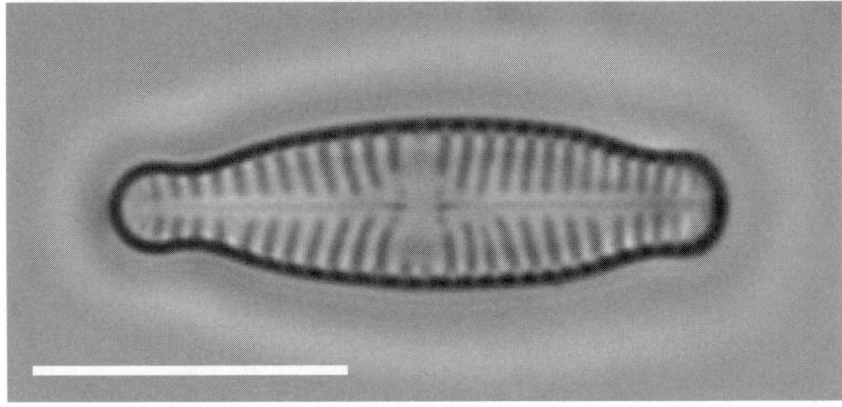

Figure 21. *Gomphonema* sp. 1

Appendix: The mixed genera data set

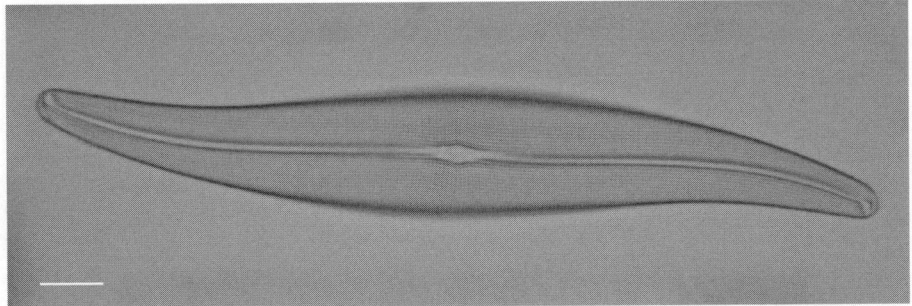

Figure 22. *Gyrosigma acuminatum* (Kützing) Rabenhorst

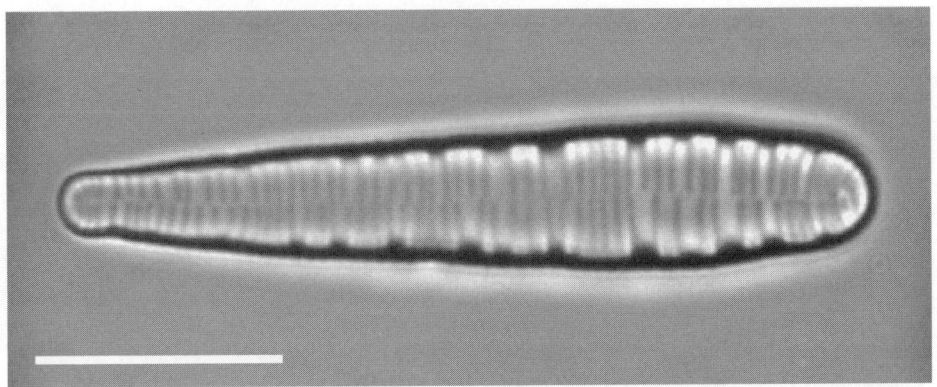

Figure 23. *Meridion circulare* var. *circulare* (Greville) Agardh

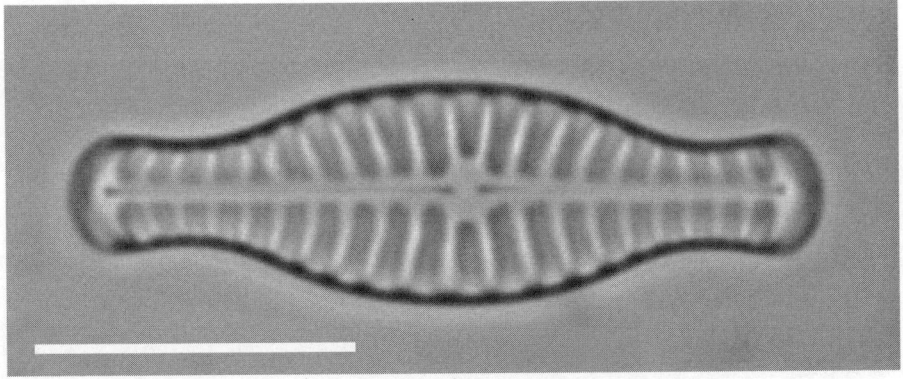

Figure 24. *Navicula capitata* var. *capitata* Ehrenberg

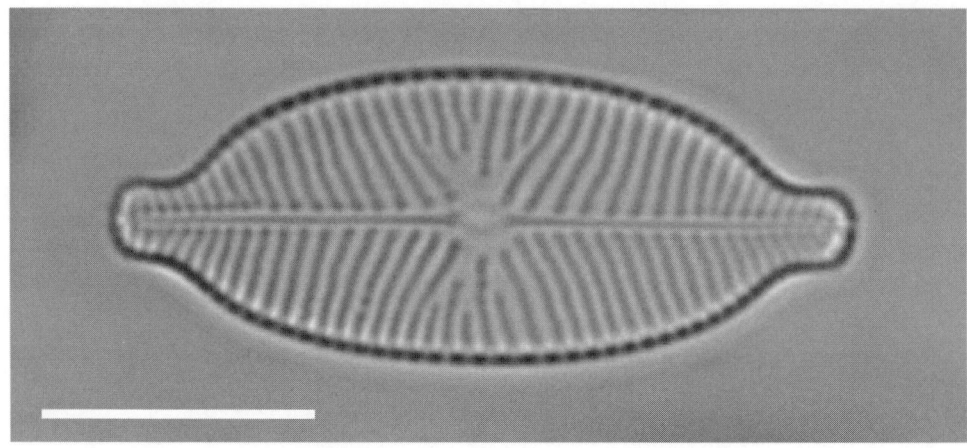

Figure 25. *Navicula constans* var. *symmetrica* Hustedt

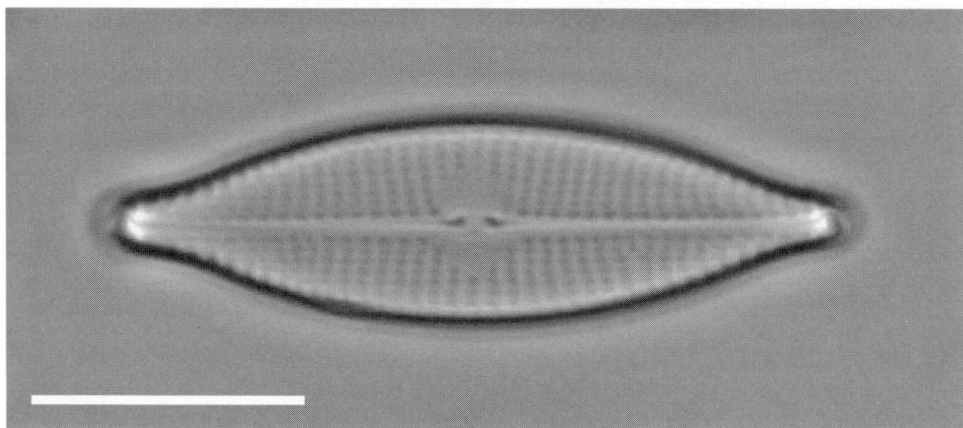

Figure 26. *Navicula gregaria* Donkin

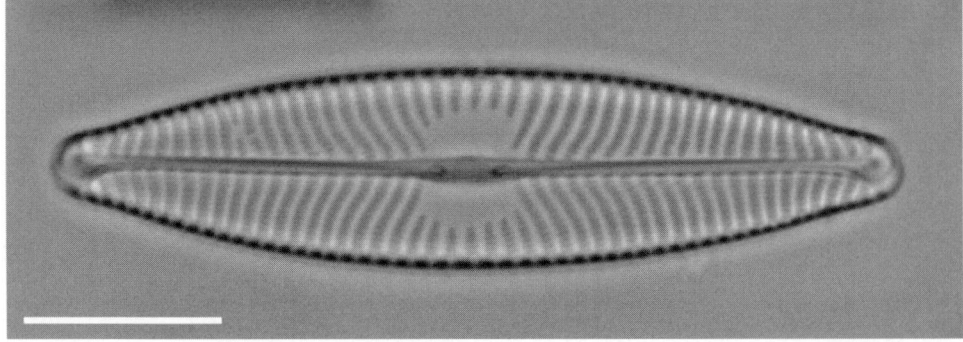

Figure 27. *Navicula lanceolata* (Agardh) Ehrenberg

Appendix: The mixed genera data set 309

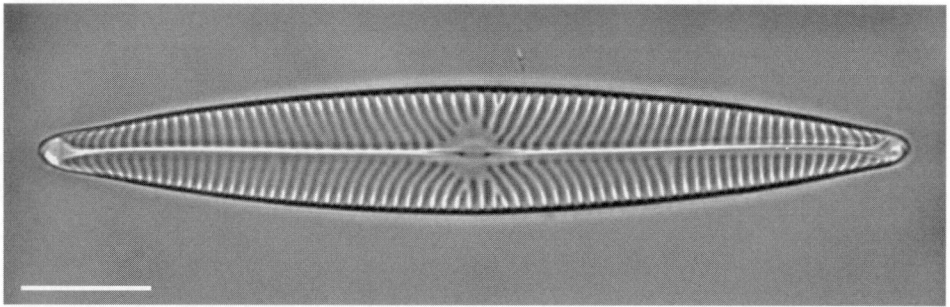

Figure 28. *Navicula radiosa* Kützing

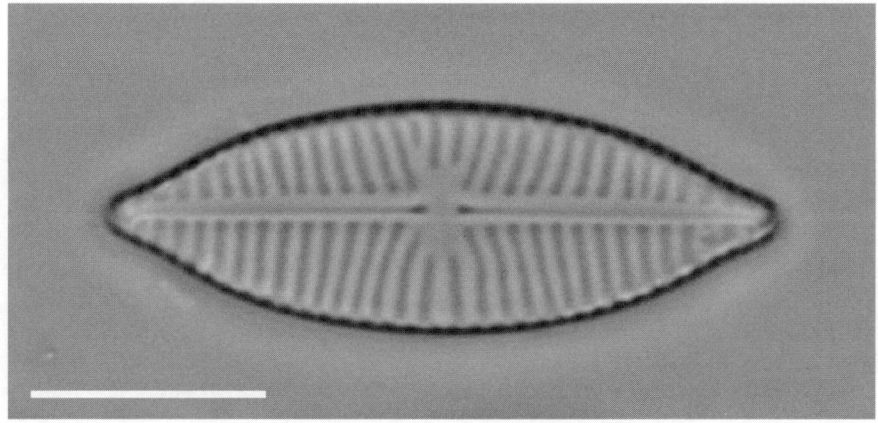

Figure 29. *Navicula menisculus* Schumann

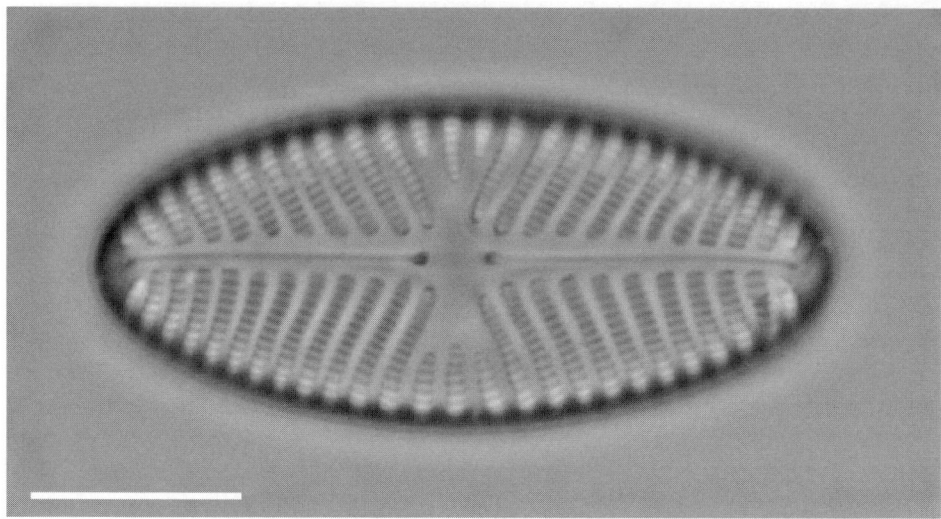

Figure 30. *Navicula reinhardtii* var. *reinhardtii* Grunow in Van Heurck

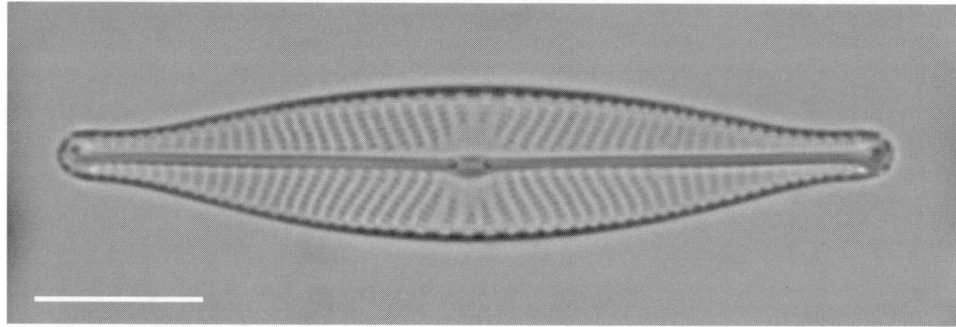

Figure 31. *Navicula rhynchocephala* Kützing

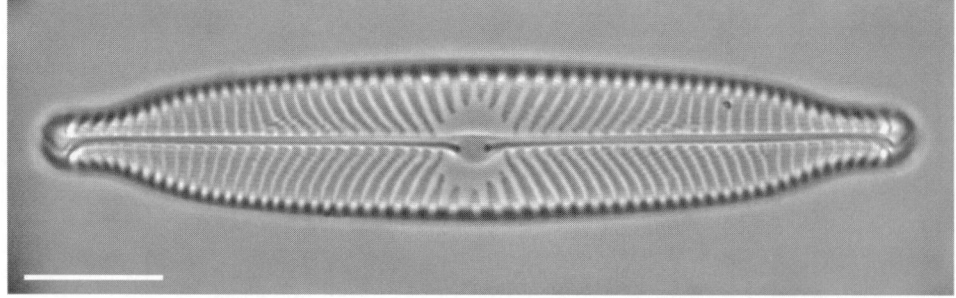

Figure 32. *Navicula viridula* var. *linearis* Hustedt

Appendix: The mixed genera data set

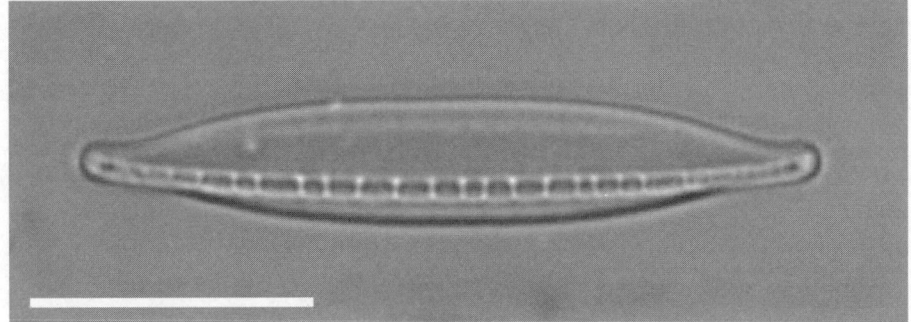

Figure 33. *Nitzschia dissipata* (Kützing) Grunow

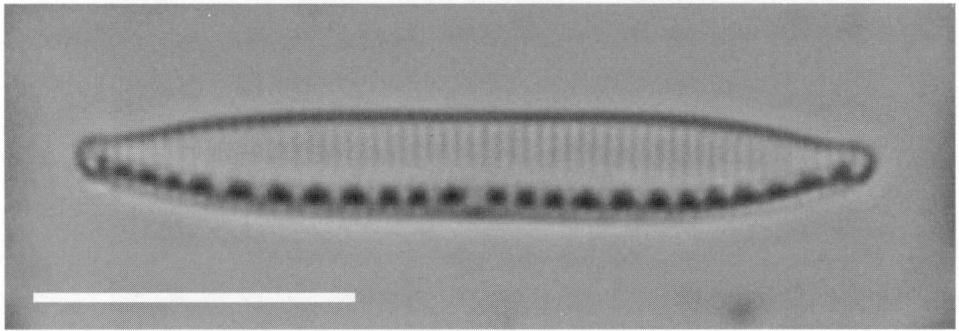

Figure 34. *Nitzschia hantzschiana* Rabenhorst

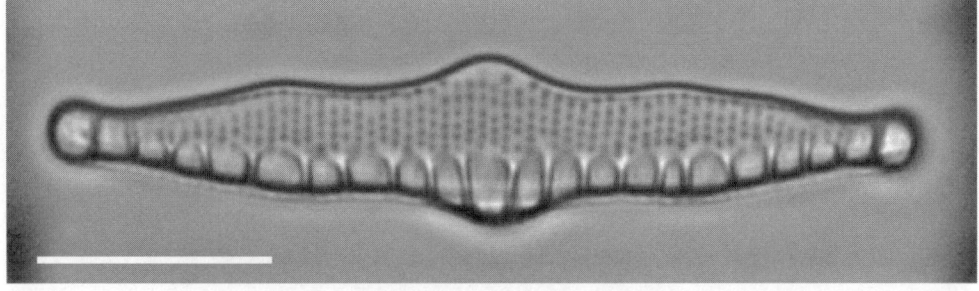

Figure 35. *Nitzschia sinuata* var. *sinuata* (Thwaites) Grunow

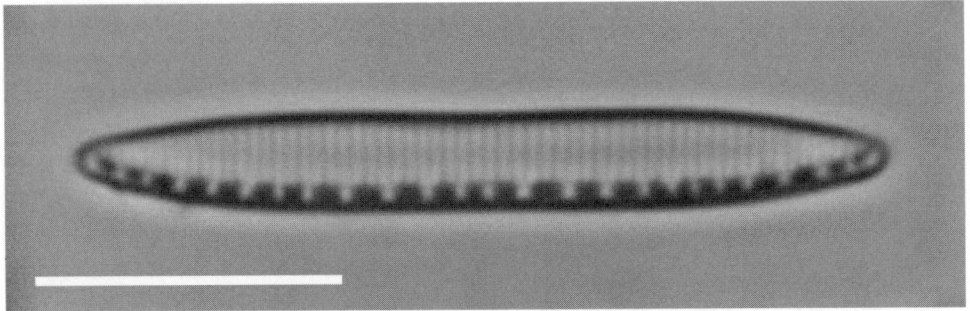

Figure 36. *Nitzschia* sp. 2

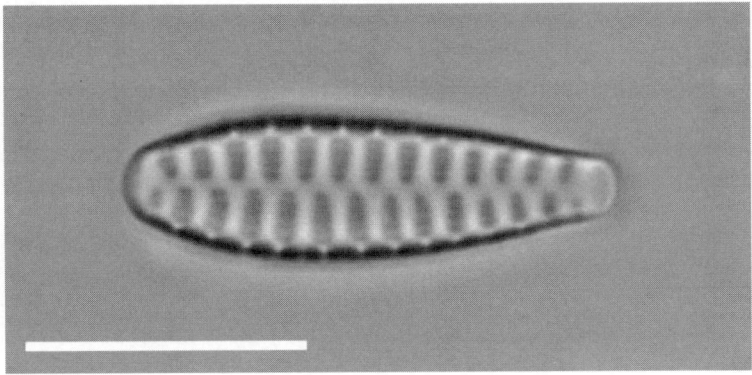

Figure 37. *Opephora olsenii* Möller

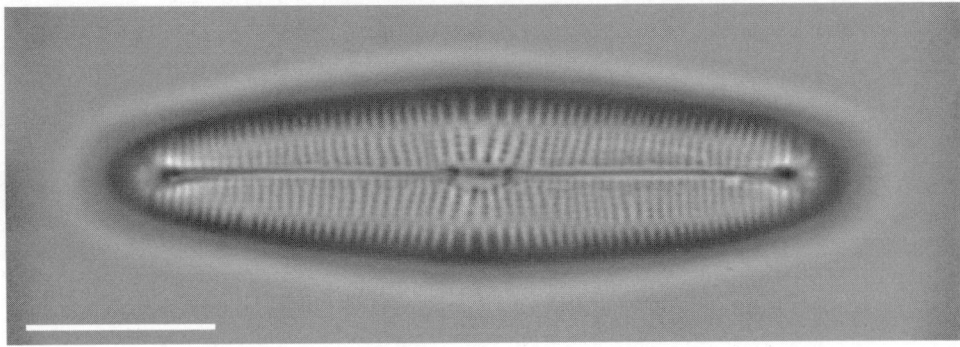

Figure 38. *Parlibellus delognei* (Van Heurck) Cox

Appendix: The mixed genera data set 313

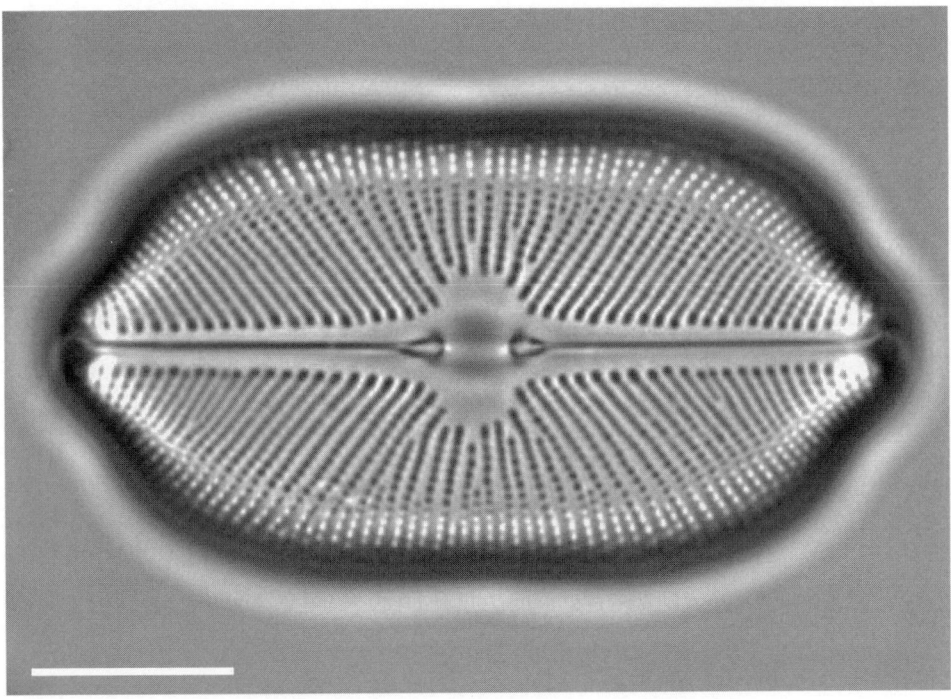

Figure 39. *Petroneis humerosa* (Brébisson ex Smith) Stickle & Mann

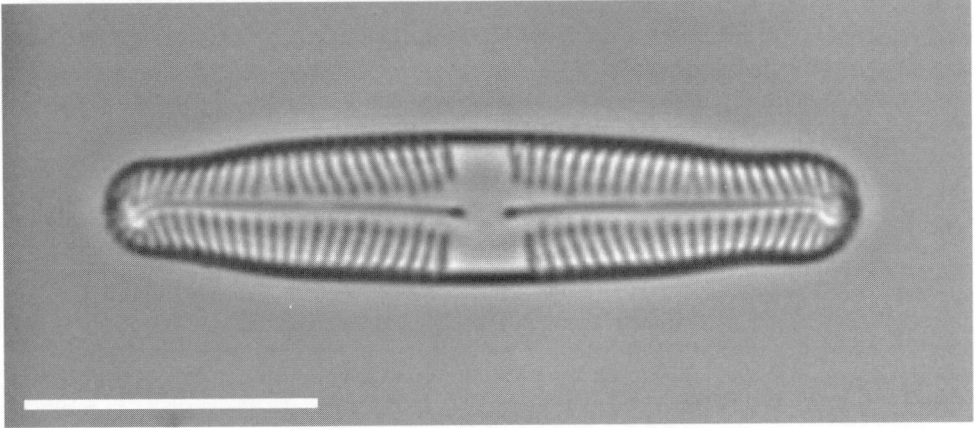

Figure 40. *Pinnularia kuetzingii* Krammer

314

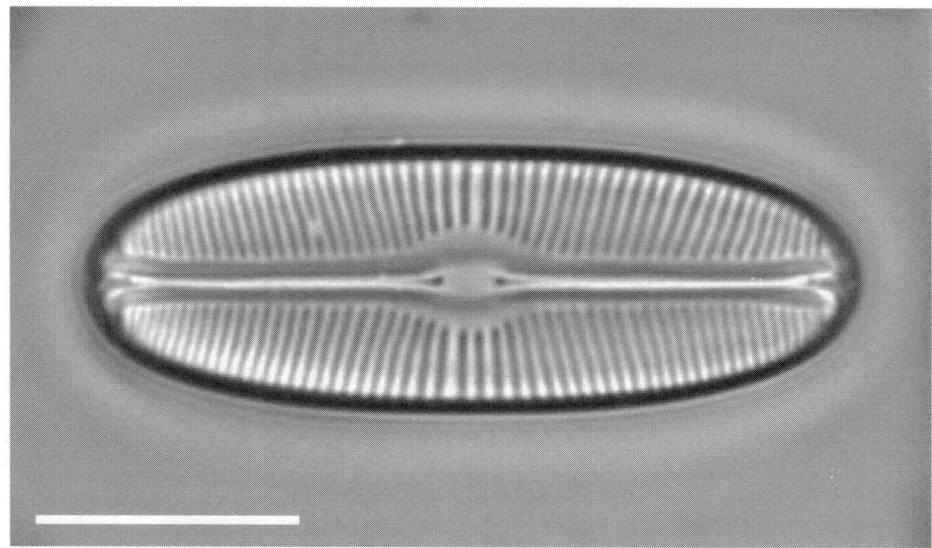

Figure 41. *Sellaphora bacillum* (Ehrenberg) D.G. Mann

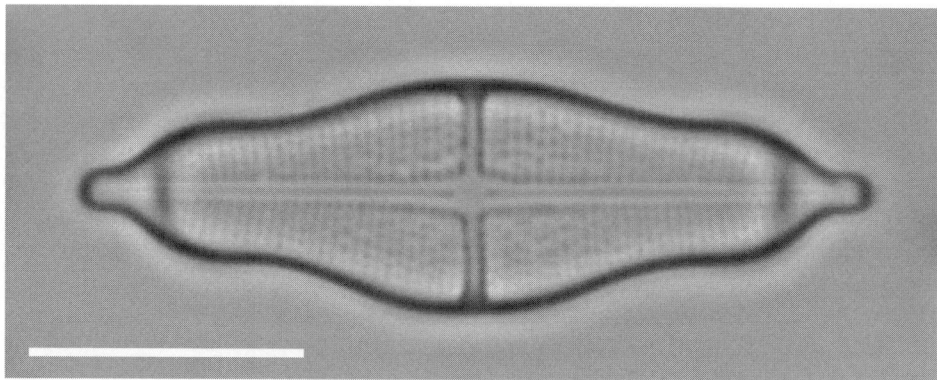

Figure 42. *Stauroneis smithii* Grunow

Appendix: The mixed genera data set 315

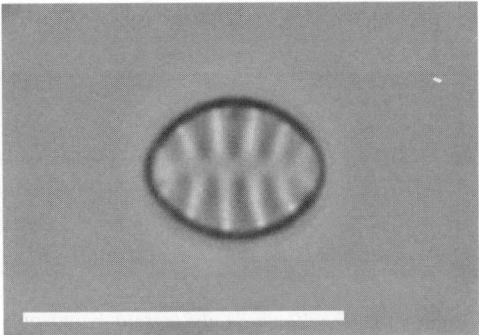

Figure 43. *Staurosirella pinnata* (Ehrenberg) Williams & Round

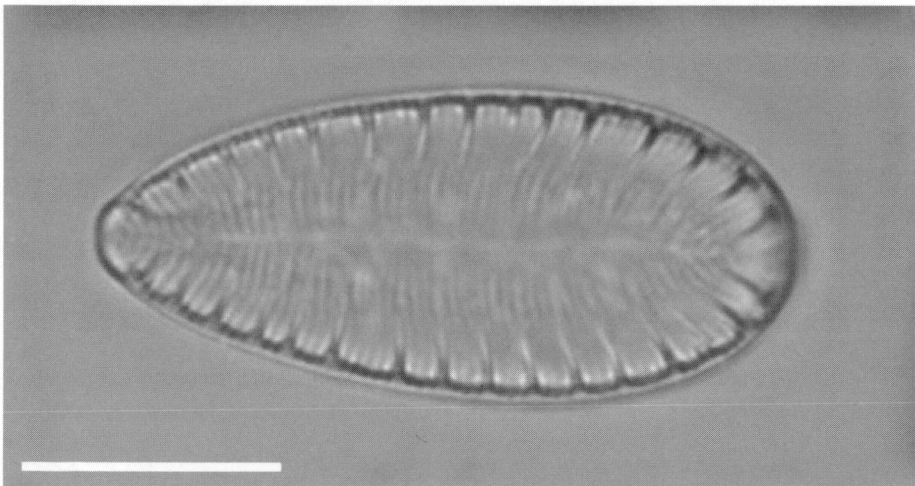

Figure 44. *Surirella brebissonii* Krammer & Lange-Bertalot

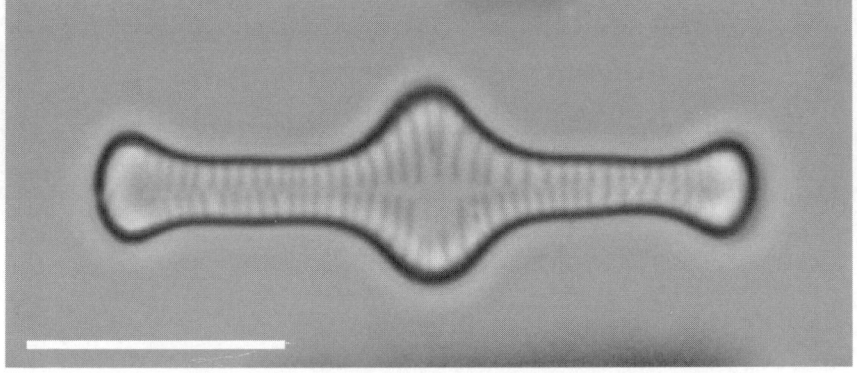

Figure 45. *Tabellaria flocculosa* (Roth) Kützing

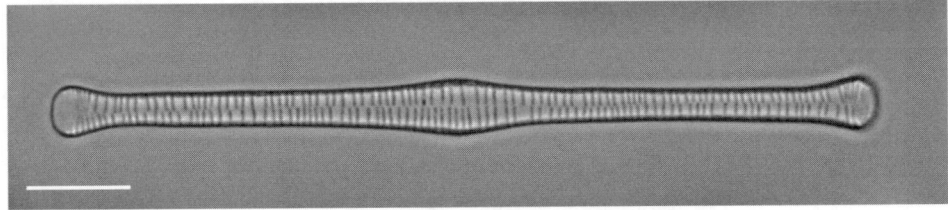

Figure 46. *Tabellaria quadriseptata* Knudson

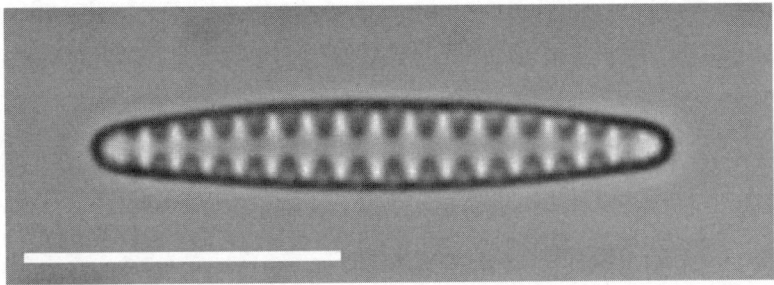

Figure 47. *Tabularia investiens* (Smith) Williams & Round

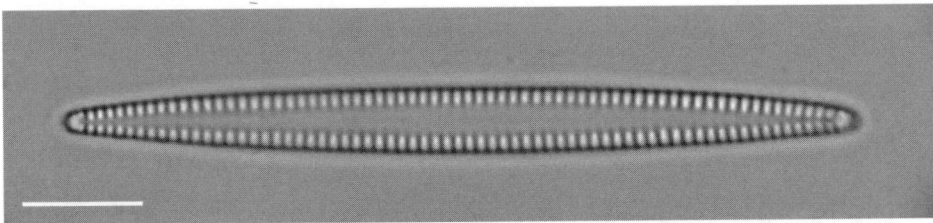

Figure 48. *Tabularia* sp. 1